Commutative Algebra

Geometric, Homological, Combinatorial and Computational Aspects

PURE AND APPLIED MATHEMATICS

A Program of Monographs, Textbooks, and Lecture Notes

LECTURE NOTES IN PURE AND APPLIED MATHEMATICS

Recent Titles

G. Lumer and L. Weis, Evolution Equations and Their Applications in Physical and Life Sciences

J. Cagnol et al., Shape Optimization and Optimal Design

J. Herzog and G. Restuccia, Geometric and Combinatorial Aspects of Commutative Algebra

G. Chen et al., Control of Nonlinear Distributed Parameter Systems

F. Ali Mehmeti et al., Partial Differential Equations on Multistructures

D. D. Anderson and I. J. Papick, Ideal Theoretic Methods in Commutative Algebra

Á. Granja et al., Ring Theory and Algebraic Geometry

A. K. Katsaras et al., p-adic Functional Analysis

R. Salvi, The Navier-Stokes Equations

F. U. Coelho and H. A. Merklen, Representations of Algebras

S. Aizicovici and N. H. Pavel, Differential Equations and Control Theory

G. Lyubeznik, Local Cohomology and Its Applications

G. Da Prato and L. Tubaro, Stochastic Partial Differential Equations and Applications

W. A. Carnielli et al., Paraconsistency

A. Benkirane and A. Touzani, Partial Differential Equations

A. Illanes et al., Continuum Theory

M. Fontana et al., Commutative Ring Theory and Applications

D. Mond and M. J. Saia, Real and Complex Singularities

V. Ancona and J. Vaillant, Hyperbolic Differential Operators and Related Problems

G. R. Goldstein et al., Evolution Equations

A. Giambruno et al., Polynomial Identities and Combinatorial Methods

A. Facchini et al., Rings, Modules, Algebras, and Abelian Groups

J. Bergen et al., Hopf Algebras

A. C. Krinik and R. J. Swift, Stochastic Processes and Functional Analysis: A Volume of Recent Advances in Honor of M. M. Rao

S. Caenepeel and F. van Oystaeyen, Hopf Algebras in Noncommutative Geometry and Physics

J. Cagnol and J.-P. Zolésio, Control and Boundary Analysis

S. T. Chapman, Arithmetical Properties of Commutative Rings and Monoids

O. Imanuvilov, et al., Control Theory of Partial Differential Equations

Corrado De Concini, et al., Noncommutative Algebra and Geometry

Alberto Corso, et al., Commutative Algebra: Geometric, Homological, Combinatorial and Computational Aspects

Giuseppe Da Prato and Luciano Tubaro, Stochastic Partial Differential Equations and Applications – VII

This collective volume is dedicated to

WOLMER V. VASCONCELOS

for being such a wonderful mathematician, mentor, colleague and friend.

Commutative Algebra

Geometric, Homological, Combinatorial and Computational Aspects

edited by

Alberto Corso
University of Kentucky
Lexington, Kentucky

Philippe Gimenez
Universidad de Valladolid
Valladolid, Spain

Maria Vaz Pinto
Instituto Superior Tecnico
Lisbon, Portugal

Santiago Zarzuela
Universitat de Barcelona
Barcelona, Spain

Chapman & Hall/CRC
Taylor & Francis Group
Boca Raton London New York Singapore

Published in 2006 by
CRC Press
Taylor & Francis Group
6000 Broken Sound Parkway NW, Suite 300
Boca Raton, FL 33487-2742

Printed in the United States of America on acid-free paper
10 9 8 7 6 5 4 3 2 1

International Standard Book Number-10: 0-8247-2335-X (Softcover)
International Standard Book Number-13: 978-0-8247-2335-4 (Softcover)
Library of Congress Card Number 2005049112

Library of Congress Cataloging-in-Publication Data

Commutative algebra : geometric, homological, combinatorial, and computational aspects / edited by Alberto Corso ...[et al.].
p. cm. – (Lecture notes in pure and applied mathematics ; v.244)
Includes bibliographical references and index.
ISBN 0-8247-2335-X (alk. Paper)
1. Commutative algebra—Congresses. I. Corso, Alberto. II. Series.

QA251.3.C6537 2005
512'.44—dc22 2005049112

T&F informa

Taylor & Francis Group
is the Academic Division of T&F Informa plc.

Visit the Taylor & Francis Web site at
http://www.taylorandfrancis.com

and the CRC Press Web site at
http://www.crcpress.com

Preface

The last decades have seen a great deal of activities and developments in several areas at the interface of commutative algebra and algebraic geometry. Exciting new results have been achieved "borrowing" methods proper to neighboring fields such as combinatorics, homological algebra, polyhedral geometry, symbolic computation, and topology. This volume – *Commutative Algebra: Geometric, Homological, Combinatorial and Computational Aspects* – would like to highlight some of these proficuous interactions by gathering a collection of refereed research papers which encompass several of those aspects. Blowup algebras, Castelnuovo-Mumford regularity, integral closure and normality, Koszul homology, liaison theory, multiplicities, polarization, and reductions of ideals are some of the topics featured in the fifteen original research articles included in this volume. Three survey articles on topics of current interest also enrich the publication: Poincaré series of singularities, uniform Artin-Rees theorems, and Gorenstein rings. Most of the results published in this volume were presented during two meetings that were held during the month of June 2003, in Spain and Portugal. Namely,

- *Commutative Algebra: Geometric, Homological, Combinatorial and Computational Aspects*, one of the Special Sessions held during the First Joint Meeting of the American Mathematical Society and the Real Sociedad Matemática Española in Sevilla, Spain, in the period June 18-21, 2003;

- *Lisbon Conference on Commutative Algebra*, held at the Instituto Superior Técnico in Lisbon, Portugal, in the period June 23-27, 2003.

These conferences are part of a long-established tradition of Western European summer meetings in commutative algebra and algebraic geometry. Besides the traditional opportunity for experts from the U.S. and Western Europe to gather and exchange the most recent discoveries, special attention was paid to allow for the participation of scholars from underrepresented countries. Overall, more than one hundred participants came together from numerous parts of the World: Brazil, Canada, France, Germany, Greece, India, Iran, Italy, Japan, Mexico, The Netherlands, Norway, Portugal, Romania, Spain, Switzerland, United Kingdom, United States of America, and Vietnam. This certainly enhanced the breath and scope of the two events.

Needless to say, we would like to express our gratitude to the contributors of this volume for their enthusiasm in the project. The anonymous referees, who worked very closely with us, also deserve a special credit for all their time spent in reading and correcting the original manuscripts: We are aware of the many demands on our time that the academic profession requires from each of us. Finally, we would also like to take the opportunity that this Preface offers us to personally acknowledge the many colleagues and institutions that contributed to the success of these two conferences. First, we would like to express our thanks to Joan Elias (Universitat de Barcelona), John Greenlees (University of Sheffield), Giuseppe Valla (Università di Genova), Wolmer V. Vasconcelos, and Charles Weibel (Rutgers University) for serving as members of the Scientific Committee of the Lisbon Conference. A special mention goes to Pedro Ferreira dos Santos, Gustavo Granja, Michael Paluch and the entire staff of the Instituto Superior Técnico for their endless help and support in planning and organizing the Lisbon Conference from the very first stage. Finally, we wish to express our heartfelt gratitude to the following institutions for their generous financial support: Banco BPI, Centro de Análise Matemática, Geometria e Sistemas Dinâmicos (Instituto Superior Técnico), Fundação Calouste Gulbenkian, Fundação Luso-Americana para o Desenvolvimento, Fundação Oriente, Fundação para a Ciência e a Tecnologia, Universitat de Barcelona and Universidad de Valladolid.

The Editors,

ALBERTO CORSO · Lexington · USA · `corso@ms.uky.edu`
PHILIPPE GIMENEZ · Valladolid · Spain · `pgimenez@agt.uva.es`
MARIA VAZ PINTO · Lisbon · Portugal · `vazpinto@math.ist.utl.pt`
SANTIAGO ZARZUELA · Barcelona · Spain · `szarzuela@ub.edu`

March 15, 2005

Contents

Participants

Ian Aberbach
University of Missouri

Alejandro Adem
University of Wisconsin-Madison

Leovigildo Alonso Tarrío
Universidade de Santiago de Compostela, Spain

Josep Àlvarez
Universitat Politècnica de Catalunya, Spain

Ricardo Andrade
Instituto Superior Técnico, Portugal

Luchezar L. Avramov
University of Nebraska

Marco Barone
Università di Bologna, Italy

Valentina Barucci
Università di Roma "La Sapienza," Italy

Isabel Bermejo
Universidad de la Laguna, Spain

Valentina Bertella
Università di Genova, Italy

Maria de Fátima Borralho
Universidade do Algarve, Portugal

Ana Bravo
Universidad Autónoma de Madrid, Spain

Massimo Brignone
Università di Genova, Italy

MARKUS BRODMANN
Universität Zürich, Switzerland

JOSÉ I. BURGOS GIL
Universitat de Barcelona, Spain

MARTA CASANELLAS
Universitat de Barcelona, Spain

GIULIO CAVIGLIA
University of Kansas

MARC CHARDIN
CNRS and Université Paris VI, France

ALDO CONCA
Università di Genova, Italy

ANA CORREIA
Universidade Aberta, Portugal

ALBERTO CORSO
University of Kentucky

STEVEN DALE CUTKOSKY
University of Missouri

MARCO D'ANNA
Università di Catania, Italy

CLARE D'CRUZ
Chennai Mathematical Institute, India

EMANUELA DE NEGRI
Università di Genova, Italy

DAVID EISENBUD
MSRI and University of California-Berkeley

SARA FARIDI
University of Ottawa, Ontario

PEDRO FERREIRA DOS SANTOS
Instituto Superior Técnico, Portugal

STEFANIA GABELLI
Università di "Roma Tre," Italy

ANTHONY GERAMITA
Queen's University at Kingston, Ontario, and Università di Genova, Italy

LAURA GHEZZI
University of Missouri

PHILIPPE GIMENEZ
Universidad de Valladolid, Spain

FLORIDA GIROLAMI
Università di "Roma Tre," Italy

Elisa Gorla
University of Notre Dame

Shiro Goto
Meiji University, Japan

Gustavo Granja
Instituto Superior Técnico, Portugal

David J. Green
Universität GH-Wuppertal, Germany

John Greenlees
University of Sheffield, United Kingdom

Futoshi Hayasaka
Meiji University, Japan

Inês Henriques
Instituto Superior Técnico, Portugal

Jürgen Herzog
Universität GH-Essen, Germany

Ricardo Inglês
Instituto Superior Técnico, Portugal

Cristodor Ionescu
Romanian Academy, Romania

Srikanth Iyengar
University of Missouri

Sean Jacques
University of Sheffield, United Kingdom

Ana Jeremías López
Universidade de Santiago de Compostela, Spain

Dikran Karagueuzian
SUNY Binghamton

Anargyros Katsabekis
University of Ioannina, Greece

Daniel Katz
University of Kansas

Mordechai Katzman
University of Sheffield, United Kingdom

Jan Kleppe
Oslo University College, Norway

Francisco Leon
Università di Roma "La Sapienza," Italy

Yves Lequain
Instituto Nacional de Matemática Pura e Aplicada (IMPA), Brazil

GENNADY LYUBEZNIK
University of Minnesota

AGUSTÍN MARCELO
Universidad de Las Palmas de Gran Canaria, Spain

NAOYUKI MATSUOKA
Meiji University, Japan

MARGARIDA MENDES LOPES
Universidade de Lisboa, Portugal

CLAUDIA MILLER
Syracuse University

EZRA MILLER
MSRI and University of Minnesota

MARCEL MORALES
Institut Fourier (Grenoble), France

YUKIO NAKAMURA
Meiji University, Japan

JORGE NEVES
University of Warwick, United Kingdom

ANNA ONETO
Università di Genova, Italy

MICHAEL PALUCH
Instituto Superior Técnico, Portugal

CHRIS PETERSON
Colorado State University

ANA PIRES
Instituto Superior Técnico, Portugal

FRANCESC PLANAS VILANOVA
Universitat Politècnica de Catalunya, Spain

CLAUDIA POLINI
University of Notre Dame

SORIN POPESCU
State University of New York at Stony Brook

HAMIDREZA RAHMATI
Universiteit Utrecht, The Netherlands

MARISA RESENDE
Instituto Superior Técnico, Portugal

MARIA EVELINA ROSSI
Università di Genova, Italy

HIDETO SAKURAI
Meiji University, Japan

Enrico Sbarra
Università di Trieste, Italy

Rodney Y. Sharp
University of Sheffield, United Kingdom

Peter Schenzel
Martin-Luther-Universität Halle-Wittenberg, Germany

Aron Simis
Universidade Federal de Pernambuco, Brazil

Nelson Sousa
Istituto Superior Técnico, Portugal

Hema Srinivasan
University of Missouri

Rosario Strano
Università di Catania, Italy

Janet Striuli
University of Kansas

Jan R. Strooker
Universiteit Utrecht, The Netherlands

Grazia Tamone
Università di Genova, Italy

Francesca Tartarone
Università di "Roma Tre," Italy

Apostolos Thoma
University of Ioannina, Greece

Ngô Viêt Trung
Institute of Mathematics (Hanoi), Vietnam

Bernd Ulrich
Purdue University

Javid Validashti
Purdue University

Giuseppe Valla
Università di Genova, Italy

Wolmer V. Vasconcelos
Rutgers University

Maria Vaz Pinto
Instituto Superior Técnico, Portugal

Joana Ventura
Instituto Superior Técnico, Portugal

Orlando Villamayor
Universidad Autónoma de Madrid, Spain

RAFAEL H. VILLARREAL
CINVESTAV, Instituto Politécnico Nacional, Mexico

NATALE PAOLO VINAI
Università di Bologna, Italy

SYLVIA WIEGAND
DMS National Science Foundation and University of Nebraska

YUJI YOSHINO
Okayama University, Japan

SANTIAGO ZARZUELA
Universitat de Barcelona, Spain

ELSA ZATINI
Università di Genova, Italy

List of Sevilla's Talks

Some invariants of local rings
JOSEP ÀLVAREZ, Universitat Politècnica de Catalunya, Spain

Saturation and Castelnuovo-Mumford regularity
ISABEL BERMEJO, Universidad de la Laguna, Spain

Multiplier ideals in exotic settings
ANA BRAVO, Universidad Autónoma de Madrid, Spain

Asymptotic behavior of cohomology
MARKUS BRODMANN, Universität Zürich, Switzerland

The structure of the Rao module and the geometry of schemes
MARTA CASANELLAS, Universitat de Barcelona, Spain

Liaison of varieties of small dimension and deficiency modules
MARC CHARDIN, CNRS and Université Paris VI, France

Deformations of monomial ideals
ALDO CONCA, Università di Genova, Italy

Poincaré series of resolution of surface singularities
STEVEN DALE CUTKOSKY, University of Missouri

Linearly presented ideals
DAVID EISENBUD, MSRI and University of California at Berkeley

List of Lisbon's Talks

The structure of F-pure rings
IAN ABERBACH, University of Missouri

Cohomological and geometric periodicity
ALEJANDRO ADEM, University of Wisconsin-Madison

Homology over local homomorphisms
LUCHEZAR AVRAMOV, University of Nebraska

A simplicial description of Kähler differentials
JOSÉ I. BURGOS GIL, Universitat de Barcelona, Spain

Gin, lex and Koszul homology
ALDO CONCA, Università di Genova, Italy

Multiplicity of the special fiber of blowups
ALBERTO CORSO, University of Kentucky

Initial ideal and its associated graded ring
EMANUELA DE NEGRI, Università di Genova, Italy

Computing the Castelnuovo-Mumford regularity of a homogeneous ideal
PHILIPPE GIMENEZ, Universidad de Valladolid, Spain

The equality $I^2 = QI$ in Buchsbaum rings
SHIRO GOTO, Meiji University, Japan

Detecting the depth of group cohomology
DAVID J. GREEN, Bergische Universität Wuppertal, Germany

Rigid resolutions and big Betti numbers
JÜRGEN HERZOG, Universität GH-Essen, Germany

Linear strands in minimal resolutions
SRIKANTH IYENGAR, University of Missouri

The module structure of a group action on a polynomial ring: A finiteness theorem
DIKRAN KARAGUEUZIAN, SUNY Binghamton

Bousfield localizations in commutative algebra
ANA JEREMÍAS LÓPEZ, Universidade de Santiago de Compostela, Spain

Fitting ideals and finite projective dimension over complete intersections
DANIEL KATZ, University of Kansas

The support of top graded local cohomology modules
MORDECHAI KATZMAN, University of Sheffield, United Kingdom

Local cohomology and its applications
GENNADY LYUBEZNIK, University of Minnesota

The canonical ring of surfaces with $p_g = 0$
MARGARIDA MENDES LOPES, Universidade de Lisboa, Portugal

On quasi-simplicial toric varieties
MARCEL MORALES, Institut Fourier, Grenoble, France

On the uniform Artin-Rees properties
FRANCESC PLANAS VILANOVA, Universitat Politècnica de Catalunya, Spain

Core of monomial ideals
CLAUDIA POLINI, University of Notre Dame

On generic initial ideals
SORIN POPESCU, State University of New York at Stony Brook

Hilbert-Samuel polynomial of a Cohen-Macaulay module and applications to the fiber cone
MARIA EVELINA ROSSI, Università di Genova, Italy

Ideals in the perfect closure, linear growth of primary decompositions and the localization problem for tight closure
RODNEY Y. SHARP, University of Sheffield, United Kingdom

Algebraic free divisors and the Jacobian ideal
ARON SIMIS, Universidade Federal de Pernambuco, Brazil

Multiplicities of monomial ideals
HEMA SRINIVASAN, University of Missouri

Stanley-Reisner rings and the radical of lattice ideals
APOSTOLOS THOMA, University of Ioannina, Greece

Mixed multiplicities of ideals and mixed volumes of polytopes
NGÔ VIÊT TRUNG, Institute of Mathematics, Hanoi, Vietnam

Linearly presented ideals and a subadditivity formula for the degrees of syzygies
BERND ULRICH, Purdue University

Big Degs and applications
WOLMER V. VASCONCELOS, Rutgers University

Normality of Rees algebras of monomial ideals
RAFAEL VILLARREAL, CINVESTAV, Instituto Politécnico Nacional, Mexico

Building interesting commutative rings
SYLVIA WIEGAND, National Science Foundation and University of Nebraska

Degeneration and G-dimension of modules
YUJI YOSHINO, Okayama University, Japan

A theorem of Eakin and Sathaye and Green's hyperplane restriction theorem

GIULIO CAVIGLIA[1]
Department of Mathematics, University of Kansas, Lawrence, KS 66045, USA
E-mail: caviglia@math.ukans.edu

Abstract: A theorem of Eakin and Sathaye relates the number of generators of a certain power of an ideal with the existence of a distinguished reduction for that ideal. We prove how this result can be obtained as a special case of Green's hyperplane restriction theorem.

1 Introduction

The purpose of these notes is to show how the following Theorem 2.1, due to Eakin and Sathaye, can be viewed, after some standard reductions, as a corollary of Green's hyperplane restriction theorem.

Theorem 2.1 (Eakin-Sathaye) *Let (R,m) be a quasi-local ring with infinite residue field. Let I be an ideal of R. Let n and r be positive integers. If the*

[1] *Current address:* The Mathematical Sciences Research Institute (MSRI), Berkeley, CA 94720. *Current E-mail*: caviglia@msri.org.

number of minimal generators of I^i, denoted by $v(I^i)$, satisfies

$$v(I^i) < \binom{i+r}{r},$$

then there are elements $h_1, \ldots, h_r$ in I such that $I^i = (h_1, \ldots, h_r)I^{i-1}$.

Before proving Theorem 2.1 we recall some general facts about Macaulay representation of integer numbers: this is needed for the understanding of Green's hyperplane restriction theorem. For more details on those topics we refer the reader to [**3**] and [**4**].

1.1 Macaulay representation of integer numbers

Let d be a positive integer. Any positive integer c can then be uniquely expressed as

$$c = \binom{k_d}{d} + \binom{k_{d-1}}{d-1} + \cdots + \binom{k_1}{1},$$

where the k_i's are non-negative and strictly increasing, i.e., $k_d > k_{d-1} > \cdots > k_1 \geq 0$. This way of writing c is called the d'th *Macaulay representation* of c, and the k_i's are called the d'th *Macaulay coefficients* of c. For instance, setting $c = 13$ and $d = 3$ we get $13 = \binom{5}{3} + \binom{3}{2} + \binom{0}{1}$.

Remark 1.1 An important property of Macaulay representation is that the usual order on the integers corresponds to the lexicographical order on the arrays of Macaulay coefficients. In other words, given two positive integer $c_1 = \binom{k_d}{d} + \binom{k_{d-1}}{d-1} + \cdots + \binom{k_1}{1}$ and $c_2 = \binom{h_d}{d} + \binom{h_{d-1}}{d-1} + \cdots + \binom{h_1}{1}$ we have $c_1 < c_2$ if and only if $(k_d, k_{d-1}, \ldots, k_1)$ is smaller lexicographically than $(h_d, h_{d-1}, \ldots, h_1)$.

Definition 1.2 Let c and d be positive integers. We define $c_{<d>}$ to be

$$c_{<d>} = \binom{k_d - 1}{d} + \binom{k_{d-1} - 1}{d-1} + \cdots + \binom{k_1 - 1}{1}$$

where $k_d, \ldots, k_1$ are d'th Macaulay coefficients of c. We use the convention that $\binom{a}{b} = 0$ whenever $a < b$.

Remark 1.3 It is easy to check that if $c_1 \leq c_2$ then $c_{1<d>} \leq c_{2<d>}$. This property, as we see in the following, allows us to iteratively apply Green's Theorem and prove Corollary 1.5.

1.2 Green's hyperplane restriction theorem

Let R be a standard graded algebra over an infinite field K. We can write R as $K[X_1,\dots,X_n]/I$ where I is an homogeneous ideal not containing any linear forms. We say that r generic homogeneous linear forms of R satisfy a certain property if there exists a Zariski open set of $(\mathbb{A}^n)^r$ whose points satisfy that property. Note that each point of $(\mathbb{A}^n)^r$ can be thought of as an r-uple of homogeneous linear forms. Given a generic homogeneous linear form L we let $R_L = K[X_1,\dots,X_{n-1}]/I_L$ denote the restriction of R to the hyperplane given by L. Note that since L is generic we can write it as $L = l_1X_1 + \cdots + l_nX_n$ where $l_n \neq 0$, therefore I_L is defined as

$$I_L = (P(X_1,\dots,X_{n-1},X_n - (L/l_n))|P \in I).$$

We will denote by R_d the d'th graded component of R. Mark Green proved the following Theorem.

Theorem 1.4 (Green's hyperplane restriction theorem) *Let R be a standard graded algebra over an infinite field K, and let L be a generic homogeneous linear form of R. Setting S to be R_L, we have*

$$\dim_K S_d \leq (\dim_K R_d)_{<d>}.$$

The hyperplane restriction theorem first appeared in [**4**], where it was proved with no assumption on the characteristic of the base field K.

A different, and more combinatorial, proof can be found in [**3**] where the characteristic zero assumption is a working hypothesis. A person interested in reading this last proof can observe that the arguments in [**3**] also work in positive characteristic with a few minor changes.

A direct corollary of Green's Theorem is the following:

Corollary 1.5 *Let R be a standard graded algebra over an infinite field K, and let $L_1,\dots,L_r$ be generic homogeneous linear forms of R. Let $\binom{k_d}{d} + \binom{k_{d-1}}{d-1} + \cdots + \binom{k_1}{1}$ be the Macaulay representation of* $\dim R_d$, *and define $S = R/(L_1,\dots,L_r)$. Then*

$$\dim_K S_d \leq \binom{k_d - r}{d} + \binom{k_{d-1} - r}{d-1} + \cdots + \binom{k_1 - r}{1}.$$

Proof: Note that the above inequality gives an open condition and therefore it is enough to show that there exist r homogeneous linear forms satisfyng it. Let L be as in Theorem 1.4. Since R_L is isomorphic to $R/(L)$ by Theorem 1.4 one deduces $\dim_K(R/(L))_d \le \binom{k_d-1}{d} + \binom{k_{d-1}-1}{d-1} + \cdots + \binom{k_1-1}{1}$. On the other hand, by Remark 1.3 we can apply Green's Theorem again and obtain the result by induction. □

2 The Eakin-Sathaye Theorem

We now prove Theorem 2.1. First of all note that since $v(I^i)$ is finite, without loss of generality we can assume that I is also finitely generated: in fact, if $J \subseteq I$ is a finitely generated ideal such that $J^i = I^i$, the result for J implies the one for I. Moreover, by the use of Nakayama's Lemma, we can replace I by the homogeneous maximal ideal of the fiber cone $S = \bigoplus_{i \ge 0} I^i/mI^i$. Note that S is a standard graded algebra finitely generated over the infinite field $R/m = K$.

Theorem 2.1 can be rephrased as:

Theorem 2.1 (Eakin-Sathaye) *Let R be a standard graded algebra finitely generated over an infinite field K. Let i and r be positive integers such that*

$$\dim_K(R_i) < \binom{i+r}{r}.$$

Then there exist homogeneous linear forms $h_1,\dots,h_r$ such that $(R/(h_1,\dots,h_r))_i$ is equal to zero.

Proof: First of all note that $\dim_K R_i \le \binom{i+r}{r} - 1 = \binom{i+r}{i} - 1 = \binom{i+r-1}{i} + \binom{i+r-2}{i-1} + \cdots + \binom{i+r-j}{i-j+1} + \cdots + \binom{r}{1}$. This well-known equality can be proven by recalling that $\binom{a+b}{a}$ is the number of monomials of degree a in $b+1$ variables. The binomial coefficient $\binom{i+r}{i}$ is interpreted as monomials in $r+1$ variables and split into parts by fixing the exponent of one of the variables from 0 up to i.

By Corollary 1.5, taking r generic homogeneous linear forms $h_1,\dots,h_r$, we have

$$\dim_K(R/(h_1,\dots,h_r))_i \le \binom{i-1}{i} + \binom{i-2}{i-1} + \cdots + \binom{0}{1}.$$

The term on the right-hand side is zero and therefore the theorem is proved. □

References

[**1**] P. Eakin and A. Sathaye, Prestable ideals, *J. Algebra* **45** (1976), 439–454.

[**2**] D. Eisenbud, *Commutative Algebra with a View Toward Algebraic Geometry*, Springer-Verlag, New York, 1995.

[**3**] M. Green, Generic initial ideals, Six lectures on commutative algebra (Bellaterra, 1996), 119–186, *Progr. Math.* **166**, Birkhäuser, Basel, 1998.

[**4**] M. Green, Restrictions of linear series to hyperplanes, and some results of Macaulay and Gotzmann, Algebraic curves and projective geometry (Trento, 1988), 76–86, *Lecture Notes in Math.* **1389**, Springer, Berlin, 1989.

[**5**] J. Sally, *Numbers of Generators of Ideals in Local Rings*, Marcel Dekker, Inc., New York-Basel, 1978.

Liaison of varieties of small dimension and deficiency modules

MARC CHARDIN
Institut de Mathématiques, CNRS & Université Paris VI
4, place Jussieu, F-75252 Paris CEDEX 05, France
E-mail: chardin@math.jussieu.fr

1 Introduction

Liaison relates the cohomology of the ideal sheaf of a scheme to the cohomology of the canonical module of its link. We here refer to Gorenstein liaison in a projective space over a field: each ideal is the residual of the other in one Gorenstein homogeneous ideal of a polynomial ring. Assuming that the linked schemes (or equivalently one of them) are Cohen-Macaulay, Serre duality expresses the cohomology of the canonical module in terms of the cohomology of the ideal sheaf. Therefore, in the case of Cohen-Macaulay linked schemes, the cohomology of ideal sheaves can be computed one from another: up to shifts in ordinary and homological degrees, they are exchanged and dualized. In terms of free resolutions this means that, up to a degree shift, they may be obtained one from another by dualizing the corresponding complexes (for instance, the generators of one cohomology module corresponds to the last syzygies of another cohomology module of the link).

If the linked schemes are not Cohen-Macaulay, this property fails. Nevertheless, experience on a computer shows that these modules are closely related. We here investigate this relation for surfaces and three-dimensionnal schemes. To describe our results, notice that the graded duals of the local cohomology modules are Ext modules (into the polynomial ring or the Gorenstein quotient that provides the linkage), let us set $—^*$ for the graded dual and $D_i(M) := H^i_{\mathfrak{m}}(M)^*$, the i-th deficiency module when $i \neq \dim M$, while $\omega_M = D_{\dim M}(M)$.

In the case of surfaces, we have to understand the behavior under liaison of D_1, the dual of the Hartshorne-Rao module and the module D_2. Note that D_2 has a finite part $H^0_{\mathfrak{m}}(D_2) \simeq D_0(D_2)^*$ and a quotient $D_2/H^0_{\mathfrak{m}}(D_2) \simeq D_1(D_1(D_2))$ that is either 0 or a Cohen-Macaulay module of dimension one. For a module M of dimension $d \geqslant 3$, there is an isomorphism $H^2_{\mathfrak{m}}(\omega_M) \simeq H^0_{\mathfrak{m}}(D_{d-1}(M))$, which gives $H^2_{\mathfrak{m}}(\omega) \simeq H^0_{\mathfrak{m}}(D_2)$ in our case, and, together with the liaison sequence, shows how these three modules behave under liaison.

In the case of dimension three, there are seven Cohen-Macaulay modules, obtained by iterating Ext's on the deficiency modules, that encode all the information on the deficiency modules. Five of these modules are permuted by liaison, up to duality and shift in degrees, but this is not the case for the two others that sit in an exact sequence with the corresponding two of the link. The key map in this four-term sequence of finite length modules generalizes the composition of the linkage map with the Serre duality map and seems an interesting map to investigate further.

We begin with a short review on duality, and present some extensions of Serre duality that we couldn't find in the literature. For instance if X is an equidimensional scheme such that

$$\operatorname{depth} \mathcal{O}_{X,x} \geqslant \dim \mathcal{O}_{X,x} - 1, \quad \forall x \in X, \tag{*}$$

then the isomorphisms given by Serre duality (in the Cohen-Macaulay case) are replaced by a long exact sequence (Corollary 2.7) in which one module over three is zero in the Cohen-Macaulay case. In the context of liaison, it turns out that condition (*) is equivalent to the fact that the S_2-ification of a direct link (or equivalently any odd link) is a Cohen-Macaulay scheme (Proposition 3.3).

2 Preliminaries on duality in a projective space

Let k be a commutative ring, n a positive integer, $A := \mathrm{Sym}_k(k^n)$, $\mathfrak{m} := A_{>0}$ and $\omega_{A/k} := A[-n]$. Let $—^*$ denote the graded dual into k :

$$—^* := \mathrm{Homgr}_A(—, k) = \bigoplus_{\mu \in \mathbf{Z}} \mathrm{Hom}_k((—)_\mu, k)$$

and $—^\vee$ the dual into $\omega_{A/k}$: $—^\vee := \mathrm{Hom}_A(—, \omega_{A/k})$.

The homology of the Čech complex $C^\bullet_\mathfrak{m}(—)$ will be denoted by $H^\bullet_\mathfrak{m}(—)$; notice that this homology is the local cohomology supported in $\mathfrak{m}$ (for any commutative ring k), as $\mathfrak{m}$ is generated by a regular sequence.

The key lemma is the following:

Lemma 2.1 *If $L_\bullet$ is a graded complex of finitely generated free A-modules, then*

(i) $H^i_\mathfrak{m}(L_\bullet) = 0$ *for* $i \neq n$,

(ii) *there is a functorial homogeneous isomorphism of degree zero:*

$$H^n_\mathfrak{m}(L_\bullet) \simeq (L^\vee_\bullet)^*.$$

Proof: Claim (i) is standard and (ii) directly follows from the classical description of $H^n_\mathfrak{m}(A)$ (see e.g. [**5**, §62] or [**1**, Ch. III, Theorem 5.1]). □

For an A-module M, we set $D_i(M) := \mathrm{Ext}^{n-i}_A(M, \omega_{A/k})$ and

$$\Gamma M := \ker(C^1_\mathfrak{m}(M) \to C^2_\mathfrak{m}(M)) = \bigoplus_{\mu \in \mathbf{Z}} H^0(\mathbf{P}^{n-1}_k, \widetilde{M}(\mu)),$$

so that we have an exact sequence

$$0 \to H^0_\mathfrak{m}(M) \to M \to \Gamma M \to H^1_\mathfrak{m}(M) \to 0.$$

We will also use the notation $\omega_M := D_{\dim M}(M)$ when M is finitely generated and k is a field.

The above lemma gives:

Proposition 2.2 *Assume that M has a graded free resolution $L_\bullet \to M \to 0$ such that each L_ℓ is finitely generated; then for all i there are functorial maps*

$$H^i_{\mathfrak{m}}(M) \xrightarrow{\phi^i} H_{n-i}((L^\vee_\bullet)^*) \xrightarrow{\psi_i} D_i(M)^*,$$

where ϕ^i is an isomorphism. Furthermore ψ_i is an isomorphism if k is a field.

Proof: The two spectral sequences arising from the double complex $C^\bullet_{\mathfrak{m}} L_\bullet$ degenerates at the second step (due to Lemma 2.1 (i) for one, and the exactness of localization for the other) and together with Lemma 2.1 (ii) provides the first isomorphism. If k is a field, $-^*$ is an exact functor, so that the natural map $H_{n-i}((L^\vee_\bullet)^*) \xrightarrow{\psi_i} H^{n-i}(L^\vee_\bullet)^* = D_i(M)^*$ is an isomorphism. □

Proposition 2.3 *Assume that M possses a graded free resolution $L_\bullet \to M \to 0$ such that each L_ℓ is finitely generated. Then there is a spectral sequence*

$$H^i_{\mathfrak{m}}(D_j(M)) \Rightarrow H_{i-j}(L^*_\bullet).$$

In particular, if k is a field. Then there is a spectral sequence

$$H^i_{\mathfrak{m}}(D_j(M)) \Rightarrow \begin{cases} M^* & \text{if } i-j=0 \\ 0 & \text{else.} \end{cases}$$

Proof: The two spectral sequences arising from the double complex $C^\bullet_{\mathfrak{m}} L^\vee_\bullet$ have as second terms $'E^{ij}_2 = H^i_{\mathfrak{m}}(\mathrm{Ext}^j_A(M, \omega_{A/k})) = H^i_{\mathfrak{m}}(D_{n-j}(M))$, $''E^{ij}_2 = 0$ for $i \neq n$ and $''E^{nj}_2 = H^j(H^n_{\mathfrak{m}}(L^\vee_\bullet)) \simeq H^j((L^{\vee\vee}_\bullet)^*) \simeq H^j(L^*_\bullet)$. This gives the first result. If k is a field, $H^j(L^*_\bullet) = H_j(L_\bullet)^*$, which gives the second claim. □

Corollary 2.4 *If k is a field, and M is a finitely generated graded A-module of dimension d, then there are functorial maps*

$$d^i_j := {'d}^{i,n-j}_2 : H^i_{\mathfrak{m}}(D_j(M)) \to H^{i+2}_{\mathfrak{m}}(D_{j+1}(M))$$

such that

(i) $d^0_{d-1} : H^0_{\mathfrak{m}}(D_{d-1}(M)) \longrightarrow H^2_{\mathfrak{m}}(\omega_M)$ *is an isomorphism if $d \geqslant 3$,*

(ii) *if $d=3$ and M is equidimensionnal and satisfies S_1, there is an exact sequence,*

$$0 \to H^1_{\mathfrak{m}}(D_2(M)) \xrightarrow{d_2^1} H^3_{\mathfrak{m}}(\omega_M) \longrightarrow \Gamma M^* \longrightarrow 0,$$

(iii) *if $d \geqslant 4$ there is an exact sequence,*

$$0 \longrightarrow H^1_{\mathfrak{m}}(D_{d-1}(M)) \xrightarrow{d_{d-1}^1} H^3_{\mathfrak{m}}(\omega_M) \xrightarrow{e} H^0_{\mathfrak{m}}(D_{d-2}(M))$$

$$\xrightarrow{d_{d-2}^0} H^2_{\mathfrak{m}}(D_{d-1}(M)) \xrightarrow{d_{d-1}^2} H^4_{\mathfrak{m}}(\omega_M) \longrightarrow C \longrightarrow 0$$

where e is the composed map

$$e\colon H^3_{\mathfrak{m}}(\omega_M) \xrightarrow{can} \operatorname{coker}(d_{d-1}^1) \xrightarrow{('d_3^{3,n-d})^{-1}} \ker(d_{d-2}^0) \xrightarrow{can} H^2_{\mathfrak{m}}(D_{d-2}(M))$$

such that

(a) *if $d=4$ and M is equidimensionnal and satisfies S_1, C sits in an exact sequence*

$$0 \to H^1_{\mathfrak{m}}(D_2(M)) \xrightarrow{'d_3^{1,n-2}} C \longrightarrow \Gamma M^* \to 0,$$

(b) *if $d \geqslant 5$, C sits in an exact sequence*

$$0 \to \ker(d_{d-2}^1) \to C \to \ker\left[\ker(d_{d-3}^0) \to \ker(d_{d-1}^3)/\operatorname{im}(d_{d-2}^1)\right] \to 0.$$

Proof: We may assume that M doesn't have associated primes of dimension $\leqslant 1$. Then these statements are direct consequences of the spectral sequence of Proposition 2.3 applied to ΓM. Note that if M is equidimensional and satisfies S_1, then $\dim D_i(M) < i$ for $i < d$ so that $H^i_{\mathfrak{m}}(D_i(M)) = 0$ for $i \neq d$. □

Remark 2.5 One may dualize all the above maps into k. By Proposition 2.2 it gives rise to natural maps $\theta_{ij} : D_{i+2}(D_{j+1}(M)) \to D_i(D_j(M))$ satisfying the "dual statments" of (i), (ii) and (iii). Note also that the map e in (iii) gives a map $\varepsilon : D_0(D_{d-2}(M)) \longrightarrow H^0_{\mathfrak{m}}(D_3(\omega_M)) \subseteq D_3(\omega_M)$. Also in (iii)(a) the graded dual K of C sits in an exact sequence

$$0 \to \Gamma M \xrightarrow{\tau} K \xrightarrow{\eta} D_1(D_2(M)) \to 0,$$

where τ is the graded dual over k of the transgression map in the spectral sequence and η is the graded dual over k of ${}'d_3^{1,n-2}$ (both composed with isomorphisms given by Proposition 2.2).

Corollary 2.6 *If k is a field and M is a finitely generated equidimensionnal graded A-module of dimension d such that $\widetilde{M}$ satisfies S_ℓ for some $\ell \geqslant 1$, then, for $1 < i \leqslant \ell$, there are functorial surjective maps*

$$f_i : H_{\mathfrak{m}}^{d+1-i}(\omega_M) \longrightarrow D_i(M) = H_{\mathfrak{m}}^0(D_i(M))$$

which are isomorphisms for $1 < i < \ell$. Moreover, if $\ell \geqslant 2$ there is an injection $D_1(M) \longrightarrow H_{\mathfrak{m}}^d(\omega_M)$ whose cokernel is $(M/H_{\mathfrak{m}}^0(M))^*$, so that in this case $H_{\mathfrak{m}}^d(\omega_M) \simeq \Gamma M^*$.

Proof: The condition on $\widetilde{M}$ implies that $H_{\mathfrak{m}}^i(D_j(M)) = 0$ if $i > \max\{0, j-\ell\}$ and $j \neq d$. These vanishings together with the spectral sequence of Proposition 2.3 gives the result. □

The next result gives a slight generalization of Serre duality to schemes (or sheaves) that are close to being Cohen-Macaulay:

Corollary 2.7 *Let k be a field and M be a finitely generated graded A-module which is unmixed of dimension $d \geqslant 3$. Set $F_M := D_{d-1}(M)$, $\mathcal{M} := \widetilde{M}$ and assume that*

$$\operatorname{depth} \mathcal{M}_x \geqslant \dim \mathcal{M}_x - 1, \quad \forall x \in Supp(\mathcal{M}).$$

Then there exists a functorial isomorphism $H_{\mathfrak{m}}^0(F_M) \simeq H_{\mathfrak{m}}^2(\omega_M)$ and a long exact sequence

$$
\begin{array}{l}
0 \longrightarrow H^1_{\mathfrak{m}}(F_M) \longrightarrow H^3_{\mathfrak{m}}(\omega_M) \longrightarrow D_{d-2}(M) \longrightarrow \\
H^2_{\mathfrak{m}}(F_M) \longrightarrow H^4_{\mathfrak{m}}(\omega_M) \longrightarrow D_{d-3}(M) \longrightarrow \\
\qquad \vdots \\
H^{d-3}_{\mathfrak{m}}(F_M) \longrightarrow H^{d-1}_{\mathfrak{m}}(\omega_M) \longrightarrow D_2(M) \longrightarrow \\
H^{d-2}_{\mathfrak{m}}(F_M) \longrightarrow H^{d}_{\mathfrak{m}}(\omega_M) \longrightarrow \Gamma M^* \longrightarrow 0.
\end{array}
$$

Proof: This is immediate from the spectral sequence of Proposition 2.3. □

Remark 2.8 The vanishing of certain collections of modules $H^i_{\mathfrak{m}}(D_j(M))$ corresponds to frequently used properties that M may have. For instance, assume that M is equidimensionnal of dimension $d > 0$ and consider all the possible non-zero modules $H^i_{\mathfrak{m}}(D_j(M))$ for $j \neq d$:

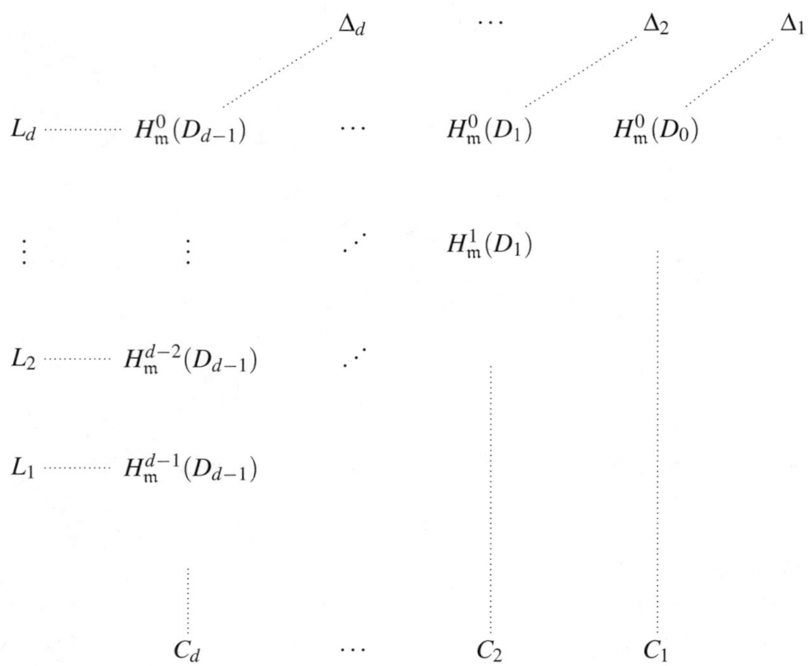

Then,

(i) The modules on lines $L_1,\ldots,L_\ell$ are zero if and only if M is Cohen-Macaulay in codimension ℓ,

(ii) The modules on columns $C_1,\ldots,C_\ell$ are zero if and only if M has depth at least ℓ,

(iii) The modules on diagonals $\Delta_1,\ldots,\Delta_\ell$ are zero if and only if M satisfies S_ℓ.

Note also that the spectral sequence shows that ω_M is Cohen-Macaulay if all non-zero modules in this diagram are located on diagonals Δ_1 and Δ_2 (even if M is not equidimensional). This somewhat generalizes the case of a sequentially Cohen-Macaulay module (due to Schenzel), where all non-zero modules are located on diagonal Δ_1.

3 The liaison sequence

In this paragraph A is a polynomial ring over a field k. Recall that a homogeneous ideal $\mathfrak{b}$ of A is Gorenstein if $B := A/\mathfrak{b}$ is Cohen-Macaulay and ω_B is a free B-module of rank 1 (i.e., B is Gorenstein), so that $\omega_B \simeq B[a]$ for some $a \in \mathbf{Z}$ called the a-invariant of B. The arithmetically Gorenstein subschemes of $\mathrm{Proj}(A)$ are the subschemes of the form $\mathrm{Proj}(A/\mathfrak{b})$ for a homogeneous Gorenstein ideal $\mathfrak{b}$; if the scheme is not empty, such a $\mathfrak{b}$ is unique.

Proposition 3.1 *Let I and $J = \mathfrak{b} : I$ be two ideals of A, linked by a Gorenstein homogeneous ideal $\mathfrak{b}$ of A (so that $I = \mathfrak{b} : J$). Set $B := A/\mathfrak{b}$, $R := A/I$, $S := A/J$, $d := \dim B$, $F_S := D_{d-1}(S)$ and $E_R := \mathrm{End}(\omega_R)$. There are natural exact sequences or isomorphisms:*

(i) $$0 \longrightarrow \omega_R \otimes \omega_B^{-1} \xrightarrow{\ \iota\ } \omega_B \otimes_B \omega_B^{-1} = B \xrightarrow{\ s\ } S \longrightarrow 0$$

where $-^{-1} := \mathrm{Hom}_B(-, B)$ *and ι and s are the canonical maps;*

(ii) $$0 \to \omega_S \xrightarrow{\mathrm{Hom}_B(s, 1_{\omega_B})} \omega_B \xrightarrow{\mathrm{Hom}_B(\iota, 1_{\omega_B})} E_R \otimes_B \omega_B \xrightarrow{\delta^r} F_S \to 0$$

where δ^r is the connecting map in the Ext sequence derived from (i), in particular $F_S \simeq (E_R/R) \otimes_B \omega_B$;

(iii) $$\lambda_i : D_i(S) \xrightarrow{\sim} D_{i+1}(\omega_R) \otimes_B \omega_B, \quad \forall i \leqslant d-2$$

given by the connecting map in the Ext sequence derived from (i).

Proof: (i) is standard (*see* e.g. [**2**] or [**3**]). Also (ii) and (iii) directly follow by the Ext sequence derived from (i). □

Remark 3.2

(1) For any B-module M, there are functorial isomorphisms:

$$\tau_i : D_i(M) \xrightarrow{\sim} \mathrm{Ext}_B^{d-i}(M, \omega_B).$$

(2) Choosing an isomomorphism $\phi : B[-a] \longrightarrow \omega_B^{-1}$, (i) gives an exact sequence,

$$0 \longrightarrow \omega_R[-a] \xrightarrow{\iota\circ(1\otimes\phi)} B \xrightarrow{s} S \longrightarrow 0.$$

Proposition 3.3 *Let X and Y be two schemes linked by an arithmetically Gorenstein projective subscheme of* $\mathrm{Proj}(A)$ *of dimension* $d \geqslant 1$ *and a-invariant a. Let* $Y_2 := \mathrm{Spec}(\mathcal{E}nd(\omega_Y))$ *be the* S_2*-ification of* Y. *The following are equivalent:*

(i) depth $\mathcal{O}_{X,x} \geqslant \dim \mathcal{O}_{X,x} - 1$ *for every* $x \in X$,

(ii) ω_Y *is Cohen-Macaulay,*

(iii) Y_2 *is Cohen-Macaulay,*

(iv) $H^i(X, \mathcal{O}_X(\mu)) \simeq H^{d-i-1}(Y, \mathcal{O}_{Y_2}(a-\mu))^*$, *for every* $0 < i < d-1$ *and every* μ, *where* $-^*$ *denotes the dual into k.*

Proof: We may assume $d \geqslant 3$. The equivalence of (i) and (ii) follows from Proposition 3.1(iii). Now $\mathcal{O}_{Y_2} = \omega_{\omega_Y}$ and $\omega_Y = \omega_{Y_2}$ which proves that (ii) and (iii) are equivalent.

On the other hand, if X satisfies (i), Corollary 2.7 and Proposition 3.1 (ii) and (iii) implies (iv).

If (iv) is satisfied, Proposition 3.1(iii) and the equality $\omega_Y = \omega_{Y_2}$ shows that $H^i(Y, \mathcal{O}_{Y_2}(\mu)) \simeq H^{d-i}(Y, \omega_{Y_2}(-\mu))^*$ for $i \neq d-1$. Applying Corollary 2.7 with $M := \Gamma\mathcal{O}_{Y_2}$ then implies that F_M has finite length, and therefore (iii) is satisfied. □

Remark 3.4 If one of the equivalent conditions of Proposition 3.3 is satisfied, and R and S are the standard graded unmixed algebras defining respectively X and Y, then $H^{d-1}(X, \mathcal{O}_X(\mu)) \simeq H^d_{\mathfrak{m}}(R)_\mu \simeq D_d(R)_{-\mu}$ and $H^0_{\mathfrak{m}}(D_d(R)) \simeq H^2_{\mathfrak{m}}(\omega_R) \simeq H^1_{\mathfrak{m}}(S)[a]$.

4 Cohomology of linked surfaces and three-folds

Let k be a field, A a polynomial ring over k and $\omega_A := A[-\dim A]$.

As in the first section, $-^*$ will denote the graded dual into k, we set $D_i(M) := \mathrm{Ext}_A^{\dim A - i}(M, \omega_A)$ and we introduce for simplicity the following abreviated notation: $D_{ij\cdots l}(M) := D_i(D_j(\cdots(D_l(M))))$.

4.1 The surface case

Let R be a homogeneous quotient of A defining a surface $X \subset \mathrm{Proj}(A)$. We will asume that $H^0_{\mathfrak{m}}(R) = 0$, as replacing R by $R/H^0_{\mathfrak{m}}(R)$ leaves both X and $H^i_{\mathfrak{m}}(R)$ for $i > 0$ unchanged. Then the three Cohen-Macaulay modules $D_1(R)$, $D_{02}(R)$ and $D_{12}(R)$ sit in the following exact sequences that show how they encode all the non-ACM characters of X:

$$0 \longrightarrow R \longrightarrow \Gamma R \longrightarrow D_1(R)^* \longrightarrow 0$$

$$0 \longrightarrow D_{02}(R)^* \longrightarrow D_2(R) \longrightarrow \Gamma D_2(R) \longrightarrow D_{12}(R)^* \longrightarrow 0$$

where $\Gamma D_2(R) = \oplus_{\mu \in \mathbf{Z}} V_\mu$ and $\delta := \dim V_\mu$ is a constant which is zero if and only if X is Cohen-Macaulay. Also $D_{02}(R)$ is a finite length module and $D_{12}(R)$ is either zero or a Cohen-Macaulay module of dimension one and degree δ.

The next result shows how these Cohen-Macaulay modules, which encode the non-ACM character of X, behave under linkage.

Proposition 4.1 *Let* $X = \mathrm{Proj}(R)$ *and* $Y = \mathrm{Proj}(S)$ *be two surfaces linked by an arithmetically Gorenstein subscheme* $\mathrm{Proj}(B) \subseteq \mathrm{Proj}(A)$ *so that* $\omega_B \simeq B[a]$. *Then there are natural degree zero isomorphisms,*

$$D_1(S) \simeq D_{02}(R) \otimes_B \omega_B, \quad D_{12}(S) \simeq D_{112}(R) \otimes_B \omega_B^{-1}.$$

Note that the roles of R and S may be reversed, so that $D_{02}(S) \simeq D_1(R) \otimes_B \omega_B^{-1}$ and $D_{112}(S) \simeq D_{12}(R) \otimes_B \omega_B$.

Proof: By Proposition 3.1 (iii), $D_1(S) \simeq D_2(\omega_R) \otimes_B \omega_B$ and $D_2(\omega_R) \simeq D_{02}(R)$ by Corollary 2.4 (i). This proves the first claim.

As for the second claim, Corollary 2.4 (ii) gives an exact sequence

$$0 \longrightarrow \Gamma S \longrightarrow D_3(\omega_S) \longrightarrow D_{12}(S) \longrightarrow 0$$

where $D_3(\omega_S) \simeq E_S$. Now $\Gamma S/S \simeq D_1(S)^*$ so that we have an exact sequence,

$$0 \longrightarrow D_1(S)^* \longrightarrow E_S/S \longrightarrow D_{12}(S) \longrightarrow 0$$

but $E_S/S \simeq D_2(R) \otimes_B \omega_B^{-1}$ by Proposition 3.1 (ii) and we know that $D_1(S)^* \simeq D_{02}(R)^* \otimes_B \omega_B^{-1}$ so that the above sequence gives

$$0 \longrightarrow D_{02}(R)^* \otimes_B \omega_B^{-1} \longrightarrow D_2(R) \otimes_B \omega_B^{-1} \longrightarrow D_{12}(S) \longrightarrow 0$$

but $D_{02}(R)^* = H^0_{\mathfrak{m}}(D_2(R))$ and $D_{112}(R) = D_2(R)/H^0_{\mathfrak{m}}(D_2(R))$. □

4.2 The three-fold case

Let R be a homogeneous quotient of A defining a three-fold $X \subset \text{Proj}(A)$. As in the surface case we will assume that $H^0_{\mathfrak{m}}(R) = 0$. Now the seven Cohen-Macaulay modules $D_1(R)$, $D_{02}(R)$, $D_{12}(R)$, $D_{03}(R)$, $D_{013}(R)$, $D_{113}(R)$ and $D_{23}(R)$ also encode all the non-ACM character of X.

The following result essentially gives the behavior of these modules under liaison.

Proposition 4.2 *Let* $X = \text{Proj}(R)$ *and* $Y = \text{Proj}(S)$ *be two three-folds linked by an arithmetically Gorenstein subscheme* $\text{Proj}(B) \subseteq \text{Proj}(A)$ *so that* $\omega_B \simeq B[a]$. *Then there are natural degree zero isomorphisms,*

$$D_1(S) \simeq D_{03}(R) \otimes_B \omega_B, D_{12}(S) \simeq D_{113}(R) \otimes_B \omega_B^{-1}, D_{23}(S) \simeq D_{223}(R) \otimes_B \omega_B^{-1},$$

and there is an exact sequence,

$$0 \longrightarrow D_{013}(R) \longrightarrow D_{02}(S) \otimes_B \omega_B \xrightarrow{\psi} D_{002}(R) \longrightarrow D_{0013}(S) \otimes_B \omega_B \longrightarrow 0,$$

where ψ *is the composition of the maps from Remark 2.5 and Proposition 3.1 (iii):*

$$D_{02}(S) \otimes_B \omega_B \xrightarrow{D_0(\lambda_2^{-1}) \otimes 1_{\omega_B}} D_{03}(\omega_R) \xrightarrow{\varepsilon^*} D_{002}(R).$$

Note that the roles of R and S may be reversed, so that we have a complete list of relations between these fourteen modules. Also the last exact sequence may be written

$$0 \longrightarrow D_{013}(R) \longrightarrow D_{02}(S) \otimes_B \omega_B \xrightarrow{\psi} D_{02}(R)^* \longrightarrow D_{013}(S)^* \otimes_B \omega_B \longrightarrow 0.$$

Proof: $D_1(S) \simeq D_2(\omega_R) \otimes_B \omega_B \simeq D_{03}(R) \otimes_B \omega_B$ by Proposition 3.1 (iii) and Corollary 2.4 (i). Also Corollary 2.4 (iii) provides exact sequences:

$$0 \longrightarrow K \xrightarrow{can} E_R \xrightarrow{\theta_{23}} D_{23}(R) \xrightarrow{\theta_{02}} D_{02}(R) \xrightarrow{e} D_3(\omega_R) \xrightarrow{\theta_{13}} D_{13}(R) \longrightarrow 0 \quad (1)$$

and

$$0 \longrightarrow \Gamma R \xrightarrow{\tau} K \xrightarrow{\eta} D_{12}(R) \longrightarrow 0\,. \qquad (2)$$

From (2) we get isomorphisms $D_i(\tau) : D_i(K) \xrightarrow{\sim} D_i(R)$ for $i \geqslant 3$ and another exact sequence

$$0 \longrightarrow D_2(K) \xrightarrow{D_2(\tau)} D_2(R) \xrightarrow{\delta} D_{112}(R) \xrightarrow{D_1(\eta)} D_1(K) \longrightarrow 0\,, \qquad (3)$$

from which it follows that $D_1(K) = 0$ (note that this vanishing also follows upon taking Ext's on sequence (1)). As $D_{02}(R)$ is of dimension zero, the right end of the first sequence then shows that $D_1(\theta_{13}) : D_{113}(R) \xrightarrow{\sim} D_{13}(\omega_R)$, but we have an isomorphism $\lambda_2^{-1} \otimes 1_{\omega_B^{-1}} : D_3(\omega_R) \xrightarrow{\sim} D_2(S) \otimes_B \omega_B^{-1}$ by Proposition 3.1 (iii) and this provides the second isomorphism. Now taking Ext's on (1) also provides two exact sequences

$$0 \longrightarrow D_{013}(R) \xrightarrow{D_0(\theta_{13})} D_{03}(\omega_R) \xrightarrow{\varepsilon^*} D_{002}(R) \xrightarrow{can} L \longrightarrow 0 \qquad (4)$$

$$0 \longrightarrow D_3(E_R) \longrightarrow D_3(R) \longrightarrow D_{223}(R) \longrightarrow D_2(E_R) \longrightarrow D_2(K) \xrightarrow{can} L' \longrightarrow 0 \qquad (5)$$

and an isomorphism $\mu : L \to L'$. Also (5) together with (3) gives us a complex

$$0 \to D_3(E_R) \to D_3(R) \to D_{223}(R) \to D_2(E_R) \to D_2(R) \to D_{112}(R) \to 0 \qquad (6)$$

whose only homology is the subquotient L of $D_2(R)$.

Taking Ext's on the exact sequence $0 \to R \to E_R \to D_3(S) \otimes_B \omega_B^{-1} \to 0$ given by Proposition 3.1 (ii), we get an exact sequence

$$\begin{aligned} 0 \to D_3(E_R) \to D_3(R) \to D_{23}(S) \otimes_B \omega_B \to D_2(E_R) \\ \to D_2(R) \to D_{13}(S) \otimes_B \omega_B \to 0. \end{aligned} \tag{7}$$

Comparing (6) and (7) proves the third isomorphism and provides the exact sequence

$$0 \to L \to D_{13}(S) \otimes_B \omega_B \to D_{112}(R) \to 0 \tag{8}$$

showing that $L \simeq D_{0013}(S) \otimes_B \omega_B$ because $D_{112}(R) \simeq D_{1113}(S) \otimes_B \omega_B$. This last identification together with (4) provides the exact sequence of the proposition and concludes the proof. □

References

[**1**] R. Hartshorne, *Algebraic Geometry*, Graduate Texts in Math. **52**, Springer-Verlag, New York-Heidelberg, 1977.

[**2**] C. Peskine and L. Szpiro, Liaison des variétés algébriques, *Invent. Math.* **26** (1974), 271–302.

[**3**] J. Migliore, *Introduction to Liaison Theory and Deficiency Modules*, *Progress in Math.* **165**, Birkhäuser, 1998.

[**4**] P. Schenzel, Notes on liaison and duality, *J. Math. Kyoto Univ.* **22** (1982/83), 485–498.

[**5**] J.-P. Serre, Faisceaux algébriques cohérents, *Ann. of Math.* **61** (1955), 197–278.

Regularity jumps for powers of ideals

ALDO CONCA
Dipartimento di Matematica, Università di Genova
Via Dodecaneso 35, I-16146 Genova, Italia
E-mail: conca@dima.unige.it

Abstract: The Castelnuovo-Mumford regularity $\mathrm{reg}(I)$ is one of the most important invariants of a homogeneous ideal I in a polynomial ring. A basic question is how the regularity behaves with respect to taking powers of ideals. It is known that in the long-run $\mathrm{reg}(I^k)$ is a linear function of k, i.e., there exist integers $a(I), b(I), c(I)$ such that $\mathrm{reg}(I^k) = a(I)k + b(I)$ for all $k \geq c(I)$. For any given integer $d > 1$ we construct an ideal J generated by $d+5$ monomials of degree $d+1$ in 4 variables such that $\mathrm{reg}(J^k) = k(d+1)$ for every $k < d$ and $\mathrm{reg}(J^d) \geq d(d+1) + d - 1$. In particular, this shows that the invariant $c(I)$ cannot be bounded above in terms of the number of variables only, not even for monomial ideals.

1 Introduction

Let K be a field. Let $R = K[x_1, \ldots, x_n]$ be the polynomial ring over K. Let $I = \oplus_{i\in\mathbb{N}} I_i$ be a homogeneous ideal. For every $i, j \in \mathbb{N}$ one defines the ijth graded Betti number of I as

$$\beta_{ij}(I) = \dim_K \mathrm{Tor}_i^R(I, K)_j$$

and set

$$t_i(I) = \max\{j \mid \beta_{ij}(I) \neq 0\}$$

with $t_i(I) = -\infty$ if it happens that $\mathrm{Tor}_i^R(I,K) = 0$. The Castelnuovo-Mumford regularity $\mathrm{reg}(I)$ of I is defined as

$$\mathrm{reg}(I) = \sup\{t_i(I) - i : i \in \mathbb{N}\}.$$

By construction $t_0(I)$ is the largest degree of a minimal generator of I. The initial degree of I is the smallest degree of a minimal generator of I, i.e., it is the least index i such that $I_i \neq 0$. The ideal I has a linear resolution if its regularity is equal to its initial degree. In other words, I has a linear resolution if its minimal generators all have the same degree and the non-zero entries of the matrices of the minimal free resolution of I all have degree 1.

Bertram, Ein and Lazarsfeld proved in [**BEL**] that if $X \subset \mathbb{P}^n$ is a smooth complex variety of codimension s which is cut out scheme-theoretically by hypersurfaces of degree $d_1 \geq \cdots \geq d_m$, then $H^i(\mathbb{P}^n, I_X^k(a)) = 0$ for $i \geq 1$ and $a \leq kd_1 + d_2 + \cdots + d_s - n$. Their result has initiated the study of the Castelnuovo-Mumford regularity of the powers of a homogeneous ideal. Note that the highest degree of a generator the k-th power I^k of I is bounded above by k times the highest degree of a generator of I, i.e., $t_0(I^k) \leq kt_0(I)$. One may wonder whether the same relation holds also for the Castelnuovo-Mumford regularity, that is, whether the inequality

$$\mathrm{reg}(I^k) \leq k\,\mathrm{reg}(I) \tag{1}$$

holds for every k. If $\dim R/I \leq 1$ then Equation (1) holds; see e.g. [**C**, **GGP**] and [**Ca**, **CH**, **Si**, **EHU**] for further generalizations. But in general (1) does not hold. We will present later some examples of ideals with linear resolution whose square does not have a linear resolution. On the other hand, Cutkosky, Herzog and Trung [**CHT**] and Kodiyalam [**K**] proved (independently) that for every homogeneous ideal I one has

$$\mathrm{reg}(I^k) = a(I)k + b(I) \qquad \text{for all } k \geq c(I) \tag{2}$$

where $a(I), b(I)$ and $c(I)$ are integers depending on I. They also showed that $a(I)$ is bounded above by the largest degree of a generator of I. Bounds for $b(I)$ and $c(I)$ are given in [**R**] in terms of invariants related to the Rees algebra of I.

Remark/Definition 1.1 Let I be an ideal generated in a single degree s. Then it is easy to see that $a(I) = s$. Hence it follows that $\mathrm{reg}(I^{k+1}) - \mathrm{reg}(I^k) = s$ for all $k \geq c(I)$. We say that the (function) regularity of the powers of I jumps at place k if $\mathrm{reg}(I^k) - \mathrm{reg}(I^{k-1}) > s$.

One of the most powerful tools in proving that a (monomial) ideal has a linear resolution is the following notion:

Definition 1.2 An ideal I generated in a single degree is said to have linear quotients if there exists a system of minimal generators $f_1, \dots, f_s$ of I such that for every $k \le s$ the colon ideal $(f_1, \dots, f_{k-1}) : f_k$ is generated by linear forms.

One has (see [**CH**]):

Lemma 1.3 (a) *If I has linear quotients then I has a linear resolution.*

(b) *If I is a monomial ideal, then the property of having linear quotients with respect to its monomial generators is independent of the characteristic of the base filed.*

We present now some (known and some new) examples of ideals with a linear resolution such that the square does have non-linear syzygies.

The first example of such an ideal was discovered by Terai. It is an ideal well-known for having another pathology: it is a square-free monomial ideal whose Betti numbers, regularity and projective dimension depend on the characteristic of the base field.

Example 1.4 Consider the ideal

$$I = (abc, abd, ace, adf, aef, bcf, bde, bef, cde, cdf)$$

of $K[a,b,c,d,e,f]$. In characteristic 0 one has $\operatorname{reg}(I) = 3$ and $\operatorname{reg}(I^2) = 7$. The only non-linear syzygy for I^2 comes at the very end of the resolution. In characteristic 2 the ideal I does not have a linear resolution.

The second example is taken from [**CH**]. It is monomial and characteristic free. The ideal is defined by 5 monomials. To the best of our knowledge, no ideal is known with the studied pathology and with less than 5 generators.

Example 1.5 Consider the ideal

$$I = (a^2b, a^2c, ac^2, bc^2, acd)$$

of $K[a,b,c,d]$. It is easy to check that I has linear quotients (with respect to the monomial generators in the given order). It follows that I has a linear resolution independently of char$K$. Furthermore I^2 has a quadratic first-syzygy in characteristic 0. But the first syzygies of a monomial ideal are independent of char$K$. So we may conclude that $\operatorname{reg}(I) = 3$ and $\operatorname{reg}(I^2) > 6$ for every base field K.

The third example is due to Sturmfels [**St**]. It is monomial, square-free and characteristic free. It is defined by 8 square-free monomials and, according to Sturmfels [**St**], there are no such examples with less than 8 generators.

Example 1.6 Consider the ideal

$$I = (def, cef, cdf, cde, bef, bcd, acf, ade)$$

of $K[a,b,c,d,e,f]$. One checks that I has linear quotients (with respect to the monomial generators in the given order) and so it has a linear resolution independently of char$K$. Furthermore I^2 has a quadratic first-syzygy. One concludes that $\operatorname{reg}(I) = 3$ and $\operatorname{reg}(I^2) > 6$ for every base field K.

So far all the examples were monomial ideals generated in degree 3. Can we find examples generated in degree 2? In view of the main result of [**HHZ**], we have to allow also non-monomial generators. One binomial generator is enough:

Example 1.7 Consider the ideal

$$I = (a^2, ab, ac, ad, b^2, ae+bd, d^2)$$

of $K[a,b,c,d,e]$. One checks that $\operatorname{reg}(I) = 2$ and $\operatorname{reg}(I^2) = 5$ in characteristic 0. The ideal has linear quotients with respect to the generators in the given order. We have not checked whether the example is characteristic free.

One may wonder whether there exists a prime ideal with this behavior. Surprisingly, one can find such an example already among the most beautiful and studied prime ideals, the generic determinantal ideals.

Example 1.8 Let I be the ideal of $K[x_{ij} : 1 \leq i \leq j \leq 4]$ generated by the 3-minors of the generic symmetric matrix (x_{ij}). It is well-known that I is a prime ideal defining a Cohen-Macaulay ring and that I has a linear resolution. One checks that I^2 does not have a linear resolution in characteristic 0.

Remark 1.9 Denote by I the ideal of 1.8 and by J the ideal of 1.4. It is interesting to note that the graded Betti numbers of I and J as well as those of I^2 and J^2 coincide. Is this just an accident? There might be some hidden relationship between the two ideals. In order to be able to reproduce the pathology it would be important to understand whether the non-linear syzygies of I^2 and J^2 have a common "explanation." We can ask whether J is an initial ideal or a specialization (or an initial ideal of a specialization) of the ideal I. Concretely, we may ask whether J can be represented as the (initial) ideal of (the ideal of) 3-minors of a 4×4 symmetric matrix of linear forms in 6 variables. We have not been able to answer this question. Note that the most natural way of filling a 4×4 symmetric matrix with 6 variables would be to put 0's on the main diagonal and to fill the remaining positions with the 6 variables. Taking 3-minors one gets an ideal, say J_1, which shares many invariants with J. For instance, we have checked that J_1 and J as well as their squares have the same graded Betti numbers (respectively of course). The ideals J and J_1 are both reduced but J has 10 components of degree 1 while J_1 has 4 components of degree 1 and 3 of degree 2. We have also checked that, in the given coordinates, J cannot be an initial ideal of J_1.

Finally, the last example shows that even among the Cohen-Macaulay ideals with minimal multiplicity one can find ideals whose squares have a non-linear resolution.

Example 1.10 Consider the ideal

$$I=(x^2,xy_1,xy_2,xy_3,y_1^2,y_2^2,y_3^2,y_1y_2-xz_{12},y_1y_3-xz_{13},y_2y_3-xz_{23})$$

of $R=K[x,y_1,y_2,y_3,z_{12},z_{13},z_{23}]$. Then R/I is Cohen-Macaulay with minimal multiplicity (and hence I has a 2-linear resolution) but I^2 does not have a linear resolution in characterisitc 0. Indeed, I^2 has a non-linear second syzygy.

More generally one can consider variables $x,y_1,\dots,y_n$ and z_{ij} with $1\le i< j\le n$. Then the ideal I_n generated by

$$x^2,xy_1,\dots,xy_n,y_1^2,\dots,y_n^2$$

and

$$y_iy_j-xz_{ij} \text{ with } 1\le i<j\le n$$

defines a Cohen-Macaualy ring with minimal multiplicity. For $n = 3$ one gets the ideal of 1.10. For $n > 3$ there is some computational evidence that I_n^2 has a non-linear first syzygy. It seems to be difficult to compute or bound the invariants $b(I_n)$ and $c(I_n)$ for this family.

2 Regularity jumps

The goal of this section is to show that the regularity of the powers of an ideal can jump for the first time at any place. This happens already in 4 variables. Let R be the polynomial ring $K[x_1,x_2,z_1,z_2]$ and for $d > 1$ define the ideal

$$J = (x_1z_1^d, x_1z_2^d, x_2z_1^{d-1}z_2) + (z_1,z_2)^{d+1}.$$

We will prove the following:

Theorem 2.1 *The ideal J^k has linear quotients for all $k < d$ and J^d has a first-syzygy of degree d. In particular* $\operatorname{reg}(J^k) = k(d+1)$ *for all $k < d$ and* $\operatorname{reg}(J^d) \geq d(d+1)+d-1$ *and this holds independently of K.*

We may deduce:

Corollary 2.2 *The invariant $c(I)$ cannot be bounded above in terms of the number of variables only, not even for monomial ideals.*

In order to prove that J^k has linear quotients we need the following technical construction.

Given three sets of variables $x = x_1,\dots,x_m$, $z = z_1,\dots,z_n$ and $t = t_1,\dots,t_k$ we consider monomials $m_1,\dots,m_k$ of degree d is the variables z. Let ϕ be a map

$$\phi : \{t_1,\dots,t_k\} \to \{x_1,\dots,x_m\}.$$

Extend its action to arbitrary monomials by setting

$$\phi(\prod t_i^{a_i}) = \prod \phi(t_i)^{a_i}.$$

Set $W = (m_1,\dots,m_k)$. Consider the bigraded presentation

$$\Phi : K[z_1,\dots,z_n,t_1,\dots,t_k] \to R(W) = K[z_1,\dots,z_n,m_1s,\dots,m_ks]$$

of the Rees algebra of the ideal W obtained by setting $\Phi(z_i) = z_i$ and $\Phi(t_i) = m_i s$ (s a new variable) and giving degree $(1,0)$ to the z's and degree $(0,1)$ to the t's. We set

$$H = \operatorname{Ker}\Phi$$

and note that H is a binomial ideal.

Definition 2.3 We say that $m_1,\dots,m_k$ are pseudo-linear of order p with respect to ϕ if for every $1 \le b \le a \le p$ and for every binomial $MA - NB$ in H with M,N monomials in the z of degree $(a-b)(d+1)$ and A,B monomials in the t of degree b such that $\phi(A) > \phi(B)$ in the lex-order there exists an element of the form $M_1 t_i - N_1 t_j$ in H where M_1, N_1 are monomials in the z such that the following conditions are satisfied:

(1) $N_1 | N$,

(2) $t_i | A$, $t_j | B$,

(3) $\phi(t_i) | \phi(A) / \mathrm{GCD}(\phi(A), \phi(B))$,

(4) $\phi(t_i) > \phi(t_j)$ in the lex-order.

The important consequence is the following:

Lemma 2.4 *Assume that $m_1 \dots, m_k$ are monomials of degree d in the z which are pseudo-linear of order p with respect to ϕ. Set*

$$J = (z_1,\dots,z_n)^{d+1} + (\phi(t_1)m_1, \phi(t_2)m_2, \dots, \phi(t_k)m_k).$$

Then J^a has linear quotients for all $a = 1,\dots,p$.

Proof: Set $Z = (z_1,\dots,z_n)^{d+1}$ and $I = (\phi(t_1)m_1, \phi(t_2)m_2, \dots, \phi(t_k)m_k)$. Take a with $1 \le a \le p$ and order the generators of J^a according to the following decomposition: $J^a = Z^a + Z^{a-1}I + \cdots + Z^b I^{a-b} + \cdots + I^a$. In the block Z^a we order the generators so that they have linear quotients; this is easy since Z^a is just a power of the $(z_1,\dots,z_n)$. In the the block $Z^b I^{a-b}$ with $b < a$ we order the generators extending (in any way) the lex-order in the x. We claim that, with this order, the ideal J^a has linear quotients. Let us check this. As long as we deal with elements of the block Z^a there is nothing to check. So let us take some monomial, say u, from the block Z^bI^{a-b} with $b < a$ and denote by V the ideal generated by the monomials which are earlier in the list. We have to show that the colon ideal $V : (u)$ is generated by variables. Note that $V : (u)$ contains $(z_1,\ldots,z_n)$ since $(z_1,\ldots,z_n)u \subset Z^{b+1}I^{a-b-1} \subset V$. Let v be a generator of V. If v comes from a block Z^cI^{a-c} with $c > b$ then we are done since $(v) : (u)$ is contained in $(z_1,\ldots,z_n)$ by degree reason. So we can assume that also v comes from the block Z^bI^{a-b}. Again, if the generator $v/\operatorname{GCD}(v,u)$ of $(v) : (u)$ involves the variables z we are done. So we are left with the case in which $v/\operatorname{GCD}(v,u)$ does not involves the variables z. It is now the time to use the assumption that the m_i's are pseudo-linear. Say $u = Nm_{s_1}\phi(t_{s_1})\cdots m_{s_a}\phi(t_{s_a})$ and $v = Mm_{r_1}\phi(t_{r_1})\cdots m_{r_a}\phi(t_{r_a})$ with M,N monomials of degree $(b-a)(d+1)$ in the z. Set $A = t_{r_1}\cdots t_{r_a}$ and $B = t_{s_1}\cdots t_{s_a}$. Since v is earlier than u in the generators of J^a we have $\phi(A) > \phi(B)$ in the lex-order. Note also that $(v) : (u)$ is generated by $\phi(A)/\operatorname{GCD}(\phi(A),\phi(B))$. Now the fact that $v/\operatorname{GCD}(v,u)$ does not involves the variables z is equivalent to saying that $MA - NB$ belongs to H. By assumption there exists $L = M_1t_i - N_1t_j$ in H such that the conditions (1)–(4) of Definition 2.3 hold. Multiplying L by $(N/N_1)(B/t_j)$, we have that $M_1(N/N_1)t_i(B/t_j) - NB$ is in H and by construction

$$v_1 = M_1(N/N_1)m_i\phi(t_i)m_{s_1}\phi(t_{s_1})\cdots m_{s_a}\phi(t_{s_a})/m_j\phi(t_j)$$

is a monomial of the block Z^bI^{a-b} which is in V by construction and such that $(v) : (u) \subseteq (v_1) : (u) = (\phi(t_i))$. This concludes the proof. □

Now we can prove:

Lemma 2.5 *For every integer $d > 1$ the monomials*

$$m_1 = z_1^d,\ m_2 = z_2^d,\ m_3 = z_1^{d-1}z_2$$

are pseudo-linear of order $(d-1)$ with respect to the map

$$\phi : \{t_1,t_2,t_3\} \to \{x_1,x_2\}$$

defined by $\phi(t_1) = x_1$, $\phi(t_2) = x_1$, $\phi(t_3) = x_2$.

Proof: It is easy to see that the defining ideal H of the Rees algebra of $W = (m_1,m_2,m_3)$ is generated by

$$(3)\ z_2t_1 - z_1t_3 \qquad (4)\ z_1^{d-1}t_2 - z_2^{d-1}t_3 \qquad (5)\ t_1^{d-1}t_2 - t_3^d.$$

Let $1 \leq b \leq a \leq d-1$ and $F = MA - NB$ a binomial of bidegree $((a-b)(d+1), b)$ in H such that $\phi(A) > \phi(B)$ in the lex-order. Denote by $v = (v_1, v_2), u = (u_1, u_2)$ the exponents of M and N and by $\alpha = (\alpha_1, \alpha_2, \alpha_3)$ and $\beta = (\beta_1, \beta_2, \beta_3)$ the exponents of A and B. We collect all the relations that hold by assumption:

$$
\begin{array}{ll}
(i) & 1 \leq b \leq a \leq d-1 \\
(ii) & \alpha_1 + \alpha_2 + \alpha_3 = \beta_1 + \beta_2 + \beta_3 = b \\
(iii) & v_1 + v_2 = u_1 + u_2 = (a-b)(d+1) \\
(iv) & v_1 + d\alpha_1 + (d-1)\alpha_3 = u_1 + d\beta_1 + (d-1)\beta_3 \\
(v) & v_2 + d\alpha_2 + \alpha_3 = u_2 + d\beta_2 + \beta_3 \\
(vi) & \alpha_1 + \alpha_2 > \beta_1 + \beta_2.
\end{array}
$$

Note that (vi) holds since $\phi(A) > \phi(B)$ in the lex-order. If

$$\alpha_1 > 0 \text{ and } \beta_3 > 0 \text{ and } u_1 > 0 \tag{6}$$

then Equation (3) does the job. If instead

$$\alpha_2 > 0 \text{ and } \beta_3 > 0 \text{ and } u_2 \geq d-1 \tag{7}$$

then Equation (4) does the job. So it is enough to show that either (6) or (7) holds. By contradiction, assume that both (6) and (7) do not hold. Note that (ii) and (vi) imply that $\beta_3 > 0$. Also note that if $a = b$ then F has bidegree $(0, a)$ and hence must be divisible by (3), which is impossible since $a < d$. So we may assume that $b < a$. But then $u_1 + u_2 = (a-b)(d+1) \geq d+1$ and hence either $u_1 > 0$ or $u_2 \geq d-1$. Summing up, if both (6) and (7) do not hold and taking into consideration that $\beta_3 > 0$, that either $u_1 > 0$ or $u_2 \geq d-1$ and that $\alpha_1 + \alpha_2 > 0$, then one of the following conditions holds:

$$\alpha_1 = 0 \text{ and } u_2 < d-1 \tag{8}$$

$$\alpha_2 = 0 \text{ and } u_1 = 0. \tag{9}$$

If (8) holds then $\alpha_2 > \beta_1 + \beta_2$. Using (v) we may write

$$u_2 = v_2 + d(\alpha_2 - \beta_1 - \beta_2) + d\beta_1 + \alpha_3 - \beta_3. \tag{10}$$

If $\beta_1 > 0$ we conclude that $u_2 \geq d + d - \beta_3$ and hence $u_2 \geq d+1$ since $\beta_3 \leq b \leq d-1$, and this is a contradiction.

If instead $\beta_1 = 0$ then $\alpha_2 + \alpha_3 = \beta_2 + \beta_3$ and hence (10) yields $u_2 = (d-1)(\alpha_2 - \beta_2) + v_2 \geq (d-1)$, a contradiction.

If (9) holds then $\alpha_1 > \beta_1 + \beta_2$. By (iv) we have

$$v_1 + d(\alpha_1 - \beta_1) + (d-1)(\alpha_3 - \beta_3) = 0.$$

Hence

$$v_1 + d(\alpha_1 - \beta_1 - \beta_2) + (d-1)(\alpha_3 - \beta_3) + d\beta_2 = 0.$$

But $\alpha_3 - \beta_3 = -\alpha_1 + \beta_1 + \beta_2$, so that

$$v_1 + (\alpha_1 - \beta_1 - \beta_2) + d\beta_2 = 0$$

which is impossible since $\alpha_1 - \beta_1 - \beta_2 > 0$ by assumption. □

Now we are ready to complete the proof of the theorem.

Proof of Theorem 2.1: Combining 2.4 and 2.5 we have that J^k has linear quotients, and hence a linear resolution, for all $k < d$. It remains to show that J^d has a first-syzygy of degree d. Denote by V the ideal generated by all the monomial generators of J^d but $u = (z_1^{d-1}z_2x_2)^d$. We claim that x_1^d is a minimal generator of $V : u$ and this is clearly enough to conclude that J^d has a first-syzygy of degree d. First note that $x_1^d u = x_2^d(z_1^d x_1)^{d-1}(z_2^d x_1) \in V$, hence $x_1^d \in V : u$. Suppose, by contradiction, x_1^d is not a minimal generator of $V : u$. Then there exists an integer $s < d$ such that $x_1^s u \in V$. In other words we may write $x_1^s u$ as the product of d generators of J, say $f_1, \dots, f_d$, not all equal to $z_1^{d-1}z_2x_2$, times a monomial m of degree s. Since the total degree in the x in $x_1^s u$ is $s + d$ at each generator of J has degree at most 1 in the x, it follows that the f_i are all of the of type $z_1^d x_1, z_2^d x_2, z_1^{d-1}z_2x_2$ and m involves only x. Since x_2 has degree d in u, $s < d$ and $z_1^{d-1}z_2x_2$ is the only generator of J containing x_2, it follows that at least one of the f_i is equal to $z_1^{d-1}z_2x_2$. Getting rid of those common factors we obtain a relation of type $x_1^s(z_1^{d-1}z_2x_2)^r = m(z_1^d x_1)^{r_1}(z_2^d x_1)^{r_2}$ with $r = r_1 + r_2 < d$. In the z-variables it gives $(z_1^{d-1}z_2)^r = (z_1^d)^{r_1}(z_2^d)^{r_2}$ with $r = r_1 + r_2 < d$ which is clearly impossible. □

What is the regularity of J^k for $k \geq d$? There is some computational evidence that the first guess, i.e., $\text{reg}(J^k) = k(d+1) + d - 1$ for $k \geq d$, might be correct.

The ideas and the strategy of the previous construction can be used, in principle, to create other kinds of "bad" behaviors. We give some hints and examples but no detailed proofs.

Hint 2.6 Given $d > 1$ consider the ideal

$$H = (x_1 z_1^d, x_1 z_2^d, x_2 z_1^{d-1} z_2) + z_1 z_2 (z_1, z_2)^{d-1}.$$

We conjecture that $\operatorname{reg}(H^k) = k(d+1)$ for all $k < d$ and $\operatorname{reg}(H^d) \geq d(d+1) + d - 1$. Note that H has two generators less than the ideal J of 2.1. In the case $d = 2$, H is exactly the ideal of 1.5.

One can ask whether there are radical ideals with a behavior as the ideal in 2.1. One would need a square-free version of the construction above. This suggests the following:

Hint 2.7 For every d consider variables $z_1, z_2, \dots, z_{2d}$ and x_1, x_2 and the ideal

$$J = \left(\begin{array}{c} x_1 z_1 z_2, x_1 z_3 z_4, \dots, x_1 z_{2d-1} z_{2d}, \\ x_2 z_2 z_3, x_2 z_4 z_5, \dots, x_2 z_{2d} z_1 \end{array} \right) + \operatorname{Sq}^3(z)$$

where $\operatorname{Sq}^3(z)$ denotes the square-free cube of $(z_1, \dots, z_{2d})$, i.e., the ideal generated by the square-free monomials of degree 3 in the z's. We conjecture that $\operatorname{reg}(J^k) = 3k$ for $k < d$ and $\operatorname{reg}(J^d) > 3d$. Note that for $d = 2$ one obtains Sturmfels' Example 1.6.

We do not know how to construct prime ideals with a behavior as the ideal in 2.1. If one wants two (or more) jumps one can try with:

Hint 2.8 Let $1 < a < b$ be integers. Define the ideal

$$I = (y_2 z_1^b, y_2 z_2^b, x z_1^{b-1} z_2) + z_1^{b-a}(y_1 z_1^a, y_1 z_2^a, x z_1^{a-1} z_2) + z_1 z_2 (z_1, z_2)^{b-1}$$

of the polynomial ring $K[x, y_1, y_2, z_1, z_2]$. We expect that $\operatorname{reg}(I) = b + 1$ and $\operatorname{reg}(I^k) - \operatorname{reg}(I^{k-1}) > (b+1)$ if $k = a$ or $k = b$.

Acknowledgments

The author wishes to thank the organizers of the Lisbon Conference on Commutative Algebra (Lisbon, June 2003) for the kind invitation and for their warm hospitality. Some parts of this research project were carried out while the author was visiting MSRI (Berkeley) within the frame of the Special Program 2002/03 on Commutative Algebra. The results and examples presented in this the paper have been inspired and suggested by computations performed by the computer algebra system CoCoA [**CNR**].

References

[BEL] A. Bertram, L. Ein and R. Lazarsfeld, Vanishing theorems, a theorem of Severi, and the equations defining projective varieties, *J. Amer. Math. Soc.* **4** (1991), 587–602.

[CNR] A. Capani, G. Niesi and L. Robbiano, *CoCoA, a System for Doing Computations in Commutative Algebra,* Available via ftp from `cocoa.dima.unige.it`.

[Ca] G. Caviglia, Bounds on the Castelnuovo-Mumford regularity of tensor products, preprint 2003, to appear in *Proc. Amer. Math. Soc.*

[C] K. Chandler, Regularity of the powers of an ideal, *Comm. Algebra* **25** (1997), 3773–3776.

[CH] A. Conca and J. Herzog, Castelnuovo-Mumford regularity of products of ideals, *Collect. Math.* **54** (2003), 137–152.

[CHT] D. Cutkosky, J. Herzog and N.V. Trung, Asymptotic behavior of the Castelnuovo-Mumford regularity, *Compositio Math.* **118** (1999), 243–261.

[EHU] D. Eisenbud, C. Huneke and B. Ulrich, Regularity of Tor and graded Betti numbers, preprint 2004, math.AC/0405373.

[GGP] A. Geramita, A. Gimigliano and Y. Pitteloud, Graded Betti numbers of some embedded rational n-folds, *Math. Ann.* **301** (1995), 363–380.

[HHZ] J. Herzog, T. Hibi and X. Zheng, Monomial ideals whose powers have a linear resolution, preprint 2003, math.AC/0307222.

[K] V. Kodiyalam, Asymptotic behavior of Castelnuovo-Mumford regularity, *Proc. Amer. Math. Soc.* **128** (2000), 407–411.

[R] T. Römer, On minimal graded free resolutions, Thesis, University Essen, 2001.

[Si] J. Sidman, On the Castelnuovo-Mumford regularity of products of ideal sheaves, *Adv. Geom.* **2** (2002), 219–229.

[St] B. Sturmfels, Four counterexamples in combinatorial algebraic geometry, *J. Algebra* **230** (2000), 282–294.

Integral closure of ideals and annihilators of homology

ALBERTO CORSO
Department of Mathematics, University of Kentucky, Lexington
KY 40506, USA
E-mail: corso@ms.uky.edu

CRAIG HUNEKE
Department of Mathematics, University of Kansas, Lawrence, KS 66045, USA
E-mail: huneke@math.ku.edu

DANIEL KATZ
Department of Mathematics, University of Kansas, Lawrence, KS 66045, USA
E-mail: dlk@math.ku.edu

WOLMER V. VASCONCELOS
Department of Mathematics, Rutgers University, Piscataway, NJ 08854, USA
E-mail: vasconce@math.rutgers.edu

1 Introduction

Let $(R, \mathfrak{m})$ be a local Noetherian ring. Given an R-ideal I of grade g, a closely related object to I is its *integral closure* $\overline{I}$. This is the set (ideal, to be precise) of all elements in R that satisfy an equation of the form

$$X^m + b_1X^{m-1} + b_2X^{m-2} + \cdots + b_{m-1}X + b_m = 0,$$

with $b_j \in I^j$ and m a non-negative integer. Clearly one has that $I \subseteq \overline{I} \subseteq \sqrt{I}$, where $\sqrt{I}$ is the *radical* of I and consists instead of the elements of R that satisfy an equation of the form $X^q - b = 0$ for some $b \in I$ and q a non-negative integer. While [**EHV**] already provides direct methods for the computation of $\sqrt{I}$, the nature of $\overline{I}$ is complex. Even the issue of validating the equality $I = \overline{I}$ is quite hard and relatively few methods are known [**CHV**]. In general, computing the integral closure of an ideal is a fundamental problem in commutative algebra. Although it is theoretically possible to compute integral closures, practical computations at present remain largely out-of-reach, except for some special ideals, such as monomial ideals in polynomial rings over a field. One known computational approach is through the theory of Rees algebras: It requires the computation of the integral closure of the Rees algebra $\mathcal{R}$ of I in $R[t]$. However, this method is potentially wasteful since the integral closures of all the powers of I are being computed at the same time. On the other hand, this method has the advantage that for the integral closure $\overline{A}$ of an affine algebra A there are readily available *conductors*: given A in terms of generators and relations (at least in characteristic zero) the Jacobian ideal Jac of A has the property that $\mathrm{Jac} \cdot \overline{A} \subseteq A$, in other words, $\overline{A} \subseteq A : \mathrm{Jac}$. This fact is the cornerstone of most current algorithms to build $\overline{A}$ [**deJ**, **V**].

On a seemingly unrelated level, let $H_i = H_i(I)$ denote the homology modules of the *Koszul complex* $\mathbb{K}_*$ built on a minimal generating set $a_1, \ldots, a_n$ of I. It is well known that all the non-zero Koszul homology modules H_i are annihilated by I, but in general their annihilators tend to be larger. To be precise, this article outgrew from an effort to understand our basic question:

> *Are the annihilators of the non-zero Koszul homology modules H_i of an unmixed ideal I of a local Cohen-Macaulay ring R contained in the integral closure $\overline{I}$ of I?*

We are particularly interested in the two most meaningful Koszul homology modules, namely H_1 and H_{n-g} — the last non-vanishing Koszul homology module. Of course the case that matters most in dealing with the annihilator of the latter module is when R is Cohen-Macaulay but not Gorenstein. Observe that the Cohen-Macaulay assumption on the ring R cannot be weakened. For instance, let R be the localized polynomial ring in the variables X and Y modulo the ideal (X^2, XY). Let us denote with x and y the images of X and Y, respectively. The principal ideal $I = (y^n)$, where $n \geq 2$, is generated by a parameter. In addition, we have that $0\colon (y^n) = (x)$ and $0\colon (x) = (x, y)$, but y is not integral over (y^n). We also stress the necessity of the unmixedness requirement on I in our question. Indeed, let $R = k[\![X, Y, Z, W]\!]$ with k a field characteristic zero. The ideal

$I=(X^2-XY,-XY+Y^2,Z^2-ZW,-ZW+W^2)$ is a height two mixed ideal with $\mathrm{Ann}(H_1)=\overline{I}=(I,XZ-YZ-XW+YW)$ and $\mathrm{Ann}(H_2)=\sqrt{I}=(X-Y,Z-W)$. It is interesting to note that this ideal has played a role in [**CHV**], where it was shown that the integral closure of a binomial ideal is not necessarily binomial, unlike the case of its radical as shown by Eisenbud and Sturmfels [**ES**]. A first approach to our question would be to decide if the annihilators of the Koszul homology modules are rigid in the sense that the annihilator of H_i is contained in the annihilator of H_{i+1}. Up to radical this is true by the well-known rigidity of the Koszul complex. If true, we could concentrate our attention on the last non-vanishing Koszul homology. Unfortunately, this rigidity is not true. Indeed, let $R=k[[X,Y,Z,W]]$ with k a field characteristic zero and consider the ideal $I=(X^3,Y^4,Z^4,W^6,XYZ-YZW,XYW-Z^2W)$. Clearly, $\mathrm{ann}(H_2)=I$ whereas an easy computation with the algebra system `Macaulay` shows that $\mathrm{ann}(H_1)=I+(Y^2ZW^2,Y^3Z^3,XZ^2W^4,Z^2W^5,X^2ZW^5,Y^3W^5)$. That is, $\mathrm{ann}(H_1)\not\subset \mathrm{ann}(H_2)$. Moreover, one has that $(\mathrm{ann}(H_1))^2=I\,\mathrm{ann}(H_1)$, which yields that $\mathrm{ann}(H_1)$ is contained in $\overline{I}$. It would be good to have an example where such behavior occurs for the Koszul homology of an ideal in a ring which is not Gorenstein. A related example was given by Aberbach: let $R=k[X,Y,Z]/(X,Y,Z)^{n+1}$ and let E be the injective hull of the residue field of R. Then Z is in the annihilator of $H_1(X,Y;E)$, but Z^n does not annihilate $H_2(X,Y;E)$.

An obvious question is: What happens when I is integrally closed? In Section 2 we provide some validation for our guiding question. Let I be an $\mathfrak{m}$-primary ideal that is not a complete intersection. In Corollary 2.4 we show that if $c\in I:\mathfrak{m}$ and $cH_1=0$, then $c\in\overline{I}$. In particular, if I is an integrally closed ideal then $\mathrm{ann}(H_1)=I$. We then proceed to study $\mathrm{ann}(H_1)$ for several classes of ideals with good structure: these include syzygetic ideals, height two perfect Cohen-Macaulay ideals, and height three perfect Gorenstein ideals. While in the case of height two perfect Cohen-Macaulay ideals the Koszul homology modules are faithful (see Proposition 2.10), in the case of syzygetic ideals we observe that $\mathrm{ann}(H_1)$ can be interpreted as $I:I_1(\varphi)$, where $I_1(\varphi)$ is the ideal generated by the entries of any matrix φ minimally presenting the ideal I (see Proposition 2.6). In the case of height three perfect Gorenstein ideals we show the weaker statement that $(\mathrm{ann}(H_1))^2\subset\overline{I}$ (see Theorem 2.12).

Section 3 contains variations on a result of Burch, which continues the theme of this paper in that they deal with annihilators of homology and integrally closed ideals. The result of Burch that we have in mind [**B**] asserts that if $\mathrm{Tor}_t^R(R/I,M)$, where M is a finitely generated R-module, vanishes for two consecutive values of t less than or equal to the projective dimension of M, then $\mathfrak{m}(I:\mathfrak{m})=\mathfrak{m}I$. This has the intriguing consequence that if I is an integrally closed ideal with finite projective dimension, then $R_{\mathfrak{p}}$ is a regular local ring for all $\mathfrak{p} \in \mathrm{Ass}(R/I)$. In particular, a local ring is regular if and only if it has an $\mathfrak{m}$-primary integrally closed ideal of finite projective dimension. A variation of Burch's theorem is given in Theorem 3.1. We then deduce a number of corollaries. For instance, we show in Corollary 3.3 that integrally closed $\mathfrak{m}$-primary ideals I can be used to test for finite projective dimension, in the sense that if $\mathrm{Tor}_i^R(M, R/I) = 0$, then the projective dimension of M is at most $i-1$. This improves Burch's result in that we do not need to assume that two consecutive Tors vanish. Recent work of Goto and Hayasaka ([**GH1**] and [**GH2**]) has many more results concerning integrally closed ideals of finite projective dimension.

The annihilator of the conormal module I/I^2 is a natural source of elements in the integral closure of I. In Section 4 we study a class of Cohen-Macaulay ideals whose conormal module is faithful. We close with a last section giving an equivalent formulation of our main question, and also include another question which came up in the course of this study.

2 Annihilators of Koszul homology

We start with some easy remarks, that are definitely not sharp exactly because of their generality. It follows from localization that $\mathrm{ann}(H_1) \subset \sqrt{I}$. Moreover, for any R-ideal I minimally presented by a matrix φ we also show that $\mathrm{ann}(H_1) \subset I : I_1(\varphi)$, where $I_1(\varphi)$ is the ideal generated by the entries of φ. Things get sharper when one focuses on the annihilator of the first Koszul homology modules of classes of ideals with good structural properties. We conclude the section with a result of Ulrich about the annihilator of the last non-vanishing Koszul homology module.

2.1 The first Koszul homology module

Our first theorem is a general result about annihilators of Koszul homology. It follows from this theorem that our basic question (even in a general Noetherian setting) has a positive answer for the first Koszul homology module in the case that I is an integrally closed $\mathfrak{m}$-primary ideal. For the reader's sake, throughout the paper, we will tacitly assume that the Koszul homology modules under consideration are non-zero.

Theorem 2.1 *Let $(R, \mathfrak{m})$ be a local Noetherian ring and let I be an $\mathfrak{m}$-primary ideal. If $c \in R$ is an element such that $cH_i(I) = 0$ then one of the following conditions hold:*

(*a*) $I : c = \mathfrak{m}I : c$.

(*b*) *There exists* $J \subseteq I$ *and* $x \in R$ *such that* $I = J + (cx)$, $\mu(I) = \mu(J) + 1$ *and* $c\,H_i(J) = c\,H_{i-1}(J) = 0$.

We will need a lemma before proving Theorem 2.1.

Lemma 2.2 *Let* $J \subseteq R$ *be an ideal and* $c, x \in R$. *Assume that* (J, cx) *is primary to the maximal ideal. Then* $\lambda(H_i(J,c)) = \lambda(\operatorname{ann}_c H_i(J,cx))$.

Proof. We use induction on i. Suppose $i = 0$. The desired equality of lengths follows immediately from the exact sequence

$$0 \to ((J,cx):c)/(J,cx) \longrightarrow R/(J,cx) \xrightarrow{\cdot c} R/(J,cx) \longrightarrow R/(J,c) \to 0.$$

Suppose $i > 0$ and the lemma holds for $i-1$. We have an exact sequence

$$0 \to H_i(J,cx)/cH_i(J,cx) \longrightarrow H_i(J,cx,c) \longrightarrow \operatorname{ann}_c(H_{i-1}(J,cx)) \to 0.$$

But $H_i(J,cx,c) = H_i(J,c) \oplus H_{i-1}(J,c)$, so

$$\lambda(H_i(J,c)) + \lambda(H_{i-1}(J,c)) = \lambda(\operatorname{ann}_c(H_{i-1}(J,c)) + \lambda(H_i(J,cx)/cH_i(J,cx)).$$

Using the induction hypothesis, we obtain $\lambda(H_i(J,c)) = \lambda(H_i(J,cx)/cH_i(J,cx)) = \lambda(\operatorname{ann}_c H_i(J,cx))$. □

Proof of Theorem 2.1. Suppose (a) does not hold. Then there exists $x \in \mathfrak{m}$ such that cx is a minimal generator of I. We can write $I = J + (cx)$ for an ideal $J \subseteq I$ satisfying $\mu(I) = \mu(J) + 1$. On the one hand, from the exact sequences

$$0 \to H_i(J)/cH_i(J) \longrightarrow H_i(J,c) \longrightarrow \operatorname{ann}_c H_{i-1}(J) \to 0$$

and

$$0 \to H_i(J)/cxH_i(J) \longrightarrow H_i(J,cx) \longrightarrow \operatorname{ann}_{cx} H_{i-1}(J) \to 0$$

we get

$$\lambda(H_i(J,c)) = \lambda(H_i(J)/cH_i(J)) + \lambda(\operatorname{ann}_c H_{i-1}(J))$$

and

$$\lambda(H_i(J,cx)) = \lambda(H_i(J)/cxH_i(J)) + \lambda(\operatorname{ann}_{cx} H_{i-1}(J)).$$

On the other hand,

$$\lambda(H_i(J)/cxH_i(J)) \geq \lambda(H_i(J)/cH_i(J)) \quad \text{and} \quad \lambda(\operatorname{ann}_{cx} H_{i-1}(J)) \geq \lambda(\operatorname{ann}_c H_i(J)).$$

Since $cH_i(J,cx) = 0$, $H_i(J,cx) = \mathrm{ann}_c H_i(J,cx)$, so $\lambda(H_i(J,cx)) = \lambda(H_i(J,c))$, by Lemma 2.2. It follows from this that $\lambda(H_i(J)/cH_i(J)) = \lambda(H_i(J)/cxH_i(J))$. Thus, $cH_i(J) = cxH_i(J)$, so $cH_i(J) = 0$, by Nakayama's lemma. Similarly, since

$$\lambda(\mathrm{ann}_c H_{i-1}(J)) = \lambda(\mathrm{ann}_{cx} H_{i-1}(J)),$$

it follows that $\lambda(H_{i-1}(J)/cH_i(J)) = \lambda(H_{i-1}(J)/cxH_i(J))$, so $cH_{i-1}(J) = 0$, as before. □

Corollary 2.3 *Let $(R,\mathfrak{m})$ be a local Noetherian ring and let I be an $\mathfrak{m}$-primary ideal. If $c \cdot H_1(I) = 0$, then $I : c = \mathfrak{m}I : c$.*

Proof. If $I : c$ properly contains $mI : c$, then by Theorem 2.1, there exists $J \subseteq I$ and $x \in \mathfrak{m}$ such that $I = J + (cx)$, $\mu(I) = \mu(J) + 1$ and $c \cdot H_0(J) = 0$. But then, $c \in J$, so $I = J$, a contradiction. □

Corollary 2.4 *Let $(R,\mathfrak{m})$ be a local Noetherian ring with* $\operatorname{depth} R > 0$ *and let I be an $\mathfrak{m}$-primary ideal. If $c \in R$ is an element such that $c \cdot H_1(I) = 0$ and $c \in I : \mathfrak{m}$, then $c \in \overline{I}$.*

Proof. Since $\mathfrak{m} \subseteq I : c$, we have $\mathfrak{m}c \subseteq \mathfrak{m}I$, by Corollary 2.3. By the determinant trick, $c \in \overline{I}$. □

Corollary 2.5 *Let $(R,\mathfrak{m})$ be a local Noetherian ring and let I be an integrally closed $\mathfrak{m}$-primary ideal. Then* $\mathrm{Ann}(H_1) = I$.

Proof. Suppose $\operatorname{ann} H_1(I)$ properly contains I. Take $c \in (\operatorname{ann} H_1(I) \backslash I) \cap (I : \mathfrak{m})$. By Corollary 2.4, $c \in \overline{I} = I$, a contradiction. Thus, $\operatorname{ann} H_1(I) = I$. □

Syzygetic ideals: Let I be an ideal containing a regular element. It follows from the determinant trick that the annihilator of I^m/I^{m+1} is contained in $\overline{I}$ for all m. Hence, another piece of evidence in support of our question is given by the close relationship between H_1 and the conormal module I/I^2. This is encoded in the exact sequence

$$0 \to \delta(I) \longrightarrow H_1 \longrightarrow (R/I)^n \longrightarrow I/I^2 \to 0,$$

where $\delta(I)$ denotes the kernel of the natural surjection from the second symmetric power $\mathrm{Sym}_2(I)$ of I onto I^2, $\mathrm{Sym}_2(I) \twoheadrightarrow I^2$; see **[SV]**. We will exploit this exact sequence in at least two places: Proposition 2.6 and Theorem 4.1. We recall that the ideal I is said to be *syzygetic* whenever $\delta(I) = 0$.

Proposition 2.6 *Let R be a Noetherian ring. For any R-ideal I minimally presented by a matrix φ,* $\operatorname{ann}(H_1) \subset I \colon I_1(\varphi)$*, where $I_1(\varphi)$ denotes the ideal generated by the entries of φ. If, in addition, I is syzygetic then* $\operatorname{ann}(H_1) = I \colon I_1(\varphi)$.

Proof. Let Z_1 and B_1 denote the modules of cycles and boundaries respectively. If $x \in \operatorname{ann}(H_1)$ one has that for $z \in Z_1$ the condition $xz \in B_1$ means that each coordinate of z is conducted into I by x. Thus $x \in I : I_1(\varphi)$. The reverse containment holds if I is syzygetic. In fact, in this situation one actually has that $H_1 \hookrightarrow (I_1(\varphi)/I)^n$. Thus $I \colon I_1(\varphi) \subset \operatorname{Ann}(H_1)$. □

Corollary 2.7 *Let R be a local Noetherian ring, and let I be an ideal of finite projective dimension n. Then $(\operatorname{ann}(H_1))^{n+1} \subseteq I$.*

Proof. Assume I is minimally presented by a matrix φ. By the above proposition, $\operatorname{ann}(H_1) \subset I \colon I_1(\varphi)$. The result then follows immediately from the following proposition of G. Levin (unpublished). The proof follows from a careful analysis of Gulliksen's Lemma, 1.3.2 in [**GL**]. □

Proposition 2.8 *Let R be a local Noetherian ring and let I be an ideal of finite projective dimension n, minimally presented by a matrix φ. Then*

$$(I \colon I_1(\varphi))^{n+1} \subseteq I.$$

Remark 2.9 In general, the ideal $I : I_1(\varphi)$ may be larger than the integral closure of I. For example the integrally closed R-ideal $I = (X,Y)^2$, where R is the localized polynomial ring $k[X,Y]_{(X,Y)}$, is such that $I \colon I_1(\varphi) = (X,Y)$. However, Levin's proposition shows that $(I \colon I_1(\varphi))^2 \subset I$.

Height two perfect ideals: The first case to tackle is the one of height two perfect ideals in local Cohen-Macaulay rings. However the Cohen-Macaulayness of the H_i's gets into the way. Indeed we have the following fact:

Proposition 2.10 *Let R be a local Cohen-Macaulay ring and let I be a height two perfect R-ideal. Then for all i (with $H_i \neq 0$) one has* $\operatorname{Ann}(H_i) = I$.

Proof. Consider the resolution of the ideal I:

$$0 \to R^{n-1} \longrightarrow R^n \longrightarrow I \to 0.$$

The submodule of 1-cycles of $\mathbb{K}_*$, Z_1, is the submodule R^{n-1} of this resolution. Also, for all i one has $Z_i = \bigwedge^i Z_1$. All these facts can be traced to [**AH**]. This implies that for any $i \leq n-2$, $H_1^i = H_i$ — this multiplication is in $H_*(\mathbb{K})$. Thus the annihilator of H_1 will also annihilate, say, H_{n-2}. But this is the canonical module of R/I, and its annihilator is I. The conclusion now easily follows. □

Gorenstein ideals: Let us consider a perfect $\mathfrak{m}$-primary Gorenstein ideal in a local Noetherian ring R. In this situation, if I is Gorenstein but not a complete intersection then $\mathrm{Ann}(H_1) \neq I$. Otherwise, R/I would be a submodule of H_1. By a theorem of Gulliksen [**GL**], if H_1 has a free summand then it must be a complete intersection. Actually, using Gulliksen's theorem one shows that if I is $\mathfrak{m}$-primary Gorenstein but not a complete intersection, then the socle annihilates H_1. Combining Proposition 2.6 and the work of [**CHV**] yields the following result:

Proposition 2.11 *Let $(R,\mathfrak{m})$ be a local Noetherian ring with embedding dimension at least 2 and let I be an $\mathfrak{m}$-primary ideal contained in $\mathfrak{m}^2$ with R/I Gorenstein. Suppose further that I is minimally presented by a matrix φ and that $I_1(\varphi) = \mathfrak{m}$, where $I_1(\varphi)$ denotes the ideal generated by the entries of φ. Then $\mathrm{ann}(H_1) \subset \overline{I}$.*

Proof. By Proposition 2.6 and our assumption we have that $\mathrm{ann}(H_1) \subset I\colon I_1(\varphi) = I\colon \mathfrak{m}$. Our assertion now follows from Lemma 3.6 in [**CHV**] since $(I\colon \mathfrak{m})^2 = I(I\colon \mathfrak{m})$. □

For a height three perfect Gorenstein ideal I which is not integrally closed (as, for example, in the $\mathfrak{m}$-primary case) we have evidence that $(\mathrm{ann}(H_1))^2 = I\cdot \mathrm{ann}(H_1)$. If this were to hold in general, it would imply that $I \subsetneq \mathrm{ann}(H_1) \subset \overline{I}$. Thus far, we can prove the weaker result that the square of the annihilator of H_1 is in the integral closure of I.

Theorem 2.12 *Let R be a local Noetherian ring with $\mathrm{char}(R) \neq 2$ and let I be a height three perfect Gorenstein ideal minimally generated by $n \geq 5$ elements. Then*

$$(\mathrm{ann}(H_1))^2 \subset \overline{I}.$$

Proof. Let $a_1, \ldots a_n$ denote a set of minimal generators of I. Notice that B_1 and Z_1 are submodules of R^n of rank $n-1$; in general, if E is a submodule of R^n of rank r, we denote by $\det(E)$ the ideal generated by the $r \times r$ minors of the matrix with any set of generators of E (as elements of R^n).

Let $c \in R$ be such that $cZ_1 \subset B_1$. It suffices to prove that $c^2 \in \overline{I}$ since the square of an ideal is always integral over the ideal generated by the squares of its generators. Note that $c^{n-1}\det(Z_1) \subset \det(B_1)$. Let V be a valuation overring of R with valuation v; the ideal IV is now principal and generated by one of the original generators, say $a_1 = a$. By the structure theorem of Buchsbaum and Eisenbud [**BE**], we may assume that a is one of the maximal Pfaffians of the matrix presenting I. Since I is generated by a, B_1V is generated by the Koszul syzygies $(a_2,-a,0,\ldots,0),(a_3,0,-a,\ldots,0),\ldots,(a_n,0,0,\ldots,-a)$. Hence $\det(B_1V) = (a^{n-1}) = I^{n-1}V$. As for Z_1V, one has that $\det(Z_1V)$ includes the determinant of the minor defining a^2 (a is the Pfaffian of the submatrix). Thus $c^{n-1}I^2V \subset I^{n-1}V$, which yields that $c^{n-1} \in I^{n-3}V$, as cancellation holds. Hence, we have that $(n-1)v(c) = v(c^{n-1}) \geq v(I^{n-3}V) = (n-3)v(IV)$. Finally, this yields

$$v(c^2) \geq 2\frac{n-3}{n-1}v(IV) \geq v(IV)$$

and, in conclusion, $c^2 \in \overline{I}$. □

Remark 2.13 It is worth remarking that the above proof shows much more. Recall that $\overline{I^{\frac{a}{b}}}$ denotes the integral closure of the ideal generated by all $x \in R$ such that $x^b \in I^a$. From [**BE**], we know that $n = 2k+1$ must be odd. Our proof shows that

$$(\operatorname{ann}(H_1)) \subset \overline{I^{\frac{k-1}{k}}}.$$

As k gets large this is very close to our main objective, proving that $(\operatorname{ann}(H_1)) \subset \overline{I}$.

2.2 Last non-vanishing Koszul homology module

Let us turn our attention towards the tail of the Koszul complex.

Proposition 2.14 *Let R be a one-dimensional Noetherian integral domain with finite integral closure. Then any integrally closed ideal is reflexive. In particular, for any ideal I its bidual $(I^{-1})^{-1}$ is contained in its integral closure $\overline{I}$.*

Proof. We may assume that R is a local ring, of integral closure B. An ideal L is integrally closed if $L = R \cap LB$. Since B is a principal ideals domain, $LB = xB$ for some x. We claim that xB is reflexive. Let $C = B^{-1} = \operatorname{Hom}_R(B,R)$ be the conductor of B/R. C is also an ideal of B, $C = yB$, and therefore $C^{-1} = y^{-1}B^{-1} = y^{-1}C$, which shows that $C^{-1} = B$. This shows that $(L^{-1})^{-1} \subset (R^{-1})^{-1} \cap ((xB)^{-1})^{-1} =$

L. The last assertion follows immediately by setting $I \subset L = \bar{I}$. □

We can interpret the above result as an annihilation of Koszul cohomology. Let $I = (a_1, \ldots, a_m)$ and let $\mathbb{K}^*$ denote the Koszul complex

$$0 \to R \longrightarrow R^m \longrightarrow \bigwedge^2 R^m \longrightarrow \cdots \longrightarrow \bigwedge^m R^m \to 0,$$

with differential $\partial(w) = z \wedge w$, where $z = a_1 e_1 + \cdots + a_n e_m$. One sees that $Z^1 = I^{-1}z$, and $B^1 = Rz$. Thus $(I^{-1})^{-1}$ is the annihilator of H^1. On the other hand $H^1 \cong H_{m-1} \cong \operatorname{Ext}^1_R(R/I, R)$. Let us raise a related issue: $(I^{-1})^{-1}$ is just the annihilator of $\operatorname{Ext}^1_R(R/I, R)$, so one might want to consider the following question, which is obviously relevant only if the ring R is not Gorenstein. Let R be a Cohen-Macaulay geometric integral domain and let I be a height unmixed ideal of codimension g. Is $\operatorname{ann}(\operatorname{Ext}^g_R(R/I, R)) = \operatorname{ann}(H_{n-g})$ always contained in $\bar{I}$? Notice that the annihilator of the last non-vanishing Koszul homology can be identified with $J : (J : I)$ for J an ideal generated by a maximal regular sequence inside I. This follows since the last non-vanishing Koszul homology is isomorphic to $(J : I)/J$.

We thank Bernd Ulrich for allowing us to reproduce the following result [**U**], which grew out of conversations at MSRI (Berkeley):

Theorem 2.15 (Ulrich) *Let $(R, \mathfrak{m})$ be a Cohen-Macaulay local ring, let I be an equimultiple ideal and let $J \subset I$ be a complete intersection with* $\operatorname{ht} J = \operatorname{ht} I$. *Then it follows that $J : (J : I) \subset \bar{I}$. In particular the annihilator of the last non-vanishing Koszul homology of I is contained in the integral closure of I.*

Proof. We may assume that I is $\mathfrak{m}$-primary since for every associated prime $\mathfrak{p}$ of $\bar{I}$, $\operatorname{ht}\mathfrak{p} = \operatorname{ht} I$. We may also assume that R has a canonical module ω. We first prove:

Lemma 2.16 *Let R be an Artinian local ring with canonical module ω and let $I \subset R$ be an ideal. Then $0 :_\omega (0 :_R I) = I\omega$.*

Proof. Note that $0 :_\omega (0 :_R I) = \omega_{R/0:I}$. To show $I\omega = \omega_{R/0:I}$ note that $I\omega$ is a module over $R/0 \colon I$ and that the socle of $I\omega$ is at most one-dimensional as it is contained in the socle of ω. Hence we only need to show that $I\omega$ is faithful over $R/0 : I$. Let $x \in \operatorname{ann}_R I\omega$, then $xI\omega = 0$, hence $xI = 0$, hence $x \in 0 : I$. □

Returning to the proof of Theorem 2.15, it suffices to show that $(J:(J:I))\omega \subset I\omega$. But $(J:(J:I))\omega \subset J\omega :_\omega (J:_R I)$. So it suffices to show $J\omega :_\omega (J:_R I) \subset I\omega$. Replacing R,ω by R/J, $\omega_{R/J} = \omega/J\omega$ we have to show $0:_\omega (0:_R I) \subset I\omega$, which holds by Lemma 2.16. □

3 Variations on a theorem of Burch

Theorem 3.1 below is a variation of Burch's theorem mentioned in the introduction, and strengthens it in the case I is integrally closed. We then deduce a number of corollaries.

Theorem 3.1 *Let $(R,\mathfrak{m})$ be a local Noetherian ring, I an integrally closed R-ideal having height greater than zero and M a finitely generated R-module. For $t \geq 1$, set $J_t := \operatorname{ann}(\operatorname{Tor}_t(R/I,M))$. Let $(\mathbb{F}_*,\varphi_i)$ be a minimal free resolution of M. If* $\operatorname{image}(\varphi_t)$ *is contained in $\overline{\mathfrak{m}J_t}F_{t-1}$, then*

$$\operatorname{image}(\varphi_t \otimes_R 1_{R/I}) \cap \operatorname{socle}(F_{t-1}/IF_{t-1}) = 0.$$

Proof. Take $u \in F_{t-1}$ such that its residue class modulo I belongs to

$$\operatorname{image}(\varphi_t \otimes_R 1_{R/I}) \cap \operatorname{socle}(F_{t-1}/IF_{t-1}).$$

Then $u = \varphi_t(v) + w$, for $v \in F_t$ and $w \in IF_{t-1}$. For all $x \in \mathfrak{m}$, $\varphi_t(xv) \equiv 0$ modulo IF_{t-1}. Thus for all $j \in J_t$, there exists $z \in F_{t+1}$ such that $\varphi_{t+1}(z) \equiv jxv$ modulo IF_t. It follows that we can write $jxv = \varphi_{t+1}(z) + w_0$, for $w_0 \in IF_t$. Therefore, $jxu = \varphi_t(jxv) + jxw = \varphi_t(w_0) + jxw$. By hypothesis, we get $jxu \in \overline{mJ_t}IF_t$, for all $j \in J_t$ and all $x \in \mathfrak{m}$. Therefore, by cancellation, $u \in I_aF_{t-1}$. But since I is integrally closed, $u \in IF_{t-1}$, which gives what we want. □

In the following corollaries, we retain the notation from Theorem 3.1.

Corollary 3.2 *Suppose I is integrally closed and $\mathfrak{m}$-primary. If* $\operatorname{image}(\varphi_t)$ *is contained in $\overline{\mathfrak{m}J_t}F_{t-1}$, then:*

(*a*) $\operatorname{image}(\varphi_t) \subseteq IF_{t-1}$.

(*b*) $J_tF_t \subseteq \operatorname{image}(\varphi_{t+1})$.

Proof. For (a), if $\operatorname{image}(\varphi_t \otimes 1_{R/I})$ were not zero, then it would contain a nonzero socle element, since I is $\mathfrak{m}$-primary. This contradicts Theorem 3.1, so

(a) holds. For (b), it follows from (a) that $\mathrm{Tor}_t(R/I, M) = F_t/(\mathrm{image}(\varphi_{t+1}) + IF_t)$, so J_tF_t is contained in $\mathrm{image}(\varphi_{t+1}) + IF_t$, and (b) follows via Nakayama's Lemma. □

The next corollary shows that integrally closed $\mathfrak{m}$-primary ideals can be used to test for finite projective dimension.

Corollary 3.3 *Suppose that I is integrally closed and $\mathfrak{m}$-primary. Then M has projective dimension less than t if and only if* $\mathrm{Tor}_t(R/I, M) = 0$.

Proof. The hypothesis implies that $J_t = R$. Therefore, $\mathrm{image}(\varphi_t)$ is automatically contained in $\overline{\mathfrak{m}J_t}F_{t-1}$. By part (b) of Corollary 3.2, $F_t \subseteq \mathrm{image}(\varphi_{t+1})$, so $F_t = 0$, by Nakayama's Lemma. □

Corollary 3.4 *Let $J \subseteq R$ be an ideal and I an integrally closed $\mathfrak{m}$-primary ideal. If $J \subseteq \overline{\mathfrak{m}(IJ : I \cap J)}$, then $J \subseteq I$.*

Proof. Apply Corollary 3.2 with $M = R/J$ and $t = 1$. □

Corollary 3.5 *Suppose that R is reduced and M has infinite projective dimension over R. Then for all $t \geq 1$, the entries of φ_t do not belong to $\overline{\mathfrak{m} \cdot \mathrm{ann}(M)}$. In particular, each map in the minimal resolution of k has an entry not belonging to $\overline{\mathfrak{m}^2}$.*

Proof. Let I be any $\mathfrak{m}$-primary integrally closed ideal. If the entries of φ_t belong to $\overline{\mathfrak{m} \cdot \mathrm{ann}(M)}$, then $\mathrm{image}(\varphi_t)$ is contained in $\overline{\mathfrak{m}J_t}F_{t-1}$. By Corollary 3.2, $\mathrm{image}(\varphi_t)$ is contained in IF_{t-1}. But the intersection of the integrally closed $\mathfrak{m}$-primary ideals is zero, therefore, $\mathrm{image}(\varphi_t) = 0$, contrary to the hypothesis on M. Thus, the conclusion of the corollary holds.

The last statement follows in the case where R is regular from the fact that the Koszul complex on a minimal set of generators of the maximal ideal gives a resolution of k. If R is not regular, then k has infinite projective dimension, and the result follows at once from the first statement. □

In regard to the above corollary, it is well-known that the Koszul complex of a minimal set of generators of the maximal ideal is part of a minimal resolution of k in all cases, so for the maps occuring in the minimal resolution up to the dimension of the ring, the last statement is clear. The new content of the last statement is for the maps past the dimension of the ring.

Corollary 3.6 *Suppose I is integrally closed and $\mathfrak{m} \in \mathrm{Ass}(R/I)$. If* $\mathrm{image}(\varphi_t)$ *is contained in $\overline{\mathfrak{m}J_t}F_{t-1}$, e.g., $\mathrm{Tor}_t(R/I,M)=0$, then either M has projective dimension less than $t-1$ or $\mathfrak{m} \in \mathrm{Ass}(\mathrm{Tor}_{t-1}(R/I,M))$.*

Proof. Suppose M has projective dimension greater than or equal to $t-1$. Then $F_{t-1} \neq 0$. By hypothesis, the socle of F_{t-1}/IF_{t-1} is non-zero, so a non-zero element u in this socle goes to zero under $\varphi_{t-1} \otimes 1_{R/I}$. But the theorem implies that the image of u in $\mathrm{Tor}_{t-1}(R/I,M)$ remains non-zero, so the result holds. □

4 The conormal module

We end the article with a result in the spirit of our investigation. More precisely we show that the conormal module I/I^2 is faithful for a special class of Cohen-Macaulay ideals.

Theorem 4.1 *Let $(R,\mathfrak{m})$ be a Gorenstein local ring and I a Cohen-Macaulay almost complete intersection. Let φ be a matrix minimally presenting I. If $I_1(\varphi)$ is a complete intersection, then I/I^2 is a faithful R/I-module.*

Proof. Let g denote the height of I, and write $n=g+1$ for the minimal number of generators of $I=(a_1,\ldots,a_n)$. We may assume that the ideals generated by any g of the a_i's are complete intersection ideals. Let e_i, with $1 \leq i \leq n$, denote the n-tuple $(0,\ldots,0,1,0,\ldots,0)$ where 1 is in the i-th position. Finally, note that H_1 is the canonical module of R/I.

Let us consider the exact sequence

$$0 \to \delta(I) \longrightarrow H_1 \xrightarrow{\theta} (R/I)^n \xrightarrow{\pi} I/I^2 \to 0,$$

where $\delta(I)$ is the kernel of the natural surjection $\mathrm{Sym}_2(I) \twoheadrightarrow I^2$; see [**SV**]. Notice that for any $\varepsilon' = \sum r'_j e_j + B_1 \in H_1$, where $\sum r'_j a_j = 0$, one has $\theta(\varepsilon') = (r'_1+I)e_1 + \ldots + (r'_n+I)e_n$ while for any element in $(R/I)^n$ one has $\pi((r_1+I)e_1+\ldots+(r_n+I)e_n) = r_1a_1+\ldots+r_na_n+I^2$. Apply $(\_)^\vee = \mathrm{Hom}_{R/I}(\_,H_1)$ to the above exact sequence. We obtain

$$0 \to (I/I^2)^\vee \xrightarrow{\pi^\vee} \mathrm{Hom}((R/I)^n,H_1) \xrightarrow{\theta^\vee} \mathrm{Hom}(H_1,H_1) = R/I \longrightarrow \delta(I)^\vee \to 0.$$

To conclude it will be enough to show that $(I/I^2)^\vee$ is faithful.

First, we claim that the image of $\theta^\vee$ belongs to $I_1(\varphi)/I$. In fact, any element of $\mathrm{Hom}((R/I)^n,H_1)$ can be written as a combination of elementary homomorphism of the form

$$\xi_i((1+I)e_i) = \varepsilon, \qquad \xi_i((1+I)e_j) = 0, \quad \text{if } i \neq j,$$

with $\varepsilon = \sum r_j e_j + B_1 \in H_1$, where $\sum r_j a_j = 0$. Thus, for any $\varepsilon' \in H_1$ we have

$$(\theta^\vee(\xi_i))(\varepsilon') = \xi_i(\theta(\varepsilon')) = \xi_i(\sum(r'_j + I)e_j) = (r'_i + I)\varepsilon.$$

Observe that $(r'_i + I)\varepsilon = (r_i + I)\varepsilon'$ in H_1. Indeed, $r'_i\varepsilon - r_i\varepsilon' = \sum(r'_i r_j - r_i r'_j)e_j + B_1$. But $\sum(r'_i r_j - r_i r'_j)e_j$ is a syzygy of the complete intersection $(a_1, \ldots, \widehat{a_i}, \ldots, a_n)$ and thus it is a Koszul syzygy of the smaller ideal, hence it is in B_1. In conclusion, $\theta^\vee(\xi_i)$ is nothing but multiplication by $r_i + I \in I_1(\varphi)/I$. Given that ε and i were chosen arbitrarily one has that the image of $\theta^\vee$ is $I_1(\varphi)/I$.

Notice that the number of generators of $I_1(\varphi)$ is strictly smaller than n. So we can say that the image of $\theta^\vee$ is given say by $(\theta^\vee(\xi_2), \ldots, \theta^\vee(\xi_n))$. Write, for some $c_i \in R/I$,

$$\theta^\vee(\xi_1) = \sum_{i \geq 2} c_i \theta^\vee(\xi_i).$$

Hence $\xi_1 - \sum_{i \geq 2} c_i \xi_i \in \text{Ker}(\theta^\vee) = \text{Im}(\pi^\vee)$ so that we can find $\gamma \in (I/I^2)^\vee$ such that

$$\xi_1 - \sum_{i \geq 2} c_i \xi_i = \pi^\vee(\gamma) = \gamma \circ \pi.$$

Restricting these homomorphisms to the first component of $\text{Hom}((R/I)^n, H_1)$ gives a homomorphism from R/I to H_1. Now, something that annihilates γ would also annihilate the restriction, but that restriction is faithful. □

Remark 4.2 From the proof of Theorem 4.1 we also obtain that $\text{Hom}(\delta(I), H_1) = R/I_1(\varphi)$. In addition, if $I_1(\varphi)$ is Cohen-Macaulay of codimension g then by the theorem of Hartshorne-Ogus we have that $\delta(I)$ (which is S_2) is Cohen-Macaulay and therefore $\text{depth}\, I/I^2 \geq d - g - 2$.

Unfortunately, there is not much hope to stretch the proof of Theorem 4.1 as the following example shows.

Example 4.3 Let R be the localized polynomial ring $k[x,y]_{(x,y)}$. The ideal $I = (x^5 - y^5, x^4y, xy^4)$ is such that $I^2 : I = (I, x^3y^3)$. In this case $I_1(\varphi) = (x,y)^2$ so that $\mu(I) = \mu(I_1(\varphi)) = 3$.

5 More Questions

We end by considering some other closely related questions which came up during the course of this investigation. We let I be an $\mathfrak{m}$-primary ideal of the local ring R minimally generated by n elements, and let J_i be the annihilator of the ith Koszul cohomology of I with respect to a minimal generating set of I.

> *Set d equal to the dimension of R. Is $J_1 \cdot J_2 \cdots J_{n-d}$ contained in $\overline{I^{n-d}}$?*

Notice that the Koszul homology of I vanishes for values larger than $n-d$, so that the product above represents all the interesting annihilators of the Koszul homology of I. Furthermore, a postive answer to this question gives a positive answer to our main question. This follows since each J_i contains I. Along any discrete valuation v, this means that $v(I) \geq v(J_i)$ for all i. A positive answer to the question above implies that

$$\sum_{i=1}^{n-d} v(J_i) \geq (n-d)v(I) \geq \sum_{i=1}^{n-d} v(J_i).$$

It would follow that $v(J_i) = v(I)$ for all i, implying that $J_i \subseteq \overline{I}$ for all $1 \leq i \leq n-d$. Conversely, if $J_i \subseteq \overline{I}$ for all $1 \leq i \leq n-d$, then clearly $J_1 \cdot J_2 \cdots J_{n-d}$ is contained in $\overline{I^{n-d}}$, so the above question is equivalent to saying that $J_i \subseteq \overline{I}$ for all $1 \leq i \leq n-d$. This form of the question suggests using homotopies to compare the Koszul complex of a set of generators of I with the free resolution of I. However, we have not been able to use this idea to settle the question.

Another question which arose during our work is the following:

> *Let n be the number of minimal generators of an $\mathfrak{m}$-primary ideal I in a Cohen-Macaulay local ring R with infinite residue field, and let d be the dimension of the ring. For every j, $d \leq j \leq n-1$, choose j general minimal generators of I, and let J_j be the ideal they generate. Let H_{n-j} denote the $(n-j)$th Koszul homology of a minimal set of generators of I. Is*
>
> $$\operatorname{Ann}(H_{n-j}) \subseteq J_j : (J_j : I) \subseteq \overline{I}?$$

We have positive answers to this question for the two extremes: $j = d$ and $j = n-1$, in the latter case assuming I is integrally closed.

Acknowledgments

C. Huneke and W.V. Vasconcelos gratefully acknowledge partial support from the National Science Foundation through the grants DMS-0244405 and DMS-0097093, respectively. The authors would also like to express their gratitude to the referee of the paper for a very careful reading and for the useful suggestions.

References

[AH] L. Avramov and J. Herzog, The Koszul algebra of a codimension 2 embedding, *Math. Z.* **175** (1980), 249–260.

[BE] D. Buchsbaum and D. Eisenbud, Algebra structures for finite free resolutions, and some structure theorems for ideals of codimension 3, *Amer. J. Math.* **99** (1977), 447–485.

[B] L. Burch, On ideals of finite homological dimension in local rings, *Math. Proc. Camb. Phil. Soc.* **74** (1968), 941–948.

[CHV] A. Corso, C. Huneke and W.V. Vasconcelos, On the integral closure of ideals, *Manuscripta Math.* **95** (1998), 331–347.

[deJ] T. de Jong, An algorithm for computing the integral closure, *J. Symbolic Comput.* **26** (1998), 273–277.

[EHV] D. Eisenbud, C. Huneke and W.V. Vasconcelos, Direct methods for primary decomposition, *Invent. Math.* **110** (1992), 207–235.

[ES] D. Eisenbud and B. Sturmfels, Binomial ideals, *Duke Math. J.* **84** (1996), 1–45.

[GH1] S. Goto and F. Hayasaka, Finite homological dimension and primes associated to integrally closed ideals, *Proc. Amer. Math. Soc.* **130** (2002), 3159–3164.

[GH2] S. Goto and F. Hayasaka, Finite homological dimension and primes associated to integrally closed ideals II, *J. Math. Kyoto Univ.* **42** (2002), 631–639.

[GL] T.H. Gulliksen and G. Levin, *Homology of Local Rings*, Queen's Papers in Pure and Applied Math. **200**, Queen's University, Kingston, 1967.

[SV] A. Simis and W.V. Vasconcelos, The syzygies of the conormal module, *Amer. J. Math.* **103** (1981), 203–224.

[U] B. Ulrich, Personal communication.

[V] W.V. Vasconcelos, Computing the integral closure of an affine domain, *Proc. Amer. Math. Soc.* **113** (1991), 633–638.

Poincaré series of surface singularities

STEVEN DALE CUTKOSKY[1]
Department of Mathematics, University of Missouri, Columbia, MO 65211, USA
E-mail: cutkoskys@missouri.edu

1 Introduction

In this article we summarize the results of the papers [**7**] and [**6**]. In Section 2 we discuss the results of [**7**] on surface singularities, and in Section 3 we discuss results on the corresponding problem on varieties, which appear in [**6**].

2 Surface singularities

The material in this section is a survey of the paper [**7**], which is a joint work with Juergen Herzog and Ana Reguera.

Suppose that k is an algebraically closed field of characteristic zero, and (R,m,k) is a normal complete local ring of dimension 2. We restrict to characteristic zero as most of the results in this article are false in positive characteristic

[1] Partially supported by NSF.

(as is shown in [**8**], [**7**] and [**6**]). Suppose that $\pi : X \to \mathrm{spec}(R)$ is a resolution of singularities. We can write the reduced exceptional divisor as

$$\pi^{-1}(m)_{\mathrm{red}} = E_1 + \cdots + E_r$$

where $E_1, \ldots, E_r$ are the irreducible components. Since X is normal, each local ring O_{X,E_i} is a (rank 1) discrete valuation ring with valuation ν_i. To each ν_i and $n \in \mathbf{N}$ there is an associated valuation ideal in R,

$$I_n(\nu_i) = \{f \in R \mid \nu_i(f) \geq n\}.$$

The ideals $I_n(\nu_i)$ are m-primary.

To $\underline{n} = (n_1, \ldots, n_r) \in \mathbf{N}^r$ we associate a divisor with exceptional support,

$$D_{\underline{n}} = n_1E_1 + \cdots + n_rE_r.$$

Then

$$H^0(X, O_X(-D_{\underline{n}})) = \Gamma(X, O_X(-D_{\underline{n}})) = I_{n_1}(\nu_1) \cap \cdots \cap I_{n_r}(\nu_r).$$

Since this is an m-primary ideal in R, we can define

$$h_X(\underline{n}) = \ell(R/\Gamma(X, O_X(-D_{\underline{n}}))).$$

h_X is a sort of Hilbert function, so we could hope that it has a nice form, at least for large $\underline{n}$.

The form of the function $h_X(n\underline{m})$ for fixed $\underline{m}$ and $n \gg 0$ is itself very interesting. Even though the ring

$$\oplus_{n \geq 0} \Gamma(X, O_X(-nD_{\underline{m}}))$$

is in general not a finitely generated R-algebra, the Hilbert function $h_X(n\underline{m})$ is always very nice, as is shown by the following theorem.

Theorem 2.1 (see [**8**, Theorem 9]) *For fixed $\underline{m}$, and $n \gg 0$,*

$$h_X(n\underline{m}) = \ell(R/\Gamma(X, O_X(-nD_{\underline{m}}))) = an^2 + bn + c(n)$$

where $a, b \in \mathbf{Q}$ and $c(n)$ is a periodic function.

The statement of this theorem with the weaker condition that $c(n)$ is bounded (but not necessarily periodic) was first proven by Morales [**12**]. This statement (with boundedness of $c(n)$) is in fact a local form of a theorem for linear systems on projective surfaces of Zariski [**13**]. The statement of periodicity of $c(n)$ was conjectured by Zariski for linear systems on surfaces in [**13**]. Zariski's conjecture is proven in Theorem 2 [**8**] for surfaces, and in the local form of Theorem 2.1.

By the local Riemann-Roch Theorem ([**10, 9**])

$$\begin{aligned} h_X(\underline{n}) &= -\tfrac{1}{2}(K_X \cdot D_{\underline{n}}) + (D_{\underline{n}})^2 + h^1(X, \mathcal{O}_X) - h^1(X, \mathcal{O}_X(-D_{\underline{n}})) \\ &= (\text{Quadratic polynomial in } \underline{n}) \; - h^1(X, \mathcal{O}_X(-D_{\underline{n}})). \end{aligned}$$

So we reduce to computing the (finite) length

$$h^1(X, \mathcal{O}_X(-D_{\underline{n}})) = \ell(H^1(X, \mathcal{O}_X(-D_{\underline{n}})).$$

Using the techniques of Zariski decomposition and Laufer decomposition we can subdivide $\mathbf{Q}^r_+$ into rational polyhedral sets where $h_X(\underline{n})$ can be better understood.

Theorem 2.2 (see [**7**, Proposition 6.3]) *There exists a subdivision of $\mathbf{Q}^{\mathbf{r}}_+$ by rational polyhedral sets C such that $\underline{n} \in C \cap \mathbf{N}^r$ implies*

$$h_X(\underline{n}) = f_C(\underline{n}) + \sigma_C(\underline{n})$$

where $f_C(\underline{n})$ is a quadratic polynomial in $\underline{n}$ with periodic coefficients and $\sigma_C(\underline{n})$ is a bounded function.

If R has rational singularities, then $\sigma_C(\underline{n}) = 0$ for all $\underline{n}$.

It is natural to consider the following series $P_X(t)$ which we call the Poincaré series of the singularity:

$$P_X(t) = \sum_{\underline{n} \in \mathbf{N}^r} h_X(\underline{n}) t^{\underline{n}}$$

where $t^{\underline{n}} = t_1^{n_1} \cdots t_r^{n_r}$. The series depends on the resolution X in an obvious way, as the number of variables in the series is the number of exceptional components of the resolution.

Related series are considered by Campillo, Delgado, Gusein-Zade [**1**, **2**], Campillo and Galindo [**3**], for plane curve singularities and for rational surface singularities.

From the series $P_X(t)$ we can recover the topology of X, as is shown in the following theorem.

Theorem 2.3 (see [**7**, Theorem 3.1]) *From $P_X(t)$ we can compute*

1. *The intersection matrix $(E_i \cdot E_j)$.*
2. *The arithmetic genus $p_a(E_i)$ of each E_i.*
3. *The arithmetic genus $h^1(X, O_X)$ of X.*

Corollary 2.4 *We can determine if R has rational singularities from $P_X(t)$ for any resolution X.*

This follows since $h^1(X, O_X)$ depends only on R, and $h^1(X, O_X) = 0$ if and only if R has a rational singularity.

We require the following combinatorial statement.

Theorem 2.5 (see [**7**, Theorem 7.5]) *Suppose that $Q \subset \mathbf{Q}_+^r$ is a rational polyhedral set, $M < \mathbf{Z}^r$ is a subgroup and $\underline{m} \in \mathbf{Z}^r$. Then*

$$\sum_{\underline{n} \in Q \cap (\underline{m}+M)} t^{\underline{n}}$$

is a rational series.

Theorem 2.6 *If R has a rational singularity then $P_X(t)$ is rational for any resolution X.*

Proof: Apply Theorem 2.5 to the conclusions of Theorem 2.2. □

Theorem 2.7 (see [**7**, the $r \leq 2$ case of Theorem 7.7]) *If $r \leq 2$ then $P(t)$ is rational.*

Proof: If $r = 1$ this is trivial, as $\pi^{-1}(m)_{red} = -E_1$ is ample.

If $r = 2$ this follows from Theorem 2.2 and the proof of Theorem 2.1 to show that h^1 has a nice form on the polyhedral sets C of Theorem 2.2. □

The proof that h^1 has a good form which ensures rationality of $P_X(t)$ does not extend beyond $r = 2$. We consider a condition on the class group $\mathrm{Cl}(R)$ of R which ensures that $P_X(t)$ is rational for all resolutions X of R.

There is an exact sequence of abelian groups

$$0 \to G \to \mathrm{Cl}(R) \to H \to 0$$

where G is a commutative algebraic group and H is a finite group [**10**]. Furthermore, there is an exact sequence of abelian groups

$$0 \to (k^+)^m \times (k^\times)^n \to G \to A \to 0$$

where A is an abelian variety. G is said to be semi-abelian if $m = 0$. We will say that $\mathrm{Cl}(R)$ is semi-abelian if G is.

It is well known that $G = 0$ if and only if R has a rational singularity [**10**]. There are many examples of non-rational singularities with semi-abelian class groups. Over any nonsingular curve C of positive genus it is possible to construct cones, with a resolution whose reduced exceptional locus is isomorphic to C, and such that the singularity has a semi-abelian class group.

Theorem 2.8 (see [**7**, Corollary 7.8]) *Suppose that $Cl(R)$ is semi-abelian. Then $P_X(t)$ is rational for any resolution X of $spec(R)$.*

If $\mathrm{Cl}(R)$ is semi-abelian, then the functions $\sigma_C(\underline{n})$ of Theorem 2.2 are determined by membership of $\underline{n}$ in translations of subgroups of $\mathbf{Z}^r$. Then we apply Theorem 2.5 to the functions

$$h_X(\underline{n}) = f_C(\underline{n}) + \sigma_C(\underline{n})$$

for $\underline{n} \in C \cap \mathbf{N}^r$.

Theorem 2.9 (see [**7**, Theorem 9.1]) *There exists R with $Cl(R) \cong \mathbf{C}^2$ and a resolution with $r = 3$ such that $P_X(t)$ is not rational.*

We can now ask if there is a necessary and sufficient condition on $\mathrm{Cl}(R)$ ensuring that $P_X(t)$ is rational for all resolutions $\pi : X \to \mathrm{spec}(R)$.

3 The geometric problem

The proofs to Theorem 2.8 and Theorem 2.9 contain the geometric problem discussed in this section. In computing the Poincaré series of surface singularities, we must consider cohomology of powers of line bundles on projective curves.

Suppose that C/k is a proper, integral curve over k (where k is an algebraically closed field of characteristic zero), and $\mathcal{L}_1, \ldots, \mathcal{L}_r$ are line bundles on C. Define

$$h(\underline{n}) = h^0(C, \mathcal{L}_1^{n_1} \otimes \cdots \otimes \mathcal{L}_r^{n_r})$$

for $\underline{n} \in \mathbf{N}^r$. Set

$$P(t) = \sum_{\underline{n} \in \mathbf{N}^r} h(\underline{n}) t^{\underline{n}}.$$

We will analyze these functions. If $r = 1$ the Riemann-Roch Theorem tells us that

$$h(n) = h^0(\mathcal{L}^n) = \begin{cases} 0 & \text{if } \deg(\mathcal{L}) < 0 \\ \text{periodic in } n & \text{if } \deg(\mathcal{L}) = 0 \\ \chi(\mathcal{L}^n) = n\deg(\mathcal{L}) + 1 - p_a(C) & \text{if } \deg(\mathcal{L}) > 0 \text{ and } n \gg 0. \end{cases}$$

In this case ($r = 1$) $P(t)$ is rational.

If $r = 2$ we easily find examples where $h(n_1, n_2)$ is more complicated. Let C be an elliptic curve, $\mathcal{L}_1$ be a degree 0 line bundle on C of infinite order and let $\mathcal{L}_2 = \mathcal{L}_1^{-1}$. Then

$$h(n_1, n_2) = h^0(C, \mathcal{L}_1^{n_1} \otimes \mathcal{L}_2^{n_2}) = \begin{cases} 1 & \text{if } n_1 = n_2 \\ 0 & \text{otherwise.} \end{cases}$$

Observe however, that the function has an essentially combinatorial nature. In fact, this is really the worst kind of behavior that can occur if $r = 2$. As such, rationality of $P(t)$ always holds if $r = 2$.

Theorem 3.1 *Suppose that $r = 2$. Then $P(t)$ is rational.*

The proof follows from [**8**] (as explained after the proof of Theorem 4.1 [**6**]). For general r, we can say the following.

Theorem 3.2 (see [**6**, Theorem 4.1 and the discussion after its proof]) *Suppose that r is arbitrary and $Pic^0(C)$ is semi-abelian (e.g., C non-singular). Then $P(t)$ is rational.*

Proof: (Sketch) We reduce to computing

$$h(\underline{n}) = h^0(\mathcal{L}_1^{n_1} \otimes \cdots \otimes \mathcal{L}_r^{n_r} \otimes \mathcal{M})$$

for $n_1 + \cdots + n_r \gg 0$, where $\deg(\mathcal{L}_i) = 0$ for $1 \le i \le r$ and $\mathcal{M}$ is a line bundle.

$$W_i = \{\mathcal{L} \in \mathrm{Pic}^0(C) \mid h^0(\mathcal{L} \otimes \mathcal{M}) \ge i\}$$

is Zariski closed.

We apply McQuillan's Theorem [**11**]: Suppose that G is a semi-abelian algebraic group, $X \subset G$ is a subvariety and $H < G$ is a finitely generated group. If $H \cap X$ is Zariski dense in X then X is a translate of a semi-abelian subvariety.

Let $G = \mathrm{Pic}^0(C)$ and $H = \mathrm{Image}(\pi)$ where $\pi : \mathbf{Z}^r \to G$ is the homomorphism

$$\pi(\underline{n}) = \mathcal{L}_1^{n_1} \otimes \cdots \otimes \mathcal{L}_r^{n_r},$$

and $X = W_i$ (which may not be irreducible). McQuillan's Theorem implies there exist $\underline{m}_1, \ldots, \underline{m}_s \in \mathbf{Z}^r$ and subgroups $G_1, \ldots, G_s$ of $\mathbf{Z}^r$ such that

$$h(\underline{n}) \ge i \text{ iff } \pi(\underline{n}) \in W_i \text{ iff } \underline{n} \in \cup_{i=1}^{s} (m_{\underline{i}} + G_i).$$

The proof of Theorem 3.2 now follows from Theorem 2.5. □

Theorem 3.3 (see [**6**, Example 4.2]) *There exists a singular curve C with*

$$\mathrm{Pic}^0(C) \cong \mathbf{C}^2$$

and line bundles $\mathcal{L}_1, \mathcal{L}_2, \mathcal{L}_3$ on C such that $P(n_1, n_2, n_3)$ is not rational.

Proof: Let C be the rational curve with isolated singularity $\tilde{p}$ such that $\mathcal{O}_{C,\tilde{p}} = \mathbf{C}[t^2, t^5]_{(t^2,t^5)}$. $\mathrm{Pic}^0(C) \cong \mathbf{C}^2$ is realized by the homomorphism for $D \in \mathrm{Div}^0(C - \tilde{p})$,

$$D \mapsto \left(\frac{d}{dt} \log(f_D) \mid_{t=0}, \frac{d^3}{dt^3} \log(f_D) \mid_{t=0} \right), \tag{1}$$

where $f_D \in \mathbf{C}(t)$ is a rational function whose divisor is D on $\mathbf{P}^1$.

Let $\infty \in C$ be a nonsingular point. Then we have the Abel-Jacobi mapping

$$AJ : C - \{\tilde{p}\} \to \mathrm{Pic}^0(C)$$

defined by $AJ(p) = \mathcal{O}_C(p - \infty)$.

It follows from (1) that

$$\mathrm{Image}(AJ) = \{y = 2x^3\}.$$

Let $W = \text{Image}(AJ)$. Suppose that $\mathcal{L} \in \text{Pic}^0(C)$ is a line bundle.

$$h^0(C, \mathcal{L} \otimes \mathcal{O}_C(\infty)) = \begin{cases} 0 & \text{if } \mathcal{L} \notin W \\ 1 & \text{if } \mathcal{L} \in W. \end{cases}$$

Set

$$\mathcal{L}_1 = (1,0), \mathcal{L}_2 = (0,1), \mathcal{L}_3 = \mathcal{O}_C(\infty).$$

Then

$$h^0(\mathcal{L}_1^{n_1} \otimes \mathcal{L}_2^{n_2} \otimes \mathcal{L}_3) = \begin{cases} 1 & \text{if } n_2 = 2n_1^3 \\ 0 & \text{otherwise.} \end{cases}$$

After a simple calculation this shows that $P(t)$ is not rational. □

References

[1] A. Campillo, F. Delgado and S.M. Gusein-Zade, The Alexander polynomial of a plane curve singularity via the ring of functions on it, *Duke Math. J.* **117** (2003), 125–156.

[2] A. Campillo, F. Delgado and S.M. Gusein-Zade, Poincaré series of rational surface singularities, *Invent. Math.* **155** (2004), 41–53.

[3] A. Campillo and C. Galindo, The Poincaré series associated with finitely many monomial valuations, *Math. Proc. Cambridge Philos. Soc.* **134** (2003), 433–443.

[4] V. Cossart, O. Piltant and A. Reguera, *Divisorial Valuations Dominating Rational Surface Singularities*, Fields Inst. Comm. **32**: "Valuation theory and its applications volume I," AMS (2002).

[5] S.D. Cutkosky, On unique and almost unique factorization of complete ideals II, *Invent. Math.* **98** (1989), 59–74.

[6] S.D. Cutkosky, Poincaré series of line bundles on varieties, in *A Tribute to C.S. Seshadri*, Perspectives in Geometry and Representation Theory, Hindustan Book Agency (2003), 220–240.

[7] S.D. Cutkosky, J. Herzog and A. Reguera, Poincaré series of surface singularities, *Trans. Amer. Math. Soc.* **356** (2003), 1833–1874.

[8] S.D. Cutkosky and V. Srinivas, On a problem of Zariski on dimensions of linear systems, *Ann. of Math.* **137** (1993), 531–559.

[9] M. Kato, Riemann-Roch theorem for strongly pseudoconvex manifolds of dimension 2, *Math. Ann.* **222** (1976), 243–250.

[10] J. Lipman, Rational singularities with applications to algebraic surfaces and unique factorization, *Publ. Math. IHES* **36** (1969), 195–279.

[**11**] M. McQuillan, Division points on semi-abelian varieties, *Invent. Math.* **120** (1995), 143–159.

[**12**] M. Morales, Calcul de quelques invariants des singularités de surface normale, in *Knots, braids and singularities* (Plans-sur-Bex, 1982), Monograph Enseign. Math. **31**, 191–203.

[**13**] O. Zariski, The Theorem of Riemann-Roch for high multiples of an effective divisor on an algebraic surface, *Ann. of Math.* **96** (1962), 560–616.

Multi-graded Hilbert coefficients

CLARE D'CRUZ
Chennai Mathematical Institute, 92, G. N. Chetty Road, Chennai 600017, India
E-mail: clare@cmi.ac.in

ANNA GUERRIERI
Dipartimento di Matematica, Universitá di L'Aquila
Via Vetoio, Loc. Coppito, 67010 L'Aquila, Italia
E-mail: guerran@univaq.it

1 Introduction

Let $(R, \mathfrak{m})$ be a local ring of dimension $d > 0$ and I an $\mathfrak{m}$-primary ideal of R. The blowup algebras $R[It] = \bigoplus_{n \geq 0} I^n t^n$ and $\mathrm{gr}_I(R) = R[It]/IR[It]$ are key ingredients in the process of resolving singularities. The numerical functions that gauge the complexity of the above graded rings are fundamental tools. The function $H(I;n) = \ell(\mathrm{gr}_I(R)_n)$ is called the Hilbert function of I; as known, it is a polynomial in n for all $n \gg 0$. The iterated Hilbert function $H^1(I;n) = \sum_{i \geq 0}^{n} H(I;i)$ is also a polynomial in n for all $n \gg 0$, denoted by $h^1(I;n)$. This can be written in the form $\sum_{q=0}^{d} (-1)^q e_q(I) \binom{n+d-q}{d-q}$. The $e_q(I)$ are the Hilbert coefficients of the ideal I.

In this paper we provide information on the coefficients of the asymptotically polynomial Hilbert-type functions associated with iterated blowups. In Section 3 we show that all these coefficients are usual Hilbert coefficients for suitably chosen ideals (see Proposition 3.1 and Corollary 3.6).

Iterated blowups give rise to multi-graded blowup algebras; here the blowup ring is $R[I_1t_1, \cdots, I_gt_g]$. There are several associated graded rings in the picture that we call *multi-form rings*. For us the *multi-graded Hilbert functions* are all the possible Hilbert functions associated to the multi-form rings. These functions are asymptotically polynomials which we call the *multi-graded Hilbert polynomials*; their coefficients are the *multi-graded Hilbert coefficients*. In Section 2 we recall many classical definitions and results.

Rees [**5**] and Teissier [**8**] showed that the multi-graded coefficients of topmost degree (*mixed-multiplicities*) are positive integers that can be expressed in terms of multiplicity of ideals in the the ring R. In this paper we concentrate on the properties of the multi-graded coefficients not appearing at the topmost degree, which is done in Section 3 (see Proposition 3.1 and Corollary 3.6).

In Section 4 we extend to the multi-graded case some well known results of Northcott [**4**], and of Narita [**3**], and we prove the non-negativity of the multi-graded coefficients corresponding to $e_1(I)$ and $e_2(I)$ (Lemma 4.1, Theorem 4.2, Theorem 4.3). We also provide an example which illustrates the situation of the multi-graded coefficients corresponding to $e_3(I)$ (Example 4.4).

2 Multi-graded Hilbert functions

In this section we define the various multi-graded structures we are interested in, their corresponding multi-form functions and we recall information about their behavior.

Definition 2.1 Let $(R, \mathfrak{m})$ be a local ring. Let $g \geq 1$, let $I_1, \ldots, I_g$ be $\mathfrak{m}$-primary ideals and let $t_1, \ldots, t_g$ be indeterminates. The *multi-Rees algebra of* $I_1, \ldots, I_g$ is

$$R[I_1t_1, \ldots, I_gt_g] := \bigoplus_{n_1, \ldots, n_g \geq 0} (I_1t_1)^{n_1} \cdots (I_gt_g)^{n_g},$$

the *multi-form ring of* $I_1,\ldots,I_g$ is

$$\begin{aligned}\mathrm{gr}_R(I_1,\ldots,I_g) &:= R[I_1t_1,\ldots,I_gt_g]\otimes_R R/I_1\cdots I_g\\ &= \bigoplus_{n_1,\ldots,n_g\geq 0} I_1^{n_1}\cdots I_g^{n_g}/I_1^{n_1+1}\cdots I_g^{n_g+1}\end{aligned}$$

and the *i-th multi-form ring of* $I_1,\ldots,I_g$ is

$$\mathrm{gr}_R(I_1,\ldots,I_g)(i) = \bigoplus_{n_1,\ldots,n_g\geq 0} I_1^{n_1}\cdots I_g^{n_g}/I_i(I_1^{n_1}\cdots I_g^{n_g}).$$

Definition 2.2 Let $I_1,\ldots,I_g$ be $\mathfrak{m}$-primary ideals in R. For $1\leq i\leq g$ the i-th *Bhattacharya function* is

$$B(\underline{n};i) := \ell(\mathrm{gr}_R(I_1,\ldots,I_g)(i)_{n_1,\ldots,n_g}) = \left(I_1^{n_1}\cdots I_g^{n_g}/I_1^{n_1}\cdots I_i^{n_i+1}\cdots I_g^{n_g}\right).$$

The *iterated Bhattacharya function in the i-th position* is

$$B^{\mathbf{1}}(\underline{n};i) := \sum_{j=0}^{n_i} B(n_1,\ldots,n_{i-1},j,n_{i+1},\ldots,n_g;i).$$

The *iterated Bhattacharya function* is

$$B^{\mathbf{1}}(\underline{n};\mathbf{I}) := B^{\mathbf{1}}(n_1;1)+\sum_{i=2}^{g} B^{\mathbf{1}}(n_1+1,\ldots,n_{i-1}+1,n_i;i) = \ell\left(\frac{R}{I_1^{n_1+1}\cdots I_g^{n_g+1}}\right).$$

Next result is a rewriting of a classical result of Bhattacharya [**1**].

Proposition 2.3 *Let* $I_1,\ldots,I_g$ *be* $\mathfrak{m}$*-primary ideals in* $(R,\mathfrak{m})$. *For all* $n_1,\ldots,n_g$ *large, the functions* $B(\underline{n};i)$ *are polynomials, say* $b(\underline{n};i)$, *of total degree* $d-1$ *in* $n_1,\ldots,n_g$ *and can be written in the form*

$$\sum_{q_1+\cdots+q_g\leq d-1} (-1)^{d-1-(q_1+\cdots+q_g)} E_{q_1,\ldots,q_i+1,\ldots q_g}(I_i)\binom{n_1+q_1}{q_1}\cdots\binom{n_g+q_g}{q_g},$$

where $E_{q_1,\ldots,q_i+1,\ldots,q_g}(I_i)\in\mathbb{Z}$.

Definition 2.4 We call the set of coefficients appearing in the i-th Bhattacharya polynomials the *i-th Bhattacharya coefficients*. The coefficients of the terms of total degree $d-1$ are the *mixed multiplicities* of the ideals $I_1,\ldots,I_g$; see [**6**] and [**8**].

Observation 2.5 Let I be an $\mathfrak{m}$-primary ideal in $(R,\mathfrak{m})$. Then we have only one i-th Bhattacharya polynomial $b(n;I)$ which is the Hilbert polynomial. Namely

$$b(n;I) \quad = \quad \sum_{q\leq d-1} (-1)^{d-1-q} E_{q+1}(I) \binom{n+q}{q},$$

and one clearly has $E_{q+1}(I) = e_{d-1-q}(I)$ for all $q = 0,\ldots,d-1$.

The function $B^{\mathbf{1}}(\underline{n};i)$ is an iteration of the i-th Bhattacharya function, therefore asymptotically it is a polynomial function of degree d. This polynomial is described below; for a proof see [**7**, Lemma 2.1.6].

Lemma 2.6 *Let $I_1,\ldots,I_g$ be $\mathfrak{m}$-primary ideals in $(R,\mathfrak{m})$. Then there exists an integer $N \geq 0$ such that for all $n_i \geq N$, $1 \leq i \leq g$, the function $B^{\mathbf{1}}(\underline{n};i)$ is a polynomial of degree d, say $b^{\mathbf{1}}(\underline{n};i)$. This polynomial $b^{\mathbf{1}}(\underline{n};i)$ can be written in the form*

$$\sum_{q_1+\cdots+q_g\leq d-1} (-1)^{d-1-(q_1+\cdots+q_g)} E_{q_1,\ldots,q_i+1,\ldots,q_g}(I_i) \binom{n_1+q_1}{q_1} \cdots \binom{n_g+q_g}{q_g}$$
$$+(-1)^d E_{\underline{0}}(I_i),$$

where

$$(-1)^d E_{\underline{0}}(I_i) := \sum_{j=0}^{N-1} [B(\ldots,n_{i-1},j,n_{i+1},\ldots;i) - b(\ldots,n_{i-1},j,n_{i+1},\ldots;i)].$$

Definition 2.7 For $1 \leq i \leq g$ we call the polynomial $b^{\mathbf{1}}(\underline{n};i)$ the *iterated Bhattacharya polynomial in the i-th position* of the ideals $I_1,\ldots,I_g$. We call the set of its coefficients the *iterated Bhattacharya coefficients in the i-th position.*

Observation 2.8 Except for $E_{\underline{0}}(I_i)$, the coefficients of the polynomials $b^{\mathbf{1}}(\underline{n};i)$ and $b(\underline{n};i)$ coincide.

3 The iterated Bhattacharya polynomial

In this section we analyze the structure of $B^{\mathbf{1}}(\underline{n};\mathbf{I})$, which is classically the most interesting object. For the sake of simplicity we work with two ideals, but this can be done for any number of ideals; see Remark 3.2.

Proposition 3.1 *Let I_1 and I_2 be $\mathfrak{m}$-primary ideals in a local ring of positive dimension d. Then for $n_1, n_2 \gg 0$, the function $B^1(n_1,n_2;\mathbf{I})$ is a polynomial in n_1 and n_2 of degree d, say $b^1(n_1,n_2;\mathbf{I})$, that can be written in the form*

$$b^1(n_1,n_2;\mathbf{I}) = \sum_{q_1+q_2\leq d} (-1)^{d-(q_1+q_2)} A_{q_1,q_2}(\mathbf{I}) \binom{n_1+q_1}{q_1}\binom{n_2+q_2}{q_2}$$

where

$$\begin{aligned} A_{q_1,0}(\mathbf{I}) &= E_{q_1}(I_1) = e_{d-q_1}(I_1) = \sum_{j=0}^{d-q_1} (-1)^j E_{q_1,j}(I_1) \qquad q_1 = 1,\ldots,d, \\ A_{0,q_2}(\mathbf{I}) &= E_{q_2}(I_2) = e_{d-q_2}(I_2) = \sum_{j=0}^{d-q_2} (-1)^j E_{j,q_2}(I_2) \qquad q_2 = 1,\ldots,d, \end{aligned}$$

$$\begin{aligned} A_{q_1,q_2}(\mathbf{I}) &= \sum_{j=0}^{d-(q_1+q_2)} (-1)^j E_{q_1,q_2+j}(I_1) \\ &= \sum_{j=0}^{d-(q_1+q_2)} (-1)^j E_{q_1+j,q_2}(I_2) \qquad q_1,q_2 \neq 0, \\ A_{0,0}(\mathbf{I}) &= E_0(I_2) + E_{0,0}(I_1) = e_d(I_2) + E_{0,0}(I_1) \\ &= E_0(I_1) + E_{0,0}(I_2) = e_d(I_1) + E_{0,0}(I_2) \\ &= e_d(I_1I_2). \end{aligned}$$

Proof: By definition, for $n_1, n_2 \gg 0$,

$$\begin{aligned} b^1(n_1\,, n_2;\mathbf{I}) &= \\ &= b^1(n_1;1) + b^1(n_1+1,n_2;2) = \sum_{q=0}^{d} (-1)^{d-q} E_q(I_1)\binom{n_1+q}{q} \\ &\quad + \sum_{q_1+q_2\leq d-1} (-1)^{d-1-(q_1+q_2)} E_{q_1,q_2+1}(I_2)\binom{n_1+q_1+1}{q_1}\binom{n_2+q_2+1}{q_2+1} \\ &\quad + (-1)^d E_{0,0}(I_2). \end{aligned}$$

Substituting the binomial identity $\binom{n_1+q_1+1}{q_1} = \sum_{j=0}^{q_1}\binom{n_1+j}{j}$ in the second sum we see that the coefficient of

$$(-1)^{d-1-(q_1+q_2)}\binom{n_1+q_1}{q_1}\binom{n_2+q_2+1}{q_2+1}$$

is $\sum_{j=0}^{d-1-(q_1+q_2)} (-1)^j E_{q_1+j,q_2+1}(I_2)$ and this is $A_{q_1,q_2+1}(\mathbf{I})$. Hence

$$A_{q_1,q_2}(\mathbf{I}) = \begin{cases} E_{q_1}(I_1) & \text{if } q_1 \neq 0, q_2 = 0, \\ \sum_{j=0}^{d-(q_1+q_2)} (-1)^j E_{q_1+j,q_2}(I_2) & \text{if } q_2 \neq 0, \\ E_0(I_1) + E_{0,0}(I_2) & \text{if } q_1, q_2 = 0. \end{cases} \quad (1)$$

If we write

$$b^{\mathbf{1}}(n_1,n_2;\mathbf{I}) = b^1_{e_1}(n_2;2) + b^1_{e_2}(n_1,n_2+1;1)$$

we get

$$A_{q_1,q_2}(\mathbf{I}) = \begin{cases} E_{q_2}(I_2) & \text{if } q_2 \neq 0, q_1 = 0, \\ \sum_{j=0}^{d-(q_1+q_2)} (-1)^j E_{q_1,q_2+j}(I_1) & \text{if } q_1 \neq 0, \\ E_0(I_2) + E_{0,0}(I_1) & \text{if} q_1, q_2 = 0. \end{cases} \quad (2)$$

Comparing (1) and (2) we get our result. Also $A_{0,0}(\mathbf{I}) = e_d(I_1 I_2)$ follows from the fact that $\ell(R/I_1^n I_2^n) = \ell(R/(I_1 I_2)^n)$. □

We used the functions $B^{\mathbf{1}}(\underline{n};i)$ in order to show that the edge coefficients in $B^{\mathbf{1}}(\underline{n};\mathbf{I})$ are Hilbert coefficients.

Remark 3.2 In general, for $n_1,\ldots,n_g \gg 0$ the function $B^{\mathbf{1}}(\underline{n};\mathbf{I})$ is a polynomial of degree d:

$$b^{\mathbf{1}}(\underline{n};\mathbf{I}) = \sum_{q_1+\cdots+q_g \le d} (-1)^{d-(q_1+\cdots+q_g)} A_{q_1,\ldots,q_g}(\mathbf{I}) \binom{n_1+q_1}{q_1} \cdots \binom{n_g+q_g}{q_g}$$

where the topmost degree coefficients are the mixed multiplicity. For any $q_i \ge 1$, $A_{0,\ldots,q_i,\ldots,0}(\mathbf{I}) = e_{d-q_i}(I_i)$ and $A_{0,\ldots,0}(\mathbf{I}) = e_d(I_1 \cdots I_g)$.

Definition 3.3 We call the polynomial $b^{\mathbf{1}}(\underline{n};\mathbf{I})$ the *iterated Bhattacharya polynomial* of the ideals $I_1,\ldots,I_g$. We call the set of coefficients appearing in it the *iterated Bhattacharya coefficients*.

Definition 3.4 Let $(R,\mathfrak{m})$ be a local ring and (I_1,I_2) be a set of ideals of R. Then we say that an element $x_1 \in I_1$ is superficial for I_1 and I_2 if there exists an integer s_1 such that for $r_1 \geq s_1$ and all nonnegative integers r_2,

$$(x_1)\cap I_1^{r_1}I_2^{r_2} = x_1 I_1^{r_1-1}I_2^{r_2}.$$

Such elements exist when the residue field is infinite.

Lemma 3.5 *Let I_1 and I_2 be $\mathfrak{m}$-primary ideals in a local ring of dimension $d \geq 2$. Let $x \in I_1$ be a superficial element for I_1 and I_2. Let $'$ denote the image in R/xR. Then $A'_{q_1,q_2}(\mathbf{I}) = A_{q_1+1,q_2}(\mathbf{I})$.*

Proof: Without loss of generality we can assume that x is a non-zero divisor. Then for $n_1,n_2 \gg 0$,

$$\begin{aligned}
\ell\left(\frac{R'}{(I_1')^{n_1+1}(I_2')^{n_2+1}}\right) &= \ell\left(\frac{R}{I_1^{n_1+1}I_2^{n_2+1}+xR}\right)\\
&= \ell\left(\frac{R}{I_1^{n_1+1}I_2^{n_2+1}}\right) - \ell\left(\frac{R}{I_1^{n_1}I_2^{n_2+1}}\right)\\
&= \sum_{q_1+q_2\leq d}(-1)^{d-(q_1+q_2)}A_{q_1,q_2}(\mathbf{I})\binom{n_1+q_1-1}{q_1-1}\binom{n_2+q_2}{q_2}.
\end{aligned}$$

On the other hand we can also write

$$\ell\left(\frac{R'}{(I_1')^{n_1+1}(I_2')^{n_2+1}}\right) = \sum_{q_1+q_2\leq d-1}(-1)^{d-1-(q_1+q_2)}A'_{q_1,q_2}(\mathbf{I})\binom{n_1+q_1}{q_1}\binom{n_2+q_2}{q_2}.$$

Comparing the above equations we get the desired result. □

Corollary 3.6 *The iterated Bhattacharya coefficients are Hilbert coefficients of ideals in suitable quotients of R.*

Proof: Choosing a superficial sequence of q_1 elements, say $x_1,\ldots,x_{q_1}$, from I_1 and q_2 elements, say $y_1,\ldots,y_{q_2}$, from I_2 and using Lemma 3.5 repeatedly, one obtains $A_{q_1,q_2} = e_{d-q_1}(I_1') = e_{d-q_2}(I_2'')$ where $'$ denotes going modulo $x_1,\ldots,x_{q_1}$ and $''$ denotes going modulo $y_1,\ldots,y_{q_2}$. □

4 Non-negativity of the coefficients

In [**4**, Theorem 1, p. 212], Northcott proved that if I is an $\mathfrak{m}$-primary ideal in a Cohen-Macaulay local ring of dimension d, then $\ell\left(\frac{R}{I}\right) \geq e_0(I) - e_1(I)$. In the case of many ideals one can say the following.

Lemma 4.1 *Let $(R,\mathfrak{m})$ be a Cohen-Macaulay local ring of dimension d. Let I_1 and I_2 be $\mathfrak{m}$-primary ideals in R. Then $\ell(R/I_1) \geq A_{d,0}(\mathbf{I}) - A_{d-1,0}(\mathbf{I})$ and $\ell(R/I_2) \geq A_{0,d}(\mathbf{I}) - A_{0,d-1}(\mathbf{I})$.*

Proof: By Northcott's result $\ell(R/I_i) \geq e_0(I_i) - e_1(I_i)$ for $i = 1,2$. Now by Proposition 3.1 we have that $e_0(I_1) = A_{d,0}(\mathbf{I})$, $e_1(I_1) = A_{d-1,0}(\mathbf{I})$, $e_0(I_2) = A_{0,d}(\mathbf{I})$, $e_1(I_2) = A_{0,d-1}(\mathbf{I})$. □

The above lemma was proved in [**2**, Lemma 6.2], for two $\mathfrak{m}$-primary ideals in a Cohen-Macaulay local ring of dimension two using a different argument.

We now extend, to the iterated Bhattacharya coefficients and the i-th Bhattacharya coefficients, the results of Northcott [**4**], and Narita [**3**], on the non-negativity of the first two Hilbert coefficients.

Theorem 4.2 *Let I_1 and I_2 be $\mathfrak{m}$-primary ideals in a Cohen-Macaulay local ring of dimension $d \geq 1$. Then for all $q_1 + q_2 \geq d-2$, $A_{q_1,q_2}(\mathbf{I})$ are non-negative.*

Proof: Let $q_1 + q_2 = d$. Then $A_{q_1,q_2}(\mathbf{I})$ is the multiplicity of the ideal generated by q_1 elements from I_1 and $d - q_1$ elements from I_2 chosen sufficiently general, by Rees and Teissier, and hence are positive integers. Now let $q_1 + q_2 \in \{d-1, d-2\}$. We prove by induction on dimension. If $d = 1$, then $A_{0,0}(\mathbf{I}) = e_1(I_1I_2)$. If $d = 2$, then $A_{1,0}(\mathbf{I}) = e_1(I_1)$, $A_{0,1}(\mathbf{I}) = e_1(I_2)$ and $A_{0,0}(\mathbf{I}) = e_2(I_1I_2)$ by Proposition 3.1. By the results of Northcott [**4**], and of Narita [**3**], all the above four coefficients are non-negative. Let $d \geq 3$ and $q_1 + q_2 \in \{d-1, d-2\}$. By Proposition 3.1, $A_{q_1,0}(\mathbf{I})$ and $A_{0,q_2}(\mathbf{I})$, are the Hilbert coefficients and are non-negative; see [**4**] and [**3**]. If q_1 and q_2 are non-zero then by Lemma 3.5, we get $A_{q_1,q_2}(\mathbf{I}) = A_{q_1-1,q_2}(\mathbf{I}')$ where $'$ denotes the image modulo a superficial element $x \in I_1$, and this gives the non-negativity by induction on the dimension of R. □

Theorem 4.3 *Let I_1 and I_2 be $\mathfrak{m}$-primary ideals in a Cohen-Macaulay local ring of dimension $d \geq 1$. Then for all $q_1 + q_2 \geq d-3$, $E_{q_1+1,q_2}(I_1)$ and $E_{q_1,q_2+1}(I_2)$ are non-negative.*

Proof: From Proposition 3.1, we have $E_{q_1+1,0}(I_1) = e_{d-q_1-1}(I_1)$ and hence are non-negative for $q_1+1 \in \{d, d-1, d-2\}$. Similarly, $E_{0,q_2+1}(I_2)$ are non-negative for $q_2+1 \in \{d, d-1, d-2\}$. If $q_1, q_2 \neq 0$, then by Proposition 3.1, $E_{q_1+1,q_2}(I_1) = A_{q_1+1,q_2}(\mathbf{I}) + A_{q_1+1,q_2+1}(\mathbf{I})$ and $E_{q_1,q_2+1}(I_2) = A_{q_1,q_2+1}(\mathbf{I}) + A_{q_1+1,q_2+1}(\mathbf{I})$, which are non-negative since $A_{q_1+1,q_2}(\mathbf{I})$, $A_{q_1,q_2+1}(\mathbf{I})$ and $A_{q_1+1,q_2+1}(\mathbf{I})$ are non-negative by Theorem 4.2. □

Modifying slightly an example from Narita [**3**], one sees that the higher coefficients need not be positive.

Example 4.4 Let $R = k[x_1, x_2, x_3, x_4]_{\mathfrak{m}}/(x_4^3)$ where k is a field and x_1, x_2, x_3, x_4 are variables and $\mathfrak{m} = (x_1, x_2, x_3, x_4)$. Put $I_1 = I_2 = I = (x_1, x_2^2, x_3^2, x_2x_4, x_3x_4)$. Then $\dim R = 3$ and $A_{00}(\mathbf{I}) = e_3(I^2) = e_3(I) < 0$.

Acknowledgments

The first author thanks the Università di L'Aquila, the Institut de Mathématiques CNRS & Université Paris 6 and Progetto INDAM-GNSAGA. The second author thanks the Institut de Mathématiques CNRS & Université Paris 6 and MURST-GNSAGA.

References

[**1**] P.B. Bhattacharya, The Hilbert function of two ideals, *Proc. Cambridge Philos. Soc.* **53** (1957), 568–575.

[**2**] A.V. Jayanthan and J.K. Verma, Grothendieck-Serre formula and bigraded Cohen-Macaulay Rees algebras, *J. Algebra* **254** (2002), 1–20.

[**3**] M. Narita, A note on the coefficients of Hilbert characteristic functions in semi-regular local rings, *Proc. Cambridge Philos. Soc.* **59** (1963), 269–275.

[**4**] D.G. Northcott, A note on the coefficients of the abstract Hilbert function, *J. London Math. Soc.* **35** (1960), 209–214.

[**5**] D. Rees, $\mathcal{A}$-transforms of local rings and a theorem on multiplicities of ideals, *Proc. Cambridge Philos. Soc.* **57** (1961), 8–17.

[**6**] D. Rees, Generalizations of reductions and mixed multiplicities, *J. London Math. Soc.* **29** (1984), 397–414.

[**7**] P.C. Roberts, *Multiplicities and Chern Classes in Local Algebra*, Cambridge University Press, 1998.

[**8**] B. Teissier, Cycles évanescents, sections planes, et conditions de Whitney, Singularités à Cargèse, 1972, *Astérisque* **7-8** (1973), 285–362.

Torsion freeness and normality of blowup rings of monomial ideals

CESAR A. ESCOBAR
Departamento de Matemáticas
Escuela Superior de Física y Matemáticas, Instituto Politécnico Nacional
Unidad Adolfo López Mateos, 07738 Mexico City, D.F.
E-mail: escobar@esfm.ipn.mx

RAFAEL H. VILLARREAL
Departamento de Matemáticas
Centro de Investigación y de Estudios Avanzados del IPN
Apartado Postal 14–740, 07000 México City, D.F.
E-mail: vila@esfm.ipn.mx

YUJI YOSHINO
Department of Mathematics, Okayama University, Okayama 700-8530, Japan
E-mail: yoshino@math.okayama-u.ac.jp

Abstract: Let $I = (x^{v_1}, \ldots, x^{v_q})$ be a monomial ideal of a polynomial ring $K[\mathbf{x}]$ over a field K and let A be the matrix with column vectors $v_1, \ldots, v_q$. We show that the associated graded ring of I is reduced if A is totally unimodular. Then we express the symbolic Rees algebra of I in terms of Hilbert bases provided that I is square-free. A normality criterion is presented which is adequate for monomial ideals.

1 Introduction

Let $R = K[x_1,\ldots,x_n]$ be a polynomial ring over a field K and let I be a monomial ideal of R generated by a finite set of monomials $\{x^{v_1},\ldots,x^{v_q}\}$. As usual we use x^a as an abbreviation for $x_1^{a_1}\cdots x_n^{a_n}$, where $a = (a_1,\ldots,a_n)$ is in $\mathbb{N}^n$. The three central objects of study here are the following blowup algebras: (a) the *Rees algebra*

$$R[It] := R \oplus It \oplus \cdots \oplus I^i t^i \oplus \cdots \subset R[t],$$

where t is a new variable, (b) the *associated graded ring*

$$\mathrm{gr}_I(R) := R/I \oplus I/I^2 \oplus \cdots \oplus I^i/I^{i+1} \oplus \cdots \simeq R[It] \otimes_R (R/I),$$

with multiplication

$$(a + I^{i+1})(b + I^{j+1}) = ab + I^{i+j+1} \qquad (a \in I^i,\ b \in I^j),$$

and (c) the *symbolic Rees algebra*

$$R_s(I) := R + I^{(1)}t + I^{(2)}t^2 + \cdots + I^{(i)}t^i + \cdots \subset R[t],$$

where $I^{(i)}$ is the *ith* symbolic power of I.

It is well known [**20**, p. 168] that the integral closure of $R[It]$ in its field of fractions can be expressed as

$$\overline{R[It]} = R \oplus \overline{I}t \oplus \cdots \oplus \overline{I^i}t^i \oplus \cdots,$$

where $\overline{I^i}$ is the integral closure of I^i. The integral closure of the *ith* power of I is again a monomial ideal [**20**, p. 169] given by:

$$\overline{I^i} = (\{x^a \in R \mid (x^a)^p \in I^{pi} \text{ for some } p \geq 1\}). \tag{1}$$

If $I^i = \overline{I^i}$ for $i \geq 1$, the ideal I is called *normal*. The ring $R[It]$ is said to be *normal* if $R[It]$ is equal to its integral closure. Thus $R[It]$ is normal if and only if I is normal. Recall that the ring $\mathrm{gr}_I(R)$ is called *reduced* if its nilradical is zero.

The content of the paper is as follows. Our first result proves that if A is the matrix with column vectors $v_1,\ldots,v_q$ and each $i \times i$ minor of A is 0 or ± 1 for all

$i \geq 1$, then $\mathrm{gr}_I(R)$ is reduced (Theorem 2.3). As a consequence it is recovered that $R[It]$ is normal [**2**]; see Remark 2.5.

We point out an effective criterion to determine when $\mathrm{gr}_I(R)$ is reduced (Proposition 3.4). If I is square-free, we give a description of its symbolic Rees algebra in terms of Hilbert bases (Theorem 3.5). As a consequence we recover that $R_s(I)$ is finitely generated [**12**].

In Section 4 we present a normality criterion which is adequate for monomial ideals (Theorem 4.4). To illustrate the practical use of this criterion it is shown that ideals of Veronese type are normal (Proposition 4.9). As a byproduct, using a result of [**8**], we derive that a monomial subring of Veronese type is the Ehrhart ring of a lattice polytope [**7**].

The main tools to show those results are commutative algebra and linear optimization techniques as described in [**11, 19, 20, 23**].

2 Torsion freeness of the associated graded ring

We begin with a basic result about the nilpotent elements of the associated graded ring.

Lemma 2.1 *If I is a monomial ideal of R, then the nilradical of the associated graded ring of I is given by*

$$\mathrm{nil}(\mathrm{gr}_I(R)) = (\{\overline{x^\alpha} \in I^i/I^{i+1} | x^{s\alpha} \in I^{si+1}; i \geq 0; s \geq 1\}).$$

Proof: The nilradical of $\mathrm{gr}_I(R)$ is graded with respect to the fine grading, and thus it is generated by homogeneous elements. □

Definition 2.2 An integral matrix A is called *totally unimodular* if each $i \times i$ minor of A is 0 or ± 1 for all $i \geq 1$.

Theorem 2.3 *Let $I = (x^{v_1}, \ldots, x^{v_q})$ be a monomial ideal of R and let A be the $n \times q$ matrix with column vectors $v_1, \ldots, v_q$. If A is totally unimodular, then $\mathrm{gr}_I(R)$ is reduced.*

Proof: Let $m = x^\alpha$ be a monomial such that: (i) $\overline{m} \in I^i/I^{i+1}$ and (ii) $m^s \in I^{is+1}$, for some $0 \neq s \in \mathbb{N}$. By Lemma 2.1 it suffices to prove that $m \in I^{i+1}$. From (ii) there are $a_1, \ldots, a_q$ in $\mathbb{N}$ and $\delta \in \mathbb{N}^n$ such that

$$x^{s\alpha} = (x^{v_1})^{a_1} \cdots (x^{v_q})^{a_q} x^\delta \text{ and } \textstyle\sum_{j=1}^q a_j = is + 1.$$

Hence the linear program

$$\text{Maximize } f(y) = \textstyle\sum_{i=1}^q y_i \qquad (*)$$

$$\text{Subject to } Ay \leq \alpha \text{ and } y \geq 0$$

has an optimal value greater than or equal to $z_0 = i + 1/s$ because the rational vector $y_0 = (a_1/s, \ldots, a_q/s)$ satisfies $Ay_0 \leq \alpha$, $y_0 \geq 0$ and the sum of its entries is z_0. Note that the polyhedron $Q = \{y | Ay \leq \alpha; y \geq 0\}$ is bounded. Indeed any $y = (y_1, \ldots, y_q)$ in Q must satisfy $y_j \leq \max\{\alpha_1, \ldots, \alpha_q\}$ for all j, where α_j is the *jth* entry of α. By a result of Hoffman and Kruskal (see [**11**, Theorem 5.19 and Corollary 5.20]) the optimal value of the linear program $(*)$ is attained by an integral vector $b = (b_1, \ldots, b_q)$. Thus $b_1 + \cdots + b_q \geq i + 1$. As $b \in Q$, we can write

$$x^\alpha = (x^{v_1})^{b_1} \cdots (x^{v_q})^{b_q} x^\gamma,$$

for some $\gamma \in \mathbb{N}^n$. This proves $x^\alpha \in I^{i+1}$, as required. □

Definition 2.4 An ideal I of a ring R is called *normally torsion free* if $\mathrm{Ass}(R/I^i)$ is contained in $\mathrm{Ass}(R/I)$ for all $i \geq 1$ and $I \neq R$.

Remark 2.5 Assume that I is a square-free monomial ideal of R. Then, according to [**10**, Corollary 1.10], I is normally torsion free $\Leftrightarrow$ $\mathrm{gr}_I(R)$ is torsion-free over R/I $\Leftrightarrow$ $\mathrm{gr}_I(R)$ is reduced. If $\mathrm{gr}_I(R)$ is reduced, then $R[It]$ is normal [**1**]. Hence it is seen that Theorem 2.3 is a generalization of both [**2**, Theorem 3.2] and [**17**, Theorem 5.9].

3 Rees cones and symbolic Rees algebras

Let $I = (x^{v_1}, \ldots, x^{v_q})$ be a square-free monomial ideal of $R = K[x_1, \ldots, x_n]$ of height $g \geq 2$ such that $\deg(x^{v_i}) \geq 2$ for all i and let

$$\mathcal{A}' = \{e_1, \ldots, e_n, (v_1, 1), \ldots, (v_q, 1)\} \subset \mathbb{R}^{n+1},$$

where e_i is the *ith* unit vector in $\mathbb{R}^{n+1}$.

The *Rees cone* of the point configuration $\mathcal{A} = \{v_1, \ldots, v_q\}$ is the rational polyhedral cone, denoted by $\mathbb{R}_+\mathcal{A}'$, consisting of the linear combinations of $\mathcal{A}'$ with non-negative coefficients. Notice that $\dim(\mathbb{R}_+\mathcal{A}') = n+1$ and also that $H_{e_i} \cap \mathbb{R}_+\mathcal{A}'$ is a facet of $\mathbb{R}_+\mathcal{A}'$ for $i = 1, \ldots, n+1$. Thus according to [**23**] there is a unique irreducible representation

$$\mathbb{R}_+\mathcal{A}' = H_{e_1}^+ \cap \cdots \cap H_{e_{n+1}}^+ \cap H_{\ell_1}^+ \cap \cdots \cap H_{\ell_r}^+ \tag{2}$$

such that each ℓ_i has relatively prime integral entries. As usual H_a^+ denotes the closed halfspace

$$H_a^+ = \{\alpha \in \mathbb{R}^{n+1} \,|\, \langle \alpha, a \rangle \geq 0\}$$

and H_a is the hyperplane through the origin with normal vector a. Let $\mathfrak{p}_1, \ldots, \mathfrak{p}_s$ be the minimal primes of I. For each $\mathfrak{p}_k$ set $C_k = \{x_i | x_i \in \mathfrak{p}_k\}$ and define the vector ℓ_k as:

$$\ell_k = -e_{n+1} + \sum_{x_i \in C_k} e_i \qquad (k = 1, \ldots, s). \tag{3}$$

This definition of ℓ_k is consistent with Eq. (2) because of the following lemma. In the sequel we assume that $\ell_1, \ldots, \ell_s$ are the vectors given by Eq. (3).

Lemma 3.1 *$H_{\ell_1}^+, \ldots, H_{\ell_s}^+$ occur in the irreducible representation of $\mathbb{R}_+\mathcal{A}'$.*

Proof: Let $1 \leq k \leq s$. It suffices to verify that $F = \mathbb{R}_+\mathcal{A}' \cap H_{\ell_k}$ is a facet of the Rees cone. Note that $\mathbb{R}_+\mathcal{A}' \subset H_{\ell_k}^+$ and $\mathbb{R}_+\mathcal{A}' \not\subset H_{\ell_k}$ because $I \subset \mathfrak{p}_k$ and $\mathfrak{p}_k \subsetneq (x_1, \ldots, x_n)$. From the minimality of $\mathfrak{p}_k$ it follows that there is a linearly independent set $\mathcal{F} \subset \mathcal{A}' \cap H_{\ell_k}$ with $n = |\mathcal{F}|$. $\square$

Lemma 3.2 *If $1 \leq j \leq r$ and $\langle \ell_j, e_{n+1} \rangle = -1$, then $1 \leq j \leq s$.*

Proof: Let $\ell_j = (a_1, \ldots, a_n, -1)$. Since $\mathbb{R}_+\mathcal{A}' \subset H^+_{\ell_j}$ we get $a_i \geq 0$ for $i = 1, \ldots, n$. Consider the ideal $\mathfrak{p}$ of R generated by the set of x_i such that $a_i > 0$. Note that $I \subset \mathfrak{p}$ because $\langle \ell_j, (v_i, 1) \rangle \geq 0$ for all $1 \leq i \leq n$. For simplicity of notation assume that $\ell_j = (a_1, \ldots, a_m, 0, \ldots, 0, -1)$, where $a_i > 0$ for all i. Take an arbitrary vector $v \in \mathcal{A}'$ such that $v \in H_{\ell_j}$, then v satisfies the equation

$$a_1x_1 + \cdots + a_mx_m = x_{n+1}.$$

Observe that v also satisfies the equation

$$x_1 + \cdots + x_m = x_{n+1}.$$

Because H_{ℓ_j} contains n linearly independent vectors from $\mathcal{A}'$ we conclude that $H_{\ell_j} = H_b$, where $b = e_1 + \cdots + e_m - e_{n+1}$. Hence $\ell_j = b$ and consequently $a_i = 1$ for all i. It remains to show that $\mathfrak{p}$ is a minimal prime of I, and this follows from the irreducibility of Eq. (2). □

Next we present two results that highlight the importance of the irreducible representation of the Rees cone.

Proposition 3.3 *If $B = R[It]$ is normal and $\mathrm{Cl}(B)$ is its divisor class group, then $\mathrm{Cl}(B) \simeq \mathbb{Z}^r$ and $r \geq s$.*

Proof: By [**16**, Theorem 1.1] the divisor class group of B is a free abelian group of finite rank. Using [**5**, Theorem 2] it is seen that r is the rank of the divisor class group of B. □

Since the program *Normaliz* [**3**] computes the irreducible representation of the Rees cone and the integral closure of $R[It]$ the following is an effective criterion for the reducedness of the associated graded ring.

Proposition 3.4 (Effective criterion) *$\mathrm{gr}_I(R)$ is reduced if and only if $R[It]$ is normal and $\langle \ell_i, e_{n+1} \rangle = -1$ for $i = 1, \ldots, r$.*

Proof: By [**10**, Theorem 1.11] and Proposition 3.3 we get that $R[It]$ is normal and $r = s$. Thus $\langle \ell_i, e_{n+1} \rangle = -1$ for all $i = 1, \ldots, r$.

Using Lemma 3.2 we obtain $r = s$. Therefore the associated graded ring is reduced by [**10**, Theorem 1.11]. □

Symbolic Rees algebra Next we introduce one of the main objects of this section. The *symbolic Rees cone* or *Simis cone* of the point configuration $\mathcal{A}$ is the rational polyhedral cone defined as:

$$\mathrm{Cn}(\mathcal{A}) := H_{e_1}^+ \cap \cdots \cap H_{e_{n+1}}^+ \cap H_{\ell_1}^+ \cap \cdots \cap H_{\ell_s}^+. \tag{4}$$

By [**11**, Lemma 5.4] there exists a finite set $\mathcal{H} \subset \mathbb{N}^{n+1}$ such that

(a) $\mathrm{Cn}(\mathcal{A}) = \mathbb{R}_+\mathcal{H}$, and

(b) $\mathbb{Z}^{n+1} \cap \mathbb{R}_+\mathcal{H} = \mathbb{N}\mathcal{H}$,

where $\mathbb{N}\mathcal{H}$ is the additive subsemigroup of $\mathbb{N}^{n+1}$ generated by $\mathcal{H}$. The set $\mathcal{H}$ is called an integral *Hilbert basis* of $\mathrm{Cn}(\mathcal{A})$. If we require $\mathcal{H}$ to be minimal (with respect to containment), then $\mathcal{H}$ is unique [**15**].

For use below recall that the *ith* symbolic power of the monomial ideal I is given by $I^{(i)} = \mathfrak{p}_1^i \cap \cdots \cap \mathfrak{p}_s^i$.

Theorem 3.5 *If $K[\mathbb{N}\mathcal{H}]$ is the semigroup ring of $\mathbb{N}\mathcal{H}$, then*

$$R_s(I) = K[\mathbb{N}\mathcal{H}].$$

Proof: Recall that $K[\mathbb{N}\mathcal{H}] = K[\{x^at^b \mid (a,b) \in \mathbb{N}\mathcal{H}\}]$. Take $x^at^b \in R_s(I)$, that is, $x^a \in \mathfrak{p}_i^b$ for all i. Hence $\langle (a,b), \ell_i \rangle \geq 0$ for all i or equivalently $(a,b) \in \mathrm{Cn}(\mathcal{A})$. Thus $(a,b) \in \mathbb{N}\mathcal{H}$ and $x^at^b \in K[\mathbb{N}\mathcal{H}]$. Conversely take $x^at^b \in K[\mathbb{N}\mathcal{H}]$. Then (a,b) is in $\mathrm{Cn}(\mathcal{A})$ and $\langle (a,b), \ell_i \rangle \geq 0$ for all i. Hence $x^a \in \mathfrak{p}_i^b$ for all i and $x^a \in I^{(b)}$, as required. □

Corollary 3.6 ([12]) *$R_s(I)$ is a finitely generated K-algebra.*

Proof: It follows at once from Theorem 3.5. □

Remark 3.7 The Rees algebra of I can be written as

$$R[It] = K[\{x^at^b \mid (a,b) \in \mathbb{N}\mathcal{A}'\}].$$

On the other hand, according to [**22**, Theorem 7.2.28] one has

$$\overline{R[It]} = K[\{x^at^b \mid (a,b) \in \mathbb{Z}^{n+1} \cap \mathbb{R}_+\mathcal{A}'\}].$$

Hence $R[It]$ is normal if and only if $\mathbb{N}\mathcal{A}' = \mathbb{Z}^{n+1} \cap \mathbb{R}_+\mathcal{A}'$.

Corollary 3.8 $\overline{R[It]} = R_s(I) \iff \langle \ell_i, e_{n+1}\rangle = -1$ *for* $i = 1,\dots,r$.

Proof: The equality $\overline{R[It]} = R_s(I)$ holds if and only if

$$\mathbb{Z}^{n+1} \cap \mathbb{R}_+\mathcal{A}' = \mathbb{Z}^{n+1} \cap \mathrm{Cn}(\mathcal{A}).$$

Hence the asserted equality holds if and only if the Rees cone and the Simis cone are equal. By the irreducibility of the representations of both cones it follows that $\overline{R[It]} = R_s(I)$ if and only if $r = s$. To finish the proof note that $r = s$ if and only if $\langle \ell_i, e_{n+1}\rangle = -1$ for $i = 1,\dots,r$. □

Hilbert bases of Rees cones For the remaining of the section we do not assume that I is generated by square-free monomials. Here we will show a degree bound for the minimal integral Hilbert basis of $\mathbb{R}_+\mathcal{A}'$.

Remark 3.9 Let a be a point of a set A in $\mathbb{R}^n$ and let x be a point in aff(A) not lying in A. If $\lambda_0 = \sup\{\lambda \in [0,1] \mid (1-\lambda)a + \lambda x \in A\}$, then $x_0 = (1-\lambda_0)a + \lambda_0 x$ is a relative boundary point of A.

A monomial $x^\alpha t^b$ in $\overline{R[It]}$ correspond to a vector (α, b) in $\mathbb{Z}^{n+1} \cap \mathbb{R}_+\mathcal{A}'$. Thus it is natural to assign degree b to $\beta = (\alpha, b)$ and write $\deg_{n+1}(\beta) = b$.

Proposition 3.10 *Let* $\mathbb{R}_+\mathcal{A}'$ *be the Rees cone of* $\mathcal{A}$ *and let* $\mathcal{B}$ *be the its minimal integral Hilbert basis. If* $\beta \in \mathcal{B}$*, then* $\deg_{n+1}(\beta) < n$.

Proof: For convenience of notation we write $\beta = (\alpha, b)$, where α is a vector in $\mathbb{N}^n$ and $b = \deg_{n+1}(\beta)$ is in $\mathbb{N}$. Note that $\mathbb{R}_+\mathcal{A}' \cap \mathbb{Z}^{n+1}$ is equal to $\mathbb{Q}_+\mathcal{A}' \cap \mathbb{Z}^{n+1}$ and $\dim(\mathbb{R}_+\mathcal{A}') = n+1$. Thus, since $(\alpha, b) \in \mathbb{R}_+\mathcal{A}'$, by Carathéodory's theorem [**9**, Theorem 2.3, p. 10] we can write

$$(\alpha, b) = \lambda_1(v_{i_1}, 1) + \cdots + \lambda_r(v_{i_r}, 1) + \mu_1 e_{j_1} + \cdots + \mu_s e_{j_s} \quad (\lambda_i, \mu_k \in \mathbb{Q}_+), \qquad (*)$$

where $\{(v_{i_1}, 1), \dots, (v_{i_r}, 1), e_{j_1}, \dots, e_{j_s}\}$ is a linearly independent set of vectors contained in $\mathcal{A}'$. By the minimality of $\mathcal{B}$ it is seen that $0 \le \lambda_i < 1$ and $0 \le \mu_k < 1$ for all i,k. From Eq. $(*)$ we get

$$b = \lambda_1 + \cdots + \lambda_r < r \implies b \le r - 1.$$

If $r \leq n$, then $b \leq n-1$. Thus from now on we may assume $r = n+1$ and Eq. $(*)$ takes the simpler form

$$(\alpha, b) = \lambda_1(v_{i_1}, 1) + \cdots + \lambda_{n+1}(v_{i_{n+1}}, 1).$$

Consider the cone C generated by $\mathcal{A}'' = \{(v_{i_1}, 1), \ldots, (v_{i_{n+1}}, 1)\}$. Since $-e_1$ is not in C, using Remark 3.9 we obtain a point

$$x_0 = (1-\lambda_0)(\alpha, b) + \lambda_0(-e_1) \quad (0 \leq \lambda_0 < 1)$$

in the relative boundary of C. According to [**23**, Theorem 3.2.1] the relative boundary of C is the union of its facets. Therefore using that any facet of C is an n-dimensional cone generated by a subset of $\mathcal{A}''$ (see [**22**, Proposition 7.2.21]) together with the minimality of $\mathcal{B}$ it follows that we can write the vector (α, b) as:

$$(\alpha, b) = \rho_0 e_1 + \rho_1(v_{j_1}, 1) + \cdots + \rho_n(v_{j_n}, 1) \qquad (0 \leq \rho_i < 1 \, \forall i),$$

and consequently $b \leq n-1$, as required. □

As a consequence we obtain (cf. [**4**, Theorem 3.3]):

Corollary 3.11 *$\overline{R[It]}$ is generated by monomials of t-degree at most $n-1$.*

Proof: If $\mathcal{B}$ is the minimal integral Hilbert basis of the Rees cone, then $\overline{R[It]} = K[\mathbb{N}\mathcal{B}]$ and we may apply Proposition 3.10. □

4 Normality criterion

In this section we present a simple normality criterion that is suitable to study the normality of monomial ideals. Then, as an application, we show that ideals of Veronese type are normal.

To prove the normality of a monomial ideal I it suffices to show that the power I^i is integrally closed for $i = 1, \ldots, n-1$, where n is the number of variables [**14**]; see [**13**] for a generalization of this fact. If I has finite co-length, then its integral closure can be constructed [**6**, Proposition 3.4].

Proposition 4.1 *Let $(R,\mathfrak{m})$ be a Noetherian local ring and let I be a proper ideal of R. Then I is integrally closed if and only if $IR_\mathfrak{p}$ is integrally closed for any prime ideal $I \subset \mathfrak{p} \neq \mathfrak{m}$ and $\overline{I} \cap (I\colon \mathfrak{m}) = I$.*

Proof: It follows from the proof of Proposition 4.2. □

Proposition 4.2 *Let $(R,\mathfrak{m})$ be a Noetherian local ring and let I be a proper ideal of R. Then I is normal if and only if*

(a) *$IR_\mathfrak{p}$ is normal for any prime ideal $I \subset \mathfrak{p} \neq \mathfrak{m}$, and*

(b) *$\overline{I^r} \cap (I^r\colon \mathfrak{m}) = I^r$ for all $r \geq 1$.*

Proof: That (a) is satisfied follows from the fact that integral closures and powers of ideals commute with localizations [**22**, Proposition 3.3.4]. Part (b) is clearly satisfied.

Assume $I^r \neq \overline{I^r}$ for some $r \geq 1$. Take $\mathfrak{p} \in \mathrm{Ass}_R(M)$, where $M = \overline{I^r}/I^r$. Since $\mathfrak{p}$ is in the support of M we get $M_\mathfrak{p} \neq (0)$, and consequently by (a) we must have $\mathfrak{p} = \mathfrak{m}$. Hence there is an embedding

$$R/\mathfrak{m} \hookrightarrow M \quad (\overline{1} \mapsto \overline{x_0}),$$

where $x_0 \in \overline{I^r} \setminus I^r$ and $\mathfrak{m} = \mathrm{ann}(\overline{x_0})$. From (b) one concludes that $x_0 \in I^r$, a contradiction. Thus $I^r = \overline{I^r}$ for all $r \geq 1$. □

The following result asserts that a normal monomial ideal stays normal if we make a variable, say x_n, equal to 1 or 0.

Proposition 4.3 *Let $I = (x^{v_1},\ldots,x^{v_q})$ be a normal ideal. If each x^{v_i} is written as $x_n^{a_i} g_i$, where g_i is a monomial in $R' = K[x_1,\ldots,x_{n-1}]$, then*

(a) *$J = (g_1,\ldots,g_q) \subset R'$ is a normal ideal, and*

(b) *$L = (\{g_i \mid a_i = 0\}) \subset R'$ is a normal ideal.*

Proof: (a) We will show that $J^r = \overline{J^r}$ for all $r \geq 1$. Take a monomial x^α of R' that belongs to $\overline{J^r}$. Then $(x^\alpha)^p \in J^{rp}$ for some $p \geq 1$ and we can write

$$(x^\alpha)^p = g_1^{b_1} \cdots g_q^{b_q} x^\beta,$$

where $\sum_{i=1}^{q} b_i = rp$ and $x^\beta \in R'$. Multiplying both sides of the equation above by x_n^{sp} with $s = \sum_{i=1}^{q} a_i b_i$, we get

$$(x_n^s x^\alpha)^p = (x_n^{a_1} g_1)^{b_1} \cdots (x_n^{a_q} g_q)^{b_q} x^\delta \in I^{rp}.$$

Hence $x_n^s x^\alpha \in I^r$ and we can write

$$x_n^s x^\alpha = (x_n^{a_1} g_1)^{c_1} \cdots (x_n^{a_q} g_q)^{c_q} x^\gamma,$$

where $\sum_{i=1}^{q} c_i = r$. Evaluating the last equality at $x_n = 1$ gives that $x^\alpha \in J^r$, as required. Part (b) follows using similar arguments. $\Box$

Theorem 4.4 *Let $I = (x^{v_1}, \ldots, x^{v_q}) \subset R = K[x_1, \ldots, x_n]$ and let J_i be the ideal of R generated by the monomials obtained from $\{x^{v_1}, \ldots, x^{v_q}\}$ by making $x_i = 1$. Then I is normal if and only if*

(a) *J_i is normal for all i (resp. $I_\mathfrak{p}$ is normal for any prime ideal $\mathfrak{p} \neq \mathfrak{m}$), and*

(b) *$\overline{I^r} \cap (I^r \colon \mathfrak{m}) = I^r$ for all $r \geq 1$, where $\mathfrak{m} = (x_1, \ldots, x_n)$.*

Proof: Condition (a) follows from Proposition 4.3 (resp. using that integral closures and powers of ideals commute with localizations). Part (b) is clearly satisfied.

To prove that the *rth* power of I is equal to $\overline{I^r}$ set $M = \overline{I^r}/I^r$. We proceed by contradiction assuming $I^r \subsetneq \overline{I^r}$. Take an associated prime ideal $\mathfrak{p}$ of M. There are embeddings

$$R/\mathfrak{p} \overset{\varphi}{\hookrightarrow} M \hookrightarrow R/I^r.$$

Thus $\mathfrak{p}$ is an associated prime of the monomial ideal I^r. As a consequence $\mathfrak{p}$ is generated by a subset of $\{x_1, \ldots, x_n\}$ (see [**22**, Proposition 5.1.3]) and $\mathfrak{p} \subset \mathfrak{m}$. Note $\mathfrak{p} = \mathfrak{m}$, otherwise if $x_i \notin \mathfrak{p}$ for some i, then by (a), $IR_\mathfrak{p} = J_i R_\mathfrak{p}$ is normal and $M_\mathfrak{p} = (0)$, a contradiction because $\mathfrak{p}$ is in the support of M. Thus $\mathfrak{p} = \mathfrak{m}$. If $\overline{x_0} = \varphi(\overline{1})$ is the image of $\overline{1}$ under the first embedding above, we get $x_0 \in \overline{I^r} \setminus I^r$ and $x_0 \in (I^r \colon \mathfrak{m})$, which contradicts (b). $\Box$

Normality of an ideal of Veronese type Let $R = K[x_1,\ldots,x_n]$ be a ring of polynomials over an arbitrary field K. Given a sequence of integers $(s_1,s_2,\ldots,s_n;d)$ such that $1 \leq s_j \leq d \leq \sum_{i=1}^n s_i$ for all j, we define $\mathcal{A}$ as the set of partitions

$$\mathcal{A} = \{(a_1,\ldots,a_n) \in \mathbb{Z}^n \,|\, a_1+\cdots+a_n = d;\ 0 \leq a_i \leq s_i \,\forall i\},$$

and F as the set of monomials

$$F = \{x^a \,|\, a \in \mathcal{A}\} = \{f_1,\ldots,f_q\}.$$

Definition 4.5 The ideal $I = (F) \subset R$ is said to be of *Veronese type* with defining sequence $(s_1,\ldots,s_n;d)$.

The monomial subring $K[F] \subset R$ and its toric ideal have been studied in [**18**]. Here we will focus in the ideal $I = (F)$ and its Rees algebra, rather than on $K[F]$.

Lemma 4.6 *Let $\mathfrak{p} \neq \mathfrak{m}$ be a prime ideal of R and let $x_i \notin \mathfrak{p}$. For $k \neq i$ define $s_k' = \min\{s_k, d-s_i\}$. If $I = (F)$ and I' is the ideal of the ring $R' = K[x_1,\ldots,\widehat{x_i},\ldots,x_n]$ generated by all the monomials*

$$x_1^{a_1'}\cdots x_{i-1}^{a_{i-1}'} x_{i+1}^{a_{i+1}'}\cdots x_n^{a_n'}$$

of R' of degree $d-s_i$ such that $a_k' \leq s_k'$ for $k \neq i$, then $IR_\mathfrak{p} = I'R_\mathfrak{p}$.

Proof: For simplicity of notation we assume $i = 1$. Let $A = R_{x_1}$ be the localization of R at the multiplicative set $\{x_1^k\}_{k\geq 0}$. Since $A_\mathfrak{p} = R_\mathfrak{p}$ it suffices to show the equality $I'A = IA$. If $d = s_1$, then $x_1^d \in I$ and $s_k' = 0$ for $k \geq 2$. Thus in this case $I'A = A$ and $IA = A$. Assume $d > s_1$. Take a monomial

$$m = x_2^{a_2'}\cdots x_n^{a_n'}$$

of I' of degree $d-s_1$. As $a_2'+\cdots+a_n' = d-s_1$ and $a_k' \leq s_k' \leq s_k$ for $k \geq 2$, we have $x_1^{s_1}m \in I$ and $m \in IA$. Therefore $I'A \subset IA$. Conversely, take a monomial $m = x_1^{a_1}\cdots x_n^{a_n}$ of degree d in I. If $a_k \geq d-s_1$ for some $k \geq 2$, then $s_k' = d-s_1$ and $x_k^{d-s_1} \in I'$. Thus $m \in I'A$. On the contrary, if $a_k < d-s_1$ for all $k \geq 2$, then $a_k \leq s_k'$ for all $k \geq 2$. Note $a_2+\cdots+a_n = d-a_1 \geq d-s_1$. It is seen that there are integers $a_2',\ldots,a_n'$ such that $a_k' \leq a_k$ for $k \geq 2$ and $a_2'+\cdots+a_n' = d-s_1$. As a consequence the monomial

$$x_2^{a_2'}\cdots x_n^{a_n'}$$

belongs to I' and $m \in I'A$. Hence $IA \subset I'A$. □

Remark 4.7 If $d > s_i$, note that I' is an ideal of Veronese type with defining sequence $(s'_1,\dots,s'_{i-1},s'_{i+1},\dots,s'_n;d-s_i)$. If $d = s_i$, then $I'R_{\mathfrak{p}} = IR_{\mathfrak{p}} = R_{\mathfrak{p}}$.

Definition 4.8 The *support* of a monomial $f = x_1^{a_1}\cdots x_n^{a_n}$, denoted by $\mathrm{supp}(f)$, is a subset of the set of variables given by

$$\mathrm{supp}(f) = \{x_i | \, a_i > 0\}.$$

Proposition 4.9 *If $I = (F) \subset R$ is an ideal of Veronese type, then I is normal.*

Proof: For simplicity we keep the notation introduced above. The proof is by induction on d. If $d = 1$ the result is clear. Indeed in this case I is the maximal ideal $\mathfrak{m} = (x_1,\dots,x_n)$ and $\mathrm{gr}_{\mathfrak{m}}(R)$ is reduced because the identity matrix is totally unimodular. Thus $R[\mathfrak{m}t]$ is normal. Note that $\mathrm{gr}_{\mathfrak{m}}(R)$ is a polynomial ring over the field K, but we do not need this fact.

Assume $d \geq 2$ and that the result holds for ideals of Veronese type generated by monomials of degree $< d$. It suffices to prove that conditions (a) and (b) of Theorem 4.4 are satisfied. Take any prime $\mathfrak{p} \neq \mathfrak{m}$ and pick $x_i \notin \mathfrak{p}$. By Lemma 4.6 we obtain that $I'R_{\mathfrak{p}} = IR_{\mathfrak{p}}$. Hence using induction hypothesis and Remark 4.7 we get that $IR_{\mathfrak{p}}$ is normal. Thus condition (a) holds. To verify condition (b) we proceed by contradiction assuming

$$\overline{I^r} \cap (I^r : \mathfrak{m}) \neq I^r \tag{5}$$

for some $r \geq 1$. This means we can choose a monomial f in $\overline{I^r} \setminus I^r$ such that $x_i f$ is in I^r for all i. Hence there are $f_i = x_1^{a_{i1}}\cdots x_n^{a_{in}}$, $i = 1,\dots,r$ of degree d such that $a_{ij} \leq s_j$ for all i, j and satisfying the equality

$$x_1 f = f_1 f_2 \cdots f_r h,$$

where h is a monomial with $\deg(h) > 0$, because $f \in \overline{I^r}$. Note $x_1 \notin \mathrm{supp}(h)$ because $f \notin I^r$. Thus we may assume $x_1 \in \mathrm{supp}(f_1)$ and $h = x_2^{c_2} x_3^{c_3} \cdots x_p^{c_p}$, where $c_i > 0$ for all i. Observe that $x_\ell^{s_\ell}$ divides f_1 for all $2 \leq \ell \leq p$, otherwise we can write

$$f = ((f_1 x_\ell)/x_1) f_2 \cdots f_r (h/x_\ell)$$

to derive $f \in I^r$, a contradiction. Thus we can write

$$f_1 = x_1^{a_{11}} x_2^{s_2} \cdots x_p^{s_p} x_{p+1}^{a_{1(p+1)}} \cdots x_n^{a_{1n}} \quad (a_{11} > 0).$$

If $r=1$, then $\deg_{x_2}(f)\geq s_2+1$ and since $x_2f\in I$ we obtain $x_2f=gh'$, with g a monomial of degree d in I. By degree considerations it follows that x_2 divides h' and $f\in I$, a contradiction. From now on we assume $r\geq 2$.

Case (I): $x_\ell^{s_\ell}$ does not divide f_k for some $2\leq\ell\leq p$ and for some $2\leq k\leq r$. Hence for each $x_i\in\operatorname{supp}(f_k)$ with $i\neq\ell$, we have that $x_i^{s_i}$ divides f_1, for otherwise if $x_i^{s_i}$ does not divide f_1 we write

$$f=h_1f_2\cdots f_{k-1}h_kf_{k+1}\cdots f_r(h/x_\ell),$$

where $h_1=(x_if_1)/x_1\in I$ and $h_k=(x_\ell f_k)/x_i\in I$. Therefore $f\in I^r$, a contradiction. Hence, since we have already seen that also $x_\ell^{s_\ell}$ divides f_1, we obtain

$$s_\ell+\sum_{x_i\in\operatorname{supp}(f_k)\setminus\{x_\ell\}} s_i\leq d=\deg(f_1)$$

and consequently

$$d-a_{k\ell}=\sum_{x_i\in\operatorname{supp}(f_k)\setminus\{x_\ell\}} a_{ki}\leq\sum_{x_i\in\operatorname{supp}(f_k)\setminus\{x_\ell\}} s_i\leq d-s_\ell.$$

Thus $a_{k\ell}\geq s_\ell$ and $x_\ell^{s_\ell}$ divides f_k, a contradiction.

Case (II): $x_\ell^{s_\ell}$ divides f_k for $2\leq\ell\leq p$ and $2\leq k\leq r$. Since we also have that $x_\ell^{s_\ell}$ divides f_1, in this case we conclude $\deg_{x_2}(f)\geq rs_2+1$. Recall that $x_2f\in I^r$, which by degree considerations readily implies $f\in I^r$, a contradiction.

Altogether we see that in both cases the inequality (5) leads to a contradiction. Hence condition (b) holds. □

Corollary 4.10 ([7]) *Let $P\subset\mathbb{R}^n$ be the lattice polytope* $\operatorname{conv}(\mathcal{A})$ *and let*

$$A(P)=K[\{x^\alpha t^i|\alpha\in\mathbb{Z}^n\cap iP\}]\subset R[t]$$

be the Ehrhart ring of P. Then $A(P)=K[Ft]$. In particular $K[Ft]$ is normal.

Proof: Since $R[Ft]$ is normal, from [**8**, Proposition 3.15] we get the equality $A(P)=K[Ft]$. As $A(P)$ is always normal, so is $K[Ft]$. □

Definition 4.11 Let $I\subset R$ be an ideal of Veronese type with defining sequence $(s_1,\ldots,s_n;d)$. I is called the *dth square-free Veronese ideal* (resp. *dth Veronese ideal*) of R if $s_i=1$ (resp. $s_i=d$) for all i.

Corollary 4.12 ([21]) *If I is the ideal of R generated by the square-free monomials of degree d, then I is a normal ideal.*

Proof: Since I is the dth square-free Veronese ideal of R, it is of Veronese type. Hence the result follows from Theorem 4.9. □

Acknowledgments

This work was partially supported by JSPS (Japan) and by CONACyT grant 45256 (Mexico).

References

[1] J. Barshay, Graded algebras of powers of ideals generated by A-sequences, *J. Algebra* **25** (1973), 90–99.

[2] V. Bonanzinga, C. Escobar and R. Villarreal, On the normality of Rees algebras associated to totally unimodular matrices, *Results Math.* **41** (2002), 258–264.

[3] W. Bruns and R. Koch, *Normaliz*, a program to compute normalizations of affine semigroups. Available via anonymous ftp from: ftp.mathematik.Uni-Osnabrueck.DE/pub/osm/kommalg/software.

[4] W. Bruns, W.V. Vasconcelos and R. Villarreal, Degree bounds in monomial subrings, *Illinois J. Math.* **41** (1997), 341–353.

[5] L. Chouinard, Krull semigroups and divisor class groups, *Canad. J. Math.* **33** (1981), 1459–1468.

[6] D. Delfino, A. Taylor, W.V. Vasconcelos, R. Villarreal and N. Weininger, Monomial ideals and the computation of multiplicities, in *Commutative Rings Theory and Applications*, Proceedings: Fez (M. Fontana et al., Eds.), 87–106, Lecture Notes in Pure and Applied Math. **231**, Marcel Dekker, New York, 2003.

[7] E. De Negri and T. Hibi, Gorenstein algebras of Veronese type, *J. Algebra* **193** (1997), 629–639.

[8] C. Escobar, J. Martínez-Bernal and R. Villarreal, Relative volumes and minors in monomial subrings, *Linear Algebra Appl.* **374** (2003), 275–290.

[9] G. Ewald, *Combinatorial Convexity and Algebraic Geometry*, Graduate Texts in Mathematics **168**, Springer-Verlag, New York, 1996.

[10] C. Huneke, A. Simis and W.V. Vasconcelos, Reduced normal cones are domains, *Contemp. Math.* **88** (1989), 95–101.

[11] B. Korte and J. Vygen, *Combinatorial Optimization Theory and Algorithms*, Springer-Verlag, 2000.

[**12**] G. Lyubeznik, On the arithmetical rank of monomial ideals, *J. Algebra* **112** (1988), 86–89.

[**13**] C. Polini, B. Ulrich, W.V. Vasconcelos and R. Villarreal, Normalization of ideals, in preparation.

[**14**] L. Reid, L.G. Roberts and M.A. Vitulli, Some results on normal homogeneous ideals, *Comm. Algebra* **31** (2003), 4485–4506.

[**15**] A. Schrijver, On total dual integrality, *Linear Algebra Appl.* **38** (1981), 27–32.

[**16**] A. Simis and N.V. Trung, The divisor class group of ordinary and symbolic blow-ups, *Math. Z.* **198** (1988), 479–491.

[**17**] A. Simis, W.V. Vasconcelos and R. Villarreal, On the ideal theory of graphs, *J. Algebra* **167** (1994), 389–416.

[**18**] B. Sturmfels, *Gröbner Bases and Convex Polytopes*, University Lecture Series **8**, American Mathematical Society, Rhode Island, 1996.

[**19**] W.V. Vasconcelos, *Arithmetic of Blowup Algebras*, London Math. Soc., Lecture Note Series **195**, Cambridge University Press, Cambridge, 1994.

[**20**] W.V. Vasconcelos, *Computational Methods in Commutative Algebra and Algebraic Geometry*, Springer-Verlag, 1998.

[**21**] R. Villarreal, Normality of subrings generated by square free monomials, *J. Pure Appl. Algebra* **113** (1996), 91–106.

[**22**] R. Villarreal, *Monomial Algebras*, Monographs and Textbooks in Pure and Applied Mathematics **238**, Marcel Dekker, Inc., New York, 2001.

[**23**] R. Webster, *Convexity*, Oxford University Press, Oxford, 1994.

Monomial ideals via square-free monomial ideals

SARA FARIDI[1]
Mathematics Department, University of Ottawa
585 King Edward Ave. Ottawa, ON, Canada K1N 6N5
E-mail: faridi@uottawa.ca

1 Introduction

In this paper we study monomial ideals using the operation "polarization" to first turn them into square-free monomial ideals. Various forms of polarization appear throughout the literature and have been used for different purposes in algebra and algebraic combinatorics (for example, Weyman [**17**], Fröberg [**8**], Schwartau [**13**] or Rota and Stein [**11**]). One of the most useful features of polarization is that the chain of substitutions that turn a given monomial ideal into a square-free one can be described in terms of a regular sequence (Fröberg [**8**]). This fact allows many properties of a monomial ideal to transfer to its polarization. Conversely, to study a given monomial ideal, one could examine its polarization. The advantage of this latter approach is that there are many combinatorial tools dealing with square-free

[1]This research was supported by NSERC, and by Laboratoire de combinatoire et d'informatique mathématique at UQÀM, Montreal.

monomial ideals. One of these tools is Stanley-Reisner theory: Schwartau's thesis [**13**] and the book by Stückrad and Vogel [**15**] discuss how the Stanley-Reisner theory of square-free monomial ideals produces results about general monomial ideals using polarization. Another tool for studying square-free monomial ideals, which will be our focus here, is facet ideal theory, developed by the author in [**5**], [**6**] and [**7**].

The paper is organized as follows. In Section 2 we define polarization and introduce some of its basic properties. In Section 3 we introduce facet ideals and their features that are relevant to this paper. In particular, we introduce simplicial trees, which correspond to square-free monomial ideals with exceptionally strong algebraic properties. Section 4 extends the results of facet ideal theory to general monomial ideals. Here we study a monomial ideal I whose polarization is a tree, and show that many of the properties of simplicial trees hold for such ideals. This includes Cohen-Macaulayness of the Rees ring of I (Corollary 4.8), I being sequentially Cohen-Macaulay (Corollary 4.12), and several inductive tools for studying such ideals, such as localization (see Section 4.1).

Appendix A is an independent study of primary decomposition in a sequentially Cohen-Macaulay module. We demonstrate how in a sequentially Cohen-Macaulay module M, every submodule appearing in the filtration of M can be described in terms of the primary decomposition of the 0-submodule of M. This is used to prove Proposition 4.11.

2 Polarization

Definition 2.1 Let $R = k[x_1,\dots,x_n]$ be a polynomial ring over a field k. Suppose $M = {x_1}^{a_1}\dots {x_n}^{a_n}$ is a monomial in R. Then we define the *polarization* of M to be the square-free monomial

$$\mathcal{P}(M) = x_{1,1}x_{1,2}\dots x_{1,a_1}x_{2,1}\dots x_{2,a_2}\dots x_{n,1}\dots x_{n,a_n}$$

in the polynomial ring $S = k[x_{i,j} \mid 1 \leq i \leq n, 1 \leq j \leq a_i]$.

If I is an ideal of R generated by monomials $M_1,\dots,M_q$, then the *polarization* of I is defined as:

$$\mathcal{P}(I) = \big(\mathcal{P}(M_1),\dots,\mathcal{P}(M_q)\big)$$

which is a square-free monomial ideal in a polynomial ring S.

Here is an example of how polarization works.

Example 2.2 Let $J = (x_1{}^2, x_1x_2, x_2{}^3) \subseteq R = k[x_1, x_2]$. Then

$$\mathcal{P}(J) = (x_{1,1}x_{1,2}, x_{1,1}x_{2,1}, x_{2,1}x_{2,2}x_{2,3})$$

is the polarization of J in the polynomial ring

$$S = k[x_{1,1}, x_{1,2}, x_{2,1}, x_{2,2}, x_{2,3}].$$

Note that by identifying each x_i with $x_{i,1}$, one can consider S as a polynomial extension of R. Exactly how many variables S has will always depend on what we polarize. Therefore, as long as we are interested in the polarizations of finitely many monomials and ideals, S remains a finitely generated algebra.

Below we describe some basic properties of polarization, some of which appear (without proof) in [**15**]. Here we record the proofs where appropriate.

Proposition 2.3 (basic properties of polarization) *Suppose $R = k[x_1, \ldots, x_n]$ is a polynomial ring over a field k, and I and J are two monomial ideals of R.*

1. $\mathcal{P}(I + J) = \mathcal{P}(I) + \mathcal{P}(J)$;

2. *For two monomials M and N in R, $M \mid N$ if and only if $\mathcal{P}(M) \mid \mathcal{P}(N)$;*

3. $\mathcal{P}(I \cap J) = \mathcal{P}(I) \cap \mathcal{P}(J)$;

4. *If $\mathfrak{p} = (x_{i_1}, \ldots, x_{i_r})$ is a (minimal) prime containing I, then $\mathcal{P}(\mathfrak{p})$ is a (minimal) prime containing $\mathcal{P}(I)$;*

5. *If $\mathfrak{p}' = (x_{i_1,e_1}, \ldots, x_{i_r,e_r})$ is a prime over $\mathcal{P}(I)$, then $\mathfrak{p} = (x_{i_1}, \ldots, x_{i_r})$ is a prime over I. Moreover, if $\mathfrak{p}'$ has minimal height (among all primes containing $\mathcal{P}(I)$), then $\mathfrak{p}$ must have minimal height as well (among all primes containing I);*

6. height $I =$ height $\mathcal{P}(I)$.

Proof:

1. Follows directly from Definition 2.1.

2. Suppose that $M = {x_1}^{b_1} \dots {x_n}^{b_n}$ and $N = {x_1}^{c_1} \dots {x_n}^{c_n}$, and suppose that

$$\mathcal{P}(M) = x_{1,1} \dots x_{1,b_1} \dots x_{n,1} \dots x_{n,b_n}$$

and

$$\mathcal{P}(N) = x_{1,1} \dots x_{1,c_1} \dots x_{n,1} \dots x_{n,c_n}.$$

If $M \mid N$, then $b_i \leq c_i$ for all i, which implies that $\mathcal{P}(M) \mid \mathcal{P}(N)$. The converse is also clear using the same argument.

3. Suppose that $I = (M_1, \dots, M_q)$ and $J = (N_1, \dots, N_s)$ where the generators are all monomials. If $U = {x_1}^{b_1} \dots {x_n}^{b_n}$ is a monomial in $I \cap J$, then for some generator M_i of I and N_j and J, we have $M_i \mid U$ and $N_j \mid U$, hence by part 2, $\mathcal{P}(M_i) \mid \mathcal{P}(U)$ and $\mathcal{P}(N_j) \mid \mathcal{P}(U)$, which implies that $\mathcal{P}(U) \in \mathcal{P}(I) \cap \mathcal{P}(J)$.

   Conversely, if U' is a monomial in $\mathcal{P}(I) \cap \mathcal{P}(J)$, then for some generator $M_i = {x_1}^{b_1} \dots {x_n}^{b_n}$ of I and $N_j = {x_1}^{c_1} \dots {x_n}^{c_n}$ and J we have $\mathcal{P}(M_i) \mid U'$ and $\mathcal{P}(N_j) \mid U'$. This means that $\text{lcm}(\mathcal{P}(M_i), \mathcal{P}(N_j)) \mid U'$. It is easy to see (by an argument similar to the one in part 2) that $\text{lcm}(\mathcal{P}(M_i), \mathcal{P}(N_j)) = \mathcal{P}(\text{lcm}(M_i, N_j))$. Since $\text{lcm}(M_i, N_j)$ is one of the generators of $I \cap J$, it follows that $\mathcal{P}(\text{lcm}(M_i, N_j))$ is a generator of $\mathcal{P}(I \cap J)$ and hence $U' \in \mathcal{P}(I \cap J)$.

4. If $\mathfrak{p} = (x_{i_1}, \dots, x_{i_r})$ is a minimal prime over $I = (M_1, \dots, M_q)$, then for each of the x_{i_j} there is an M_t such that $x_{i_j} \mid M_t$, and no other generator of $\mathfrak{p}$ divides M_t. The same holds for the polarization of the two ideals: $\mathcal{P}(\mathfrak{p}) = (x_{i_1,1}, \dots, x_{i_r,1})$ and $\mathcal{P}(I) = (\mathcal{P}(M_1), \dots, \mathcal{P}(M_t))$, and so $\mathcal{P}(\mathfrak{p})$ is minimal over $\mathcal{P}(I)$.

5. Suppose that $\mathfrak{p}' = (x_{i_1,e_1}, \dots, x_{i_r,e_r})$ is a prime lying over $\mathcal{P}(I)$. Then for every generator M_t of I, there is an x_{i_j,e_j} in $\mathfrak{p}'$ such that $x_{i_j,e_j} \mid \mathcal{P}(M_t)$. But this implies that $x_{i_j} \mid M_t$, and therefore $I \subseteq \mathfrak{p} = (x_{i_1}, \dots, x_{i_r})$.

   Now suppose that $\mathfrak{p}'$ has minimal height r over $\mathcal{P}(I)$, and there is a prime ideal $\mathfrak{q}$ over I with height $\mathfrak{q} < r$. This implies (from part 4) that $\mathcal{P}(\mathfrak{q})$, which is a prime of height less than r, contains $\mathcal{P}(I)$, which is a contradiction.

6. This follows from parts 4 and 5.

$\square$

Example 2.4 It is not true that every minimal prime of $\mathcal{P}(I)$ comes from a minimal prime of I. For example, let $I = (x_1^2, x_1x_2^2)$. Then

$$\mathcal{P}(I) = (x_{1,1}x_{1,2}, x_{1,1}x_{2,1}x_{2,2}).$$

The ideal $(x_{1,2}, x_{2,1})$ is a minimal prime over $\mathcal{P}(I)$, but the corresponding prime (x_1, x_2) is not a minimal prime of I (however, if we had taken any minimal prime of minimal height of $\mathcal{P}(I)$, e.g. $(x_{1,1})$, then the corresponding prime over I would have been minimal; this is part 5 above).

For a monomial ideal I in a polynomial ring $R = k[x_1, \dots, x_n]$ as above, there is a unique irredundant irreducible decomposition of the form

$$I = \mathfrak{q}_1 \cap \dots \cap \mathfrak{q}_m$$

where each $\mathfrak{q}_i$ is a primary ideal generated by powers of the variables $x_1, \dots, x_n$ (see [**16**, Theorem 5.1.17]).

Proposition 2.5 (polarization and primary decomposition) *Let I be a monomial ideal in a polynomial ring $R = k[x_1, \dots, x_n]$, and let $\mathcal{P}(I)$ be the polarization of I in $S = k[x_{i,j}]$ as described in* Definition 2.1.

1. *If $I = ({x_{i_1}}^{a_1}, \dots, {x_{i_r}}^{a_r})$ where the a_j are positive integers, then*

$$\mathcal{P}(I) = \bigcap_{\substack{1 \le c_j \le a_j \\ 1 \le j \le r}} (x_{i_1,c_1}, \dots, x_{i_r,c_r}).$$

2. *If $I = (x_{i_1}, \dots, x_{i_r})^m$, where $1 \le i_1, \dots, i_r \le n$ and m is a positive integer, then $\mathcal{P}(I)$ has the following irredundant irreducible primary decomposition:*

$$\mathcal{P}(I) = \bigcap_{\substack{1 \le c_j \le m \\ \Sigma c_j \le m+r-1}} (x_{i_1,c_1}, \dots, x_{i_r,c_r}).$$

3. *Suppose that $I = \mathfrak{q}_1 \cap \dots \cap \mathfrak{q}_m$ is the unique irredundant irreducible primary decomposition of I, such that for each $i = 1, \dots, m$,*

$$\mathfrak{q}_i = ({x_1}^{a_1^i}, \dots, {x_n}^{a_n^i}),$$

where the a^i_j are nonnegative integers, and if $a^i_j = 0$ we assume that $x_j{}^{a^i_j} = 0$.

Then $\mathcal{P}(I)$ has the following irreducible primary decomposition (some primes might be repeated):

$$\mathcal{P}(I) = \bigcap_{1 \leq i \leq m} \bigcap_{\substack{1 \leq c_j \leq a^i_j \\ 1 \leq j \leq n}} (x_{1,c_1}, \ldots, x_{n,c_n})$$

where when $a^i_j = 0$, we assume that $c_j = x_{j,0} = 0$.

Proof:

1. We know that

$$\mathcal{P}(I) = (x_{i_1,1} \ldots x_{i_1,a_1}, \ldots, x_{i_1,1} \ldots x_{i_1,a_r}).$$

Clearly the minimal primes of $\mathcal{P}(I)$ are $(x_{i_1,c_1}, \ldots, x_{i_r,c_r})$ for all $c_j \leq a_j$. This settles the claim.

2. Assume, without loss of generality, that $I = (x_1, \ldots, x_r)^m$. So we can write

$$I = \left(x_1{}^{b_1} \ldots x_r{}^{b_r} \mid 0 \leq b_i \leq m,\ b_1 + \cdots + b_r = m\right)$$

so that

$$\mathcal{P}(I) = \left(x_{1,1} \ldots x_{1,b_1} \ldots x_{r,1} \ldots x_{r,b_r} \mid 0 \leq b_i \leq m,\ b_1 + \cdots + b_r = m\right).$$

We first show that $\mathcal{P}(I)$ is contained in the intersection of the ideals of the form $(x_{1,c_1}, \ldots, x_{r,c_r})$ described above. It is enough to show this for each generator of $\mathcal{P}(I)$. So we show that

$$\mathcal{U} = x_{1,1} \ldots x_{1,b_1} \ldots x_{r,1} \ldots x_{r,b_r} \in I = (x_{1,c_1}, \ldots, x_{r,c_r})$$

where $0 \leq b_i \leq m$, $b_1 + \cdots + b_r = m$, $1 \leq c_j \leq m$ and $c_1 + \cdots + c_r \leq m + r - 1$.

If for any i, $b_i \geq c_i$, then it would be clear that $\mathcal{U} \in I$.

Assume $b_i \leq c_i - 1$ for $i = 1, \ldots, r-1$. It follows that

$$\begin{aligned} m - b_r &= b_1 + \cdots + b_{r-1} \\ &\leq c_1 + \cdots + c_{r-1} - (r-1) \\ &\leq m + r - 1 - c_r - (r-1) \\ &= m - c_r \end{aligned}$$

which implies that $b_r \geq c_r$, hence $\mathcal{U} \in I$.

So far we have shown one direction of the inclusion.

To show the opposite direction, take any monomial

$$\mathcal{U} \in \bigcap (x_{1,c_1}, \ldots, x_{r,c_r})$$

where $1 \leq c_j \leq m$ and $c_1 + \cdots + c_r \leq m + r - 1$.

Notice that for some $i \leq r$, $x_{i,1} \mid \mathcal{U}$; this is because $\mathcal{U} \in (x_{1,1}, \ldots, x_{r,1})$.

We write $\mathcal{U}$ as

$$\mathcal{U} = x_{1,1} \ldots x_{1,b_1} \ldots x_{r,1} \ldots x_{r,b_r} \mathcal{U}'$$

where $\mathcal{U}'$ is a monomial, and the b_i are nonnegative integers such that for each $j < b_i$, $x_{i,j} \mid \mathcal{U}$ (if $x_{i,1} \nmid \mathcal{U}$ then set $b_i = 0$). We need to show that it is possible to find such b_i so that $b_1 + \cdots + b_r = m$.

Suppose $b_1 + \cdots + b_r \leq m - 1$, and $x_{i,b_i+1} \nmid \mathcal{U}$ for $1 \leq i \leq r$. Then

$$b_1 + \cdots + b_r + r \leq m + r - 1,$$

hence

$$\mathcal{U} \in (x_{1,b_1+1}, \ldots, x_{r,b_r+1}),$$

implying that $x_{i,b_i+1} \mid \mathcal{U}$ for some i, which is a contradiction.

Therefore $b_1, \ldots, b_r$ can be picked so that they add up to m, and hence $\mathcal{U} \in \mathcal{P}(I)$; this settles the opposite inclusion.

3. This follows from part 1 and Proposition 2.3, part 3.

□

Corollary 2.6 (polarization and associated primes) *Let I be a monomial ideal in a polynomial ring $R = k[x_1, \ldots, x_n]$, and let $\mathcal{P}(I)$ be its polarization in $S =$*

$k[x_{i,j}]$ as described in Definition 2.1. *Then $(x_{i_1},\dots,x_{i_r}) \in \mathrm{Ass}_R(R/I)$ if and only if $(x_{i_1,c_1},\dots,x_{i_r,c_r}) \in \mathrm{Ass}_S(S/\mathcal{P}(I))$ for some positive integers $c_1,\dots,c_r$. Moreover, if $(x_{i_1,c_1},\dots,x_{i_r,c_r}) \in \mathrm{Ass}_S(S/\mathcal{P}(I))$, then $(x_{i_1,b_1},\dots,x_{i_r,b_r}) \in \mathrm{Ass}_S(S/\mathcal{P}(I))$ for all b_j such that $1 \le b_j \le c_j$.*

Example 2.7 Consider the primary decomposition of $J = ({x_1}^2, {x_2}^3, x_1x_2)$:

$$J = (x_1, {x_2}^3) \cap ({x_1}^2, x_2).$$

By Proposition 2.5, $\mathcal{P}(J) = (x_{1,1}x_{1,2}, x_{2,1}x_{2,2}x_{2,3}, x_{1,1}x_{2,1})$ will have primary decomposition

$$\mathcal{P}(J) = (x_{1,1}, x_{2,1}) \cap (x_{1,1}, x_{2,2}) \cap (x_{1,1}, x_{2,3}) \cap (x_{1,2}, x_{2,1}).$$

A very useful property of polarization is that the final polarized ideal is related to the original ideal via a regular sequence. The proposition below, which looks slightly different here than the original statement in [**8**], states this fact.

Proposition 2.8 (Fröberg [8]) *Let k be a field and*

$$R = k[x_1,\dots,x_n]/(M_1,\dots,M_q),$$

where $M_1,\dots,M_q$ are monomials in the variables $x_1,\dots,x_n$, and let

$$N_1 = \mathcal{P}(M_1),\dots,N_q = \mathcal{P}(M_q)$$

be a set of square-free monomials in the polynomial ring

$$S = k[x_{i,j} \mid 1 \le i \le n, 1 \le j \le a_i]$$

such that for each i, the variable x_{i,a_i} appears in at least one of the monomials $N_1,\dots,N_q$. Then the sequence of elements

$$x_{i,1} - x_{i,j} \text{ where } 1 \le i \le n \text{ and } 1 < j \le a_i \qquad (1)$$

forms a regular sequence in the quotient ring

$$R' = S/(N_1,\dots,N_q)$$

and if J is the ideal of R' generated by the elements in (1), *then*

$$R = R'/J.$$

Moreover, R is Cohen-Macaulay (Gorenstein) if and only if R' is.

Example 2.9 Let J and R be as in Example 2.2. According to Proposition 2.8, the sequence

$$x_{1,1} - x_{1,2},\ x_{2,1} - x_{2,2},\ x_{2,1} - x_{2,3}$$

is a regular sequence in $S/\mathcal{P}(J)$, and

$$R/J = S/\big(\mathcal{P}(J) + (x_{1,1} - x_{1,2}, x_{2,1} - x_{2,2}, x_{2,1} - x_{2,3})\big).$$

3 Square-free monomial ideals as facet ideals

Now that we have introduced polarization as a method of transforming a monomial ideal into a square-free one, we can focus on square-free monomial ideals. In particular, here we are interested in properties of square-free monomial ideals that come as a result of them being considered as facet ideals of simplicial complexes. Below we review the basic definitions and notations in facet ideal theory, as well as some of the basic concepts of Stanley-Reisner theory. We refer the reader to [**2**], [**5**], [**6**], [**7**] and [**14**] for more details and proofs in each of these topics.

Definition 3.1 (simplicial complex, facet, subcollection and more) A *simplicial complex* Δ over a set of vertices $V = \{v_1, \ldots, v_n\}$ is a collection of subsets of V, with the property that $\{v_i\} \in \Delta$ for all i, and if $F \in \Delta$ then all subsets of F are also in Δ (including the empty set). An element of Δ is called a *face* of Δ, and the *dimension* of a face F of Δ is defined as $|F| - 1$, where $|F|$ is the number of vertices of F. The faces of dimensions 0 and 1 are called *vertices* and *edges*, respectively, and dim $\emptyset = -1$. The maximal faces of Δ under inclusion are called *facets* of Δ. The dimension of the simplicial complex Δ is the maximal dimension of its facets.

We denote the simplicial complex Δ with facets $F_1, \ldots, F_q$ by

$$\Delta = \langle F_1, \ldots, F_q \rangle$$

and we call $\{F_1, \ldots, F_q\}$ the *facet set* of Δ. A simplicial complex with only one facet is called a *simplex*. By a *subcollection* of Δ we mean a simplicial complex whose facet set is a subset of the facet set of Δ.

Definition 3.2 (connected simplicial complex) A simplicial complex $\Delta = \langle F_1, \ldots, F_q \rangle$ is *connected* if for every pair i, j, $1 \leq i < j \leq q$, there exists a sequence of facets $F_{t_1}, \ldots, F_{t_r}$ of Δ such that $F_{t_1} = F_i$, $F_{t_r} = F_j$ and $F_{t_s} \cap F_{t_{s+1}} \neq \emptyset$ for $s = 1, \ldots, r-1$.

Definition 3.3 (facet/non-face ideals and complexes) Consider the polynomial ring $R = k[x_1, \ldots, x_n]$ over a field k and a set of indeterminates $x_1, \ldots, x_n$. Let $I = (M_1, \ldots, M_q)$ be an ideal in R, where $M_1, \ldots, M_q$ are square-free monomials that form a minimal set of generators for I.

- The *facet complex* of I, denoted by $\delta_{\mathcal{F}}(I)$, is the simplicial complex over a set of vertices $v_1, \ldots, v_n$ with facets $F_1, \ldots, F_q$, where for each i, $F_i = \{v_j \mid x_j | M_i,\ 1 \leq j \leq n\}$. The *non-face complex* or the *Stanley-Reisner complex* of I, denoted by $\delta_{\mathcal{N}}(I)$, is be the simplicial complex over a set of vertices $v_1, \ldots, v_n$, where $\{v_{i_1}, \ldots, v_{i_s}\}$ is a face of $\delta_{\mathcal{N}}(I)$ if and only if $x_{i_1} \ldots x_{i_s} \notin I$.
- Conversely, if Δ is a simplicial complex over n vertices labeled $v_1, \ldots, v_n$, we define the *facet ideal* of Δ, denoted by $\mathcal{F}(\Delta)$, to be the ideal of R generated by square-free monomials $x_{i_1} \ldots x_{i_s}$, where $\{v_{i_1}, \ldots, v_{i_s}\}$ is a facet of Δ. The *non-face ideal* or the *Stanley-Reisner ideal* of Δ, denoted by $\mathcal{N}(\Delta)$, is the ideal of R generated by square-free monomials $x_{i_1} \ldots x_{i_s}$, where $\{v_{i_1}, \ldots, v_{i_s}\}$ is not a face of Δ.

Throughout this paper we often use the letter x to denote both a vertex of Δ and the corresponding variable appearing in $\mathcal{F}(\Delta)$, and $x_{i_1} \ldots x_{i_r}$ to denote a facet of Δ as well as a monomial generator of $\mathcal{F}(\Delta)$.

Example 3.4 If Δ is the simplicial complex $\langle xyz, yu, uvw \rangle$ drawn below,

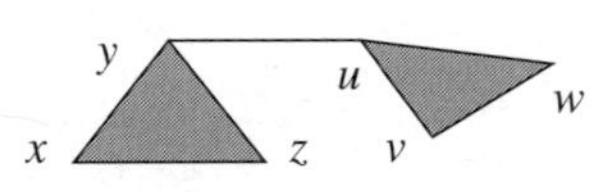

then $\mathcal{F}(\Delta) = (xyz, yu, uvw)$ and $\mathcal{N}(\Delta) = (xu, xv, xw, yv, yw, zu, zv, zw)$ are its facet ideal and nonface (Stanley-Reisner) ideal, respectively.

Facet ideals give a one-to-one correspondence between simplicial complexes and square-free monomial ideals.

Next we define the notion of a vertex cover. The combinatorial idea here comes from graph theory. In algebra, it corresponds to prime ideals lying over the facet ideal of a given simplicial complex.

Definition 3.5 (vertex covering, independence, unmixed) Let Δ be a simplicial complex with vertex set V. A *vertex cover* for Δ is a subset A of V that intersects every facet of Δ. If A is a minimal element (under inclusion) of the set of vertex covers of Δ, it is called a *minimal vertex cover*. The smallest of the cardinalities of the vertex covers of Δ is called the *vertex covering number* of Δ and is denoted by $\alpha(\Delta)$. A simplicial complex Δ is *unmixed* if all of its minimal vertex covers have the same cardinality.

A set $\{F_1,\ldots,F_u\}$ of facets of Δ is called an *independent set* if $F_i \cap F_j = \emptyset$ whenever $i \neq j$. The maximum possible cardinality of an independent set of facets in Δ, denoted by $\beta(\Delta)$, is called the *independence number* of Δ. An independent set of facets which is not a proper subset of any other independent set is called a *maximal independent set* of facets.

Example 3.6 If Δ is the simplicial complex in Example 3.4, then the vertex covers of Δ are:

$$\{\mathbf{x},\mathbf{u}\},\{\mathbf{y},\mathbf{u}\},\{\mathbf{y},\mathbf{v}\},\{\mathbf{y},\mathbf{w}\},\{\mathbf{z},\mathbf{u}\},\{x,y,u\},\{x,z,u\},\{x,y,v\},\ldots.$$

The first five vertex covers above (highlighted in bold) are the minimal vertex covers of Δ. It follows that $\alpha(\Delta)=2$, and Δ is unmixed. On the other hand, $\{xyz,uvw\}$ is the largest maximal independent set of facets that Δ contains, and so $\beta(\Delta)=2$.

Definition 3.7 (Alexander dual) Let I be a square-free monomial ideal in the polynomial ring $k[V]$ with $V=\{x_1,\ldots,x_n\}$, and let Δ_N be the non-face complex of I (i.e. $\Delta_N=\delta_{\mathcal{N}}(I)$). Then the *Alexander dual* of Δ_N is the simplicial complex

$$\Delta_N{}^{\vee}=\{F\subset V \mid F^c\notin\Delta_N\}$$

where F^c is the complement of the face F in V.

We call the nonface ideal of $\Delta_N{}^{\vee}$ the *Alexander dual* of I and denote it by $I^{\vee}$.

3.1 Simplicial trees

Considering simplicial complexes as higher dimensional graphs, one can define the notion of a *tree* by extending the same concept from graph theory. Before we define a tree, we determine what "removing a facet" from a simplicial complex means. We define this idea so that it corresponds to dropping a generator from the facet ideal of the complex.

Definition 3.8 (facet removal) Suppose Δ is a simplicial complex with facets $F_1, \dots, F_q$ and $\mathcal{F}(\Delta) = (M_1, \dots, M_q)$ is its facet ideal in $R = k[x_1, \dots, x_n]$. The simplicial complex obtained by *removing the facet* F_i from Δ is the simplicial complex

$$\Delta \setminus \langle F_i \rangle = \langle F_1, \dots, \hat{F_i}, \dots, F_q \rangle$$

and $\mathcal{F}(\Delta \setminus \langle F_i \rangle) = (M_1, \dots, \hat{M_i}, \dots, M_q)$.

The definition that we give below for a simplicial tree is one generalized from graph theory. See [**5**] and [**6**] for more on this concept.

Definition 3.9 (leaf, joint) A facet F of a simplicial complex is called a *leaf* if either F is the only facet of Δ, or for some facet $G \in \Delta \setminus \langle F \rangle$ we have

$$F \cap (\Delta \setminus \langle F \rangle) \subseteq G.$$

If $F \cap G \neq \emptyset$, the facet G above is called a *joint* of the leaf F.

Equivalently, a facet F is a leaf of Δ if $F \cap (\Delta \setminus \langle F \rangle)$ is a face of $\Delta \setminus \langle F \rangle$.

Example 3.10 Let $I = (xyz, yzu, zuv)$. Then $F = xyz$ is a leaf, but $H = yzu$ is not, as one can see in the picture below.

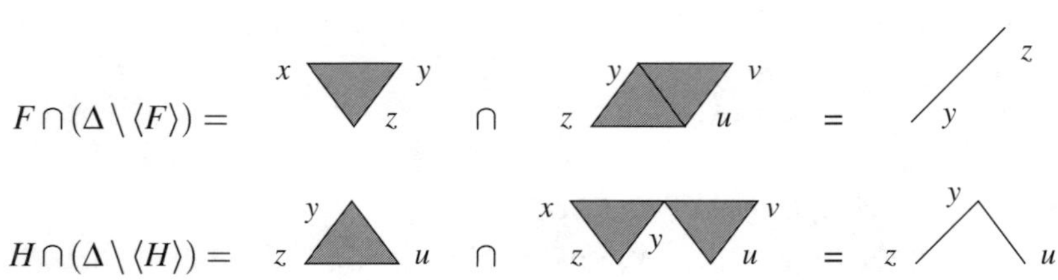

Definition 3.11 (tree, forest) A connected simplicial complex Δ is a *tree* if every nonempty subcollection of Δ has a leaf. If Δ is not necessarily connected, but every subcollection has a leaf, then Δ is called a forest.

Example 3.12 The simplicial complexes in Examples 3.4 and 3.10 are both trees, but the one below is not because it has no leaves. It is an easy exercise to see that a leaf must contain a free vertex, where a vertex is *free* if it belongs to only one facet.

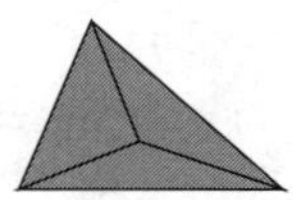

One of the most powerful properties of simplicial trees from the point of view of algebra is that they behave well under localization. This property makes it easy to use induction on the number of vertices of a tree for proving its various properties.

Lemma 3.13 (localization of a tree is a forest) *Let $I \subseteq k[x_1,\ldots,x_n]$ be the facet ideal of a simplicial tree, where k is a field. Then for any prime ideal $\mathfrak{p}$ of $k[x_1,\ldots,x_n]$, $\delta_{\mathcal{F}}(I_{\mathfrak{p}})$ is a forest.*

Proof: See [**6**, Lemma 4.5]. □

4 Properties of monomial ideals via polarization

For the purpose of all discussions in this section, unless otherwise stated, let I be a monomial ideal in the polynomial ring $R = k[x_1,\ldots,x_n]$ over a field k, whose polarization is the square-free monomial ideal $\mathcal{P}(I)$ in the polynomial ring

$$S = k[x_{i,j} \mid 1 \leq i \leq n, 1 \leq j \leq a_i].$$

We assume that the polarizing sequence (as described in (1) in Proposition 2.8) is

$$\nu = \nu_1,\ldots,\nu_v$$

which is a regular sequence in $S/\mathcal{P}(I)$ and

$$R/I = S/\big(\mathcal{P}(I) + (\mathrm{v})\big).$$

4.1 Monomial ideals whose polarization is a simplicial tree

A natural question, and one that this paper is mainly concerned with, is what properties of facet ideals of simplicial trees can be extended to general (non-square-free) monomial ideals using polarization? In other words, if for a monomial ideal I in a polynomial ring $\mathcal{P}(I)$ is the facet ideal of a tree (Definition 3.11), then what properties of $\mathcal{P}(I)$ are inherited by I?

The strongest tool when dealing with square-free monomial ideals is induction — either on the number of generators, or the number of variables in the ambient polynomial ring. This is particularly the case when the facet complex of the ideal is a tree, or in some cases when it just has a leaf. In this section we show that via polarization, one can extend these tools to monomial ideals in general. For a given monomial ideal I, we show that if $\mathcal{P}(I)$ is the facet ideal of a tree, and $\mathfrak{p}$ is a prime ideal containing I, then $\mathcal{P}(I_{\mathfrak{p}})$ is the facet ideal of a forest (Theorem 4.1); this allows induction on number of variables. Similarly, Theorem 4.3 provides us with a way to use induction on number of generators of I.

Theorem 4.1 (localization and polarization) *If $\mathcal{P}(I)$ is the facet ideal of a tree, and $\mathfrak{p}$ is a prime ideal of R containing I, then $\mathcal{P}(I_{\mathfrak{p}})$ is the facet ideal of a forest.*

Proof: The first step is to show that it is enough to prove this for prime ideals of R generated by a subset of $\{x_1,\dots,x_n\}$. To see this, assume that $\mathfrak{p}$ is a prime ideal of R containing I, and that $\mathfrak{p}'$ is another prime of R generated by all $x_i \in \{x_1,\dots,x_n\}$ such that $x_i \in \mathfrak{p}$ (recall that the minimal primes of I are generated by subsets of $\{x_1,\dots,x_n\}$; see [**16**, Corollary 5.1.5]). So $\mathfrak{p}' \subseteq \mathfrak{p}$. If $I = (M_1,\dots,M_q)$, then

$$I_{\mathfrak{p}'} = (M_1{}',\dots,M_q{}')$$

where for each i, $M_i{}'$ is the image of M_i in $I_{\mathfrak{p}'}$. In other words, $M_i{}'$ is obtained by dividing $M_i = x_1{}^{a_1}\dots x_n{}^{a_n}$ by the product of all the $x_j{}^{a_j}$ such that $x_j \notin \mathfrak{p}'$. But $x_j \notin \mathfrak{p}'$ implies that $x_j \notin \mathfrak{p}$, and so it follows that $M_i{}' \in I_{\mathfrak{p}}$. Therefore $I_{\mathfrak{p}'} \subseteq I_{\mathfrak{p}}$. On the other hand, since $\mathfrak{p}' \subseteq \mathfrak{p}$, $I_{\mathfrak{p}} \subseteq I_{\mathfrak{p}'}$, which implies that $I_{\mathfrak{p}'} = I_{\mathfrak{p}}$ (the equality and inclusion of the ideals here mean equality and inclusion of their generating sets).

Now suppose $I = (M_1, \dots, M_q)$ and $\mathfrak{p} = (x_1, \dots, x_r)$ is a prime containing I. Suppose that for each i, we write $M_i = M'_i.M''_i$, where

$$M'_i \in k[x_1, \dots, x_r] \text{ and } M''_i \in k[x_{r+1}, \dots, x_n]$$

so that

$$I_{\mathfrak{p}} = (M'_1, \dots, M'_t),$$

where without loss of generality $M'_1, \dots, M'_t$ is a minimal generating set for $I_{\mathfrak{p}}$.

We would like to show that the facet complex Δ of $\mathcal{P}(I_{\mathfrak{p}})$ is a forest. Suppose that, again without loss of generality,

$$I' = (\mathcal{P}(M'_1), \dots, \mathcal{P}(M'_s))$$

is the facet ideal of a subcollection Δ' of Δ. We need to show that Δ' has a leaf.

If $s = 1$, then there is nothing to prove. Otherwise, suppose that $\mathcal{P}(M_1)$ represents a leaf of the tree $\delta_{\mathcal{F}}(\mathcal{P}(I))$, and $\mathcal{P}(M_2)$ is a joint of $\mathcal{P}(M_1)$. Then we have

$$\mathcal{P}(M_1) \cap \mathcal{P}(M_i) \subseteq \mathcal{P}(M_2) \; for \; all \; i \in \{2, \dots, s\}.$$

Now let $x_{e,f}$ be in $\mathcal{P}(M'_1) \cap \mathcal{P}(M'_i)$ for some $i \in \{2, \dots, s\}$. This implies that

(i) $x_{e,f} \in \mathcal{P}(M_1) \cap \mathcal{P}(M_i) \subseteq \mathcal{P}(M_2)$, and

(ii) $e \in \{1 \dots, r\}$.

From *(i)* and *(ii)* we can conclude that $x_{e,f} \in \mathcal{P}(M'_2)$, which proves that $\mathcal{P}(M'_1)$ is a leaf for Δ'. □

Remark 4.2 It is not true in general that if $\mathfrak{p}$ is a (minimal) prime of I, then $\mathcal{P}(I_{\mathfrak{p}}) = \mathcal{P}(I)_{\mathcal{P}(\mathfrak{p})}$. For example, if $I = ({x_1}^3, {x_1}^2 x_2)$ and $\mathfrak{p} = (x_1)$, then $I_{\mathfrak{p}} = ({x_1}^2)$ so $\mathcal{P}(I_{\mathfrak{p}}) = (x_{1,1}x_{1,2})$, but $\mathcal{P}(I)_{\mathcal{P}(\mathfrak{p})} = (x_{1,1})$.

Another feature of simplicial trees is that they satisfy a generalization of König's theorem ([**6**, Theorem 5.3]). Below we explain how this property, and another property of trees that are very useful for induction, behave under polarization.

Recall that for a simplicial complex Δ, $\alpha(\Delta)$ and $\beta(\Delta)$ are the vertex covering number and the independence number of Δ, respectively (Definition 3.5). For simplicity of notation, if $I = (M_1, \ldots, M_q)$ is a monomial ideal, we let $\beta(I)$ denote the maximum cardinality of a subset of $\{M_1, \ldots, M_q\}$ consisting of pairwise coprime elements (so $\beta(\Delta) = \beta(\mathcal{F}(\Delta))$ for any simplicial complex Δ).

Theorem 4.3 (joint removal and polarization) *Suppose $M_1, \ldots, M_q$ are monomials that form a minimal generating set for I, and $\mathcal{P}(I)$ is the facet ideal of a simplicial complex Δ. Assume that Δ has a leaf whose joint corresponds to $\mathcal{P}(M_1)$. Then, if we let $I' = (M_2, \ldots, M_q)$, we have*

$$\text{height } I = \text{height } I'.$$

Proof: If G is the joint of Δ corresponding to $\mathcal{P}(M_1)$, then $\mathcal{P}(I') = \mathcal{F}(\Delta \setminus \langle G \rangle)$. From [**6**, Lemma 5.1] it follows that $\alpha(\Delta) = \alpha(\Delta \setminus \langle G \rangle)$, so that height $\mathcal{P}(I) =$ height $\mathcal{P}(I')$, and therefore height $I =$ height I'. □

Theorem 4.4 *Suppose $M_1, \ldots, M_q$ are monomials that form a minimal generating set for I, and $\mathcal{P}(I)$ is the facet ideal of a simplicial tree Δ. Then* height $I = \beta(I)$.

Proof: We already know that height $I =$ height $\mathcal{P}(I) = \alpha(\Delta)$. It is also clear that $\beta(I) = \beta(\mathcal{P}(I))$, since the monomials in a subset $\{M_{i_1}, \ldots, M_{i_r}\}$ of the generating set of I are pairwise coprime if and only if the monomials in $\{\mathcal{P}(M_{i_1}), \ldots, \mathcal{P}(M_{i_r})\}$ are pairwise coprime. On the other hand, from [**6**, Theorem 5.3] we know that $\alpha(\Delta) = \beta(\Delta)$. Our claim follows immediately. □

We demonstrate how to apply these theorems via an example.

Example 4.5 Suppose $I = ({x_1}^3, {x_1}^2 x_2 x_3, {x_3}^2, {x_2}^3 x_3)$. Then

$$\mathcal{P}(I) = (x_{1,1}x_{1,2}x_{1,3}, x_{1,1}x_{1,2}x_{2,1}x_{3,1}, x_{3,1}x_{3,2}, x_{2,1}x_{2,2}x_{2,3}x_{3,1})$$

is the facet ideal of the following simplicial complex (tree) Δ.

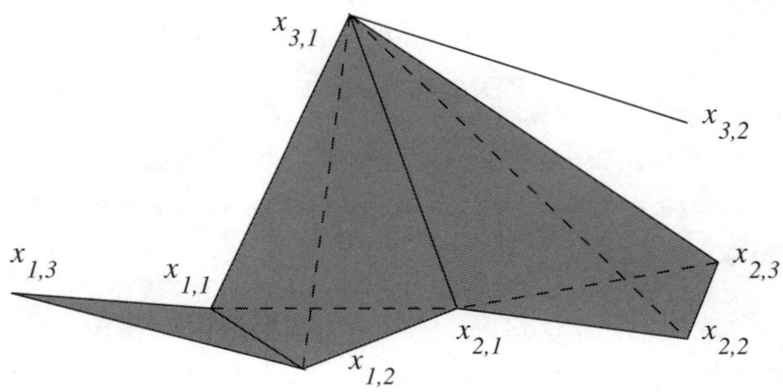

Now $\alpha(\Delta)$ = height $I = 2$ because the prime of minimal height over I is (x_1, x_3). From Theorem 4.4 it follows that $\beta(I) = 2$. This means that you can find a set of two monomials in the generating set of I that have no common variables: for example $\{x_1{}^3, x_3{}^2\}$ is such a set.

Since the monomials $x_1{}^2x_2x_3$ and $x_2{}^3x_3$ polarize into joints of Δ, by Theorem 4.3 the ideals

$$I,\ (x_1{}^3, x_3{}^2, x_2{}^3x_3),\ (x_1{}^3, x_1{}^2x_2x_3, x_3{}^2),\ \text{and}\ (x_1{}^3, x_3{}^2)$$

all have the same height.

We now focus on the Cohen-Macaulay property. In [**6**] we showed that for a simplicial tree Δ, $\mathcal{F}(\Delta)$ is a Cohen-Macaulay ideal if and only if Δ is an unmixed simplicial complex. The condition unmixed for Δ is equivalent to all minimal primes of the ideal $\mathcal{F}(\Delta)$ (which in this case are all the associated primes of $\mathcal{F}(\Delta)$) having the same height. In general, an ideal all of whose associated primes have the same height (equal to the height of the ideal) is called an *unmixed ideal*.

It now follows that:

Theorem 4.6 (Cohen-Macaulay criterion for trees) *Let $\mathcal{P}(I)$ be the facet ideal of a simplicial tree Δ. Then R/I is Cohen-Macaulay if and only if I is unmixed.*

Proof: From Proposition 2.8, R/I is Cohen-Macaulay if and only if $S/\mathcal{P}(I)$ is Cohen-Macaulay. By [**6**, Corollary 8.3], this is happens if and only if $\mathcal{P}(I)$ is unmixed. Corollary 2.6 now proves the claim. □

If $\mathcal{R}$ is a ring and J is an ideal of R, then the Rees ring of $\mathcal{R}$ along J is defined as

$$\mathcal{R}[Jt] = \oplus_{n\in\mathbb{N}} J^n t^n.$$

Rees rings come up in the algebraic process of "blowing up" ideals. One reason that trees are defined as they are is that their facet ideals produce normal and Cohen-Macaulay Rees rings ([**5**]).

Proposition 4.7 *If $S[\mathcal{P}(I)t]$ is Cohen-Macaulay, then so is $R[It]$. Conversely, if we assume that R and S are localized at their irrelevant maximal ideals, then $R[It]$ being Cohen-Macaulay implies that $S[\mathcal{P}(I)t]$ is Cohen-Macaulay.*

Proof: Suppose that $\nu_1,\dots,\nu_v$ is the polarizing sequence as described before. For $i = 1,\dots,v-1$ let

$$R_i = S/(\nu_1,\dots,\nu_i),\ I_i = \mathcal{P}(I)/(\nu_1,\dots,\nu_i),\ R_v = R \text{ and } I_v = I.$$

Notice that $S[\mathcal{P}(I)t]$ and $R[It]$ are both domains. Also note that for each i,

$$S[\mathcal{P}(I)t]/(\nu_1,\dots,\nu_i) = R_i[I_i t]$$

is the Rees ring of the monomial ideal I_i in the polynomial ring R_i, and is therefore also a domain. Therefore ν_{i+1} is a regular element in the ring $S[\mathcal{P}(I)t]/(\nu_1,\dots,\nu_i)$, which means that $\nu_1,\dots,\nu_v$ is a regular sequence in $S[\mathcal{P}(I)t]$.

Similarly, we see that

$$R[It] = S[\mathcal{P}(I)t]/(\nu_1,\dots,\nu_v).$$

[**2**, Theorem 2.1.3] now implies that if $S[\mathcal{P}(I)t]$ is Cohen-Macaulay, then so is $R[It]$. The converse follows from [**2**, Exercise 2.1.28]. □

Corollary 4.8 (Rees ring of a tree is Cohen-Macaulay) *Suppose that $\mathcal{P}(I)$ is the facet ideal of a simplicial tree. Then the Rees ring $R[It]$ of I is Cohen-Macaulay.*

Proof: This follows from the Proposition 4.7, and from [**5**, Corollary 4], which states that the Rees ring of the facet ideal of a simplicial tree is Cohen-Macaulay.

□

4.2 Polarization of sequentially Cohen-Macaulay ideals

The main result of this section is that if the polarization of a monomial ideal I is the facet ideal of a tree, then I is a sequentially Cohen-Macaulay ideal. The theorem that implies this fact (Proposition 4.11) is interesting in its own right. For a square-free monomial ideal J, Eagon and Reiner [**4**] proved that J is Cohen-Macaulay if and only if its Alexander dual $J^\vee$ has a linear resolution. Herzog and Hibi [**9**] then defined componentwise linear ideals and generalized their result, so that a square-free monomial ideal J is sequentially Cohen-Macaulay if and only if $J^\vee$ is componentwise linear (see [**9**] or [**7**]). But even though Alexander duality has been generalized to all monomial ideals from square-free ones, the criterion for sequential Cohen-Macaulayness does not generalize: it is not true that if I is any monomial ideal, then I is sequentially Cohen-Macaulay if and only of $I^\vee$ is a componentwise linear ideal; see Miller [**10**]. We show that the statement is true if $I^\vee$ is replaced by $\mathcal{P}(I)^\vee$.

Definition 4.9 ([14, Chapter III, Definition 2.9]) Let M be a finitely generated $\mathbb{Z}$-graded module over a finitely generated $\mathbb{N}$-graded k-algebra, with $R_0 = k$. We say that M is *sequentially Cohen-Macaulay* if there exists a finite filtration

$$0 = M_0 \subset M_1 \subset \ldots \subset M_r = M$$

of M by graded submodules M_i satisfying the following two conditions:

(a) Each quotient M_i/M_{i-1} is Cohen-Macaulay.

(b) $\dim\,(M_1/M_0) < \dim\,(M_2/M_1) < \ldots < \dim\,(M_r/M_{r-1})$, where dim denotes Krull dimension.

We define a componentwise linear ideal in the square-free case using [**9**, Proposition 1.5].

Definition 4.10 (componentwise linear) Let I be a square-free monomial ideal in a polynomial ring R. For a positive integer k, the k-th *square-free homogeneous component* of I, denoted by $I_{[k]}$, is the ideal generated by all square-free monomials in I of degree k. The ideal I above is said to be *componentwise linear* if for all k, the square-free homogeneous component $I_{[k]}$ has a linear resolution.

For our monomial ideal I, let $Q(I)$ denote the set of primary ideals appearing in a reduced primary decomposition of I. Suppose that $h = \text{height } I$ and $s = \max\{\text{height } \mathfrak{q} \mid \mathfrak{q} \in Q(I)\}$, and set

$$I_i = \bigcap_{\substack{\mathfrak{q}\in Q(I) \\ \text{height } \mathfrak{q} \leq s-i}} \mathfrak{q}.$$

So we have the following filtration for R/I (we assume that all inclusions in the filtration are proper; if there is an equality anywhere, we just drop all but one of the equal ideals):

$$0 = I = I_0 \subset I_1 \subset \ldots \subset I_{s-h} \subset R/I. \tag{2}$$

If R/I is sequentially Cohen-Macaulay, then by Theorem A.4, (2) is the appropriate filtration that satisfies Conditions (a) and (b) in Definition 4.9.

For the square-free monomial ideal $J = \mathcal{P}(I)$, we similarly define

$$Q(J) = \{\text{minimal primes over } J\} \text{ and } J_i = \bigcap_{\substack{\mathfrak{p}\in Q(J) \\ \text{height } \mathfrak{p} \leq s-i}} \mathfrak{p}$$

where the numbers h and s are the same as for I because of Proposition 2.5. It follows from Proposition 2.5 and Corollary 2.6 that for each i, $\mathcal{P}(I_i) = J_i$ and the polarization sequence that transforms I_i into J_i is a subsequence of $\nu = \nu_1, \ldots, \nu_\nu$.

What we have done so far is to translate, via polarization, the filtration (2) of the quotient ring R/I into one of S/J:

$$0 = J = J_0 \subset J_1 \subset \ldots \subset J_{s-h} \subset S/J. \tag{3}$$

Now note that for a given i, the sequence ν is a J_{i+1}/J_i-regular sequence in S, as ν is a regular sequence in S/J_i, which contains J_{i+1}/J_i. Also note that

$$R/I_i \simeq S/(J_i + \nu).$$

It follows that J_{i+1}/J_i is Cohen-Macaulay if and only if I_{i+1}/I_i is (see [**2**, Exercise 2.1.27(c), Exercise 2.1.28 and Theorem 2.1.3]).

Proposition 4.11 *The monomial ideal I is sequentially Cohen-Macaulay if and only if $\mathcal{P}(I)$ is sequentially Cohen-Macaulay, or equivalently, $\mathcal{P}(I)^\vee$ is a componentwise linear ideal.*

Proof: By [**7**, Proposition 4.5], $\mathcal{P}(I)^\vee$ is componentwise linear if and only if $\mathcal{P}(I)$ is sequentially Cohen-Macaulay, which by the discussion above is equivalent to I being sequentially Cohen-Macaulay. □

Corollary 4.12 (trees are sequentially Cohen-Macaulay) *Let $\mathcal{P}(I)$ be the facet ideal of a simplicial tree. Then I is sequentially Cohen-Macaulay.*

Proof: This follows from Proposition 4.11 and by [**7**, Corollary 5.5], which states that $\mathcal{P}(I)^\vee$ is a componentwise linear ideal. □

5 Further examples and remarks

To use the main results of this paper for computations on a given monomial ideal, there are two steps. One is to compute the polarization of the ideal, which as can be seen from the definition is a quick and simple procedure. This has already been implemented in Macaulay2. The second step is to determine whether the polarization is the facet ideal of a tree, or has a leaf. Algorithms that serve this purpose are under construction ([**3**]).

Remark 5.1 Let $I = (M_1, \ldots, M_q)$ be a monomial ideal in a polynomial ring R. If $\mathcal{P}(I)$ is the facet ideal of a tree, then by Corollary 4.8, $R[It]$ is Cohen-Macaulay. But more is true: if you drop any generator of I, for example if you consider $I' = (M_1, \ldots, \hat{M}_i, \ldots, M_q)$, then $R[I't]$ is still Cohen-Macaulay. This is because $\mathcal{P}(I')$ corresponds to the facet ideal of a forest, so one can apply the same result.

A natural question is whether one can say the same with the property "Cohen-Macaulay" replaced by "normal." If I is square-free, this is indeed the case. But in general, polarization does not preserve normality of ideals.

Example 5.2 (normality and polarization) A valid question is whether Proposition 4.7 holds if the word "Cohen-Macaulay" is replaced with "normal," given that simplicial trees have normal facet ideals ([**5**])?

The answer is negative. Here is an example.

Let $I = (x_1{}^3, x_1{}^2 x_2, x_2{}^3)$ be an ideal of $k[x_1, x_2]$. Then I is not normal; this is because I is not even integrally closed: $x_1 x_2{}^2 \in \bar{I}$ as $(x_1 x_2{}^2)^3 - x_1{}^3 x_2{}^6 = 0$, but $x_1 x_2{}^2 \notin I$. Now

$$\mathcal{P}(I) = (x_{1,1}x_{1,2}x_{1,3}, x_{1,1}x_{1,2}x_{2,1}, x_{2,1}x_{2,2}x_{2,3})$$

is the facet ideal of the tree

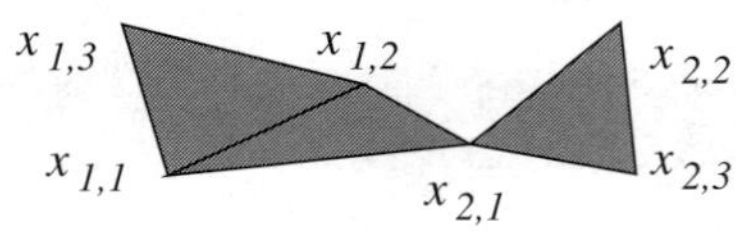

which is normal by [**5**].

The reason that normality (or integral closure in general) does not pass through polarization is much more basic: polarization does not respect multiplication of ideals, or monomials. Take, for example, two monomials M and N and two monomial ideals I and J, such that $MN \in IJ$. It is not necessarily true that $\mathcal{P}(M)\mathcal{P}(N) \in \mathcal{P}(I)\mathcal{P}(J)$.

Indeed, let $I = J = (x_1 x_2)$ and $M = x_1{}^2$ and $N = x_2{}^2$. Then $MN = x_1{}^2 x_2{}^2 \in IJ$. But

$$\mathcal{P}(M) = x_{1,1}x_{1,2},\ \mathcal{P}(N) = x_{2,1}x_{2,2},\ \text{and}\ \mathcal{P}(I) = \mathcal{P}(J) = (x_{1,1}x_{2,1})$$

and clearly $\mathcal{P}(M)\mathcal{P}(N) \notin \mathcal{P}(I)\mathcal{P}(J) = (x_{1,1}{}^2 x_{2,1}{}^2)$.

Remark 5.3 It is useful to think of polarization as a chain of substitutions. This way, as a monomial ideal I gets polarized, the ambient ring extends one variable at a time. All the in-between ideals before we hit the final square-free ideal $\mathcal{P}(I)$ have the same polarization.

For example, let $J = (x^2, xy, y^3) \subseteq k[x, y]$. We use a diagram to demonstrate the process described in the previous paragraph. Each linear form $a - b$ stands for "replacing the variable b with a," or vice versa, depending on which direction we are going.

$$J = (x^2, xy, y^3) \xrightarrow{u-x} J_1 = (xu, uy, y^3) \xrightarrow{v-y} J_2 = (xu, uv, y^2 v)$$
$$\xrightarrow{w-y} J_3 = (xu, uv, yvw)\ .$$

This final square-free monomial ideal J_3 is the polarization of J, and is isomorphic to $\mathcal{P}(J)$ as we defined it in Definition 2.1. Note that all ideals J, J_1, J_2 and J_3 have the same (isomorphic) polarization. We can classify monomial ideals according to their polarizations. An interesting question is to see what properties do ideals in the same polarization class have. A more difficult question is how far can one "depolarize" a square-free monomial ideal I, where by depolarizing I we mean finding monomial ideals whose polarization is equal to I, or equivalently, traveling the opposite direction on the above diagram.

A Appendix: Primary decomposition in a sequentially Cohen-Macaulay module

The purpose of this appendix is to study, using basic facts about primary decomposition of modules, the structure of the submodules appearing in the (unique) filtration of a sequentially Cohen-Macaulay module M. The main result (Theorem A.4) states that each submodule appearing in the filtration of M is the intersection of all primary submodules whose associated primes have a certain height and appear in an irredundant primary decomposition of the 0-submodule of M. Similar results, stated in a different language, appear in [**12**]; the author thanks Jürgen Herzog for pointing this out.

We first record two basic lemmas that we shall use later (the second one is an exercise in Bourbaki [**1**]). Throughout the discussions below, we assume that R is a finitely generated algebra over a field, and M is a finite module over R.

Lemma A.1 *Let* $\mathcal{Q}_1,\dots,\mathcal{Q}_t,\mathcal{P}$ *all be primary submodules of an* R*-module* M*, such that* $\operatorname{Ass}(M/\mathcal{Q}_i)=\{\mathfrak{q}_i\}$ *and* $\operatorname{Ass}(M/\mathcal{P})=\{\mathfrak{p}\}$*. If* $\mathcal{Q}_1\cap\dots\cap\mathcal{Q}_t\subseteq\mathcal{P}$ *and* $\mathcal{Q}_i\not\subseteq\mathcal{P}$ *for some* i*, then there is a* $j\neq i$ *such that* $\mathfrak{q}_j\subseteq\mathfrak{p}$*.*

Proof: Let $x\in\mathcal{Q}_i\setminus\mathcal{P}$. For each j not equal to i, pick the positive integer m_j such that $\mathfrak{q}_j^{m_j}x\subseteq\mathcal{Q}_j$. So we have that

$$\mathfrak{q}_1^{m_1}\dots\mathfrak{q}_{i-1}^{m_{i-1}}\mathfrak{q}_{i+1}^{m_{i+1}}\dots\mathfrak{q}_t^{m_t}x\subseteq\mathcal{Q}_1\cap\dots\cap\mathcal{Q}_t\subseteq\mathcal{P}\implies\mathfrak{q}_1^{m_1}\dots\mathfrak{q}_{i-1}^{m_{i-1}}\mathfrak{q}_{i+1}^{m_{i+1}}\dots\mathfrak{q}_t^{m_t}\subseteq\mathfrak{p}$$

where the second inclusion is because $x\notin\mathcal{P}$. Hence for some $j\neq i$, $\mathfrak{q}_j\subseteq\mathfrak{p}$. □

Lemma A.2 *Let M be an R-module and N be a submodule of M. Then for every $\mathfrak{p} \in \mathrm{Ass}(M/N)$, if $\mathfrak{p} \not\supseteq \mathrm{Ann}(N)$, then $\mathfrak{p} \in \mathrm{Ass}(M)$.*

Proof: Since $\mathfrak{p} \in \mathrm{Ass}(M/N)$, there exists $x \in M \setminus N$ such that $\mathfrak{p} = \mathrm{Ann}(x)$; in other words $\mathfrak{p}x \subseteq N$. Suppose $\mathrm{Ann}(N) \not\subseteq \mathfrak{p}$, and let $y \in \mathrm{Ann}(N) \setminus \mathfrak{p}$. Now $y\mathfrak{p}x = 0$, and so $\mathfrak{p} \subseteq \mathrm{Ann}(yx)$ in M. On the other hand, if $z \in \mathrm{Ann}(yx)$, then $zyx = 0 \subseteq N$ and so $zy \in \mathfrak{p}$. But $y \notin \mathfrak{p}$, so $z \in \mathfrak{p}$. Therefore $\mathfrak{p} \in \mathrm{Ass}(M)$.

□

Suppose M is a sequentially Cohen-Macaulay module with filtration as in Definition 4.9. We adopt the following notation. For a given integer j, we let

$$\mathrm{Ass}(M)_j = \{\mathfrak{p} \in \mathrm{Ass}(M) \mid \mathrm{height}\,\mathfrak{p} = j\}.$$

Suppose that all the j's where $\mathrm{Ass}(M)_j \neq \emptyset$ form the sequence of integers

$$0 \leq h_1 < \ldots < h_c \leq \dim R$$

so that $\mathrm{Ass}(M) = \bigcup_{1 \leq j \leq c} \mathrm{Ass}(M)_{h_j}$.

Proposition A.3 *For all $i = 0, \ldots, r-1$, we have*

1. $\mathrm{Ass}(M_{i+1}/M_i) \cap \mathrm{Ass}(M) \neq \emptyset$;
2. $\mathrm{Ass}(M)_{h_{r-i}} \subseteq \mathrm{Ass}(M_{i+1}/M_i)$ *and* $c = r$;
3. *If* $\mathfrak{p} \in \mathrm{Ass}(M_{i+1})$, *then* $\mathrm{height}\,\mathfrak{p} \geq h_{r-i}$;
4. *If* $\mathfrak{p} \in \mathrm{Ass}(M_{i+1}/M_i)$, *then* $\mathrm{Ann}(M_i) \not\subseteq \mathfrak{p}$;
5. $\mathrm{Ass}(M_{i+1}/M_i) \subseteq \mathrm{Ass}(M)$;
6. $\mathrm{Ass}(M_{i+1}/M_i) = \mathrm{Ass}(M)_{h_{r-i}}$;
7. $\mathrm{Ass}(M/M_i) = \mathrm{Ass}(M)_{\leq h_{r-i}}$;
8. $\mathrm{Ass}(M_{i+1}) = \mathrm{Ass}(M)_{\geq h_{r-i}}$.

Proof:

1. We use induction on the length r of the filtration of M. The case $r = 1$ is clear, as we have a filtration $0 \subset M$, and the assertion follows. Now suppose the statement holds for sequentially Cohen-Macaulay modules with filtrations of length less than r. Notice that M_{r-1} that appears in the filtration of M in Definition 4.9 is also sequentially Cohen-Macaulay, and so by the induction hypothesis, we have

$$\mathrm{Ass}(M_{i+1}/M_i) \cap \mathrm{Ass}(M_{r-1}) \neq \emptyset \text{ for } i = 0, \ldots, r-2$$

and since $\mathrm{Ass}(M_{r-1}) \subseteq \mathrm{Ass}(M)$ it follows that

$$\mathrm{Ass}(M_{i+1}/M_i) \cap \mathrm{Ass}(M) \neq \emptyset \text{ for } i = 0, \ldots, r-2.$$

It remains to show that $\mathrm{Ass}(M/M_{r-1}) \cap \mathrm{Ass}(M) \neq \emptyset$.

For each i, $M_{i-1} \subset M_i$, so we have ([**1**] Chapter IV)

$$\mathrm{Ass}(M_1) \subseteq \mathrm{Ass}(M_2) \subseteq \mathrm{Ass}(M_1) \cup \mathrm{Ass}(M_2/M_1). \tag{4}$$

The inclusion $M_2 \subseteq M_3$ along with the inclusions in (4) imply that

$$\begin{aligned}\mathrm{Ass}(M_2) \subseteq \mathrm{Ass}(M_3) &\subseteq \mathrm{Ass}(M_2) \cup \mathrm{Ass}(M_3/M_2)\\ &\subseteq \mathrm{Ass}(M_1) \cup \mathrm{Ass}(M_2/M_1) \cup \mathrm{Ass}(M_3/M_2).\end{aligned}$$

If we continue this process inductively, at the i-th stage we have

$$\begin{aligned}\mathrm{Ass}(M_i) &\subseteq \mathrm{Ass}(M_{i-1}) \cup \mathrm{Ass}(M_i/M_{i-1})\\ &\subseteq \mathrm{Ass}(M_1) \cup \mathrm{Ass}(M_2/M_1) \cup \mathrm{Ass}(M_3/M_2) \cup \ldots \cup \mathrm{Ass}(M_i/M_{i-1})\end{aligned}$$

and finally, when $i = r$ it gives

$$\mathrm{Ass}(M) \subseteq \mathrm{Ass}(M_1) \cup \mathrm{Ass}(M_2/M_1) \cup \ldots \cup \mathrm{Ass}(M/M_{r-1}). \tag{5}$$

Because of Condition (b) in Definition 4.9, and the fact that each M_{i+1}/M_i is Cohen-Macaulay (and hence all its associated primes have the same height; see [**2**] Chapter 2), if for every i we pick $\mathfrak{p}_i \in \mathrm{Ass}(M_{i+1}/M_i)$, then

$$h_c \geq \text{height } \mathfrak{p}_0 > \text{height } \mathfrak{p}_1 > \ldots > \text{height } \mathfrak{p}_{r-1},$$

where the left-hand-side inequality comes from the fact that $\text{Ass}(M_1) \subseteq \text{Ass}(M)$. By our induction hypothesis, $\text{Ass}(M)$ intersects $\text{Ass}(M_{i+1}/M_i)$ for all $i \leq r-2$, and so because of (5) we conclude that

$$\text{height}\, \mathfrak{p}_i = h_{c-i}, \text{ and } \text{Ass}(M)_{h_{c-i}} \subseteq \text{Ass}(M_{i+1}/M_i) \text{ for } 0 \leq i \leq r-2.$$

And now $\text{Ass}(M)_{h_0}$ has no choice but to be included in $\text{Ass}(M/M_{r-1})$, which settles our claim. It also follows that $c = r$.

2. See the proof for part 1.

3. We use induction. The case $i = 0$ is clear, since for every $\mathfrak{p} \in \text{Ass}(M_1) = \text{Ass}(M_1/M_0)$ we know from part 2 that height $\mathfrak{p} = h_r$. Suppose the statement holds for all indices up to $i-1$. Consider the inclusion

$$\text{Ass}(M_i) \subseteq \text{Ass}(M_{i+1}) \subseteq \text{Ass}(M_i) \cup \text{Ass}(M_{i+1}/M_i).$$

From part 2 and the induction hypothesis it follows that if $\mathfrak{p} \in \text{Ass}(M_{i+1})$ then height $\mathfrak{p} \geq h_{r-i}$.

4. Suppose $\text{Ann}(M_i) \subseteq \mathfrak{p}$. Since $\sqrt{\text{Ann}(M_i)} = \bigcap_{\mathfrak{p}' \in \text{Ass}(M_i)} \mathfrak{p}'$, it follows that $\bigcap_{\mathfrak{p}' \in \text{Ass}(M_i)} \mathfrak{p}' \subseteq \mathfrak{p}$, so there is a $\mathfrak{p}' \in \text{Ass}(M_i)$ such that $\mathfrak{p}' \subseteq \mathfrak{p}$. But from parts 2 and 3 above it follows that height $\mathfrak{p}' \geq h_{r-i+1}$ and height $\mathfrak{p} = h_{r-i}$, which is a contradiction.

5. From part 4 and Lemma A.2, it follows that $\text{Ass}(M_{i+1}/M_i) \subseteq \text{Ass}(M_{i+1}) \subseteq \text{Ass}(M)$.

6. This follows from parts 2 and 5, and the fact that M_{i+1}/M_i is Cohen-Macaulay, and hence all associated primes have the same height.

7. We show this by induction on $e = r - i$. The case $e = 1$ (or $i = r-1$) is clear, because by part 6 we have $\text{Ass}(M/M_{r-1}) = \text{Ass}(M)_{h_1} = \text{Ass}(M)_{\leq h_1}$.

   Now suppose the equation holds for all integers up to $e-1$ (namely $i = r-e+1$), and we would like to prove the statement for e (or $i = r-e$). Since $M_{i+1}/M_i \subseteq M/M_i$, we have

$$\text{Ass}(M_{i+1}/M_i) \subseteq \text{Ass}(M/M_i) \subseteq \text{Ass}(M_{i+1}/M_i) \cup \text{Ass}(M/M_{i+1}). \quad (6)$$

By the induction hypothesis and part 6 we know that

$$\mathrm{Ass}(M/M_{i+1}) = \mathrm{Ass}(M)_{\leq h_{r-i-1}} \text{ and } \mathrm{Ass}(M_{i+1}/M_i) = \mathrm{Ass}(M)_{h_{r-i}},$$

which put together with (6) implies that

$$\mathrm{Ass}(M)_{h_{r-i}} \subseteq \mathrm{Ass}(M/M_i) \subseteq \mathrm{Ass}(M)_{\leq h_{r-i}}.$$

We still have to show that $\mathrm{Ass}(M/M_i) \supseteq \mathrm{Ass}(M)_{\leq h_{r-i-1}}$.

Let

$$\mathfrak{p} \in \mathrm{Ass}(M)_{\leq h_{r-i-1}} = \mathrm{Ass}(M/M_{i+1}) = \mathrm{Ass}((M/M_i)/(M_{i+1}/M_i)).$$

If $\mathfrak{p} \supseteq \mathrm{Ann}(M_{i+1}/M_i)$, then (by part 6)

$$\mathfrak{p} \supseteq \bigcap_{\mathfrak{q} \in \mathrm{Ass}(M)_{h_{r-i}}} \mathfrak{q} \implies \mathfrak{p} \supseteq \mathfrak{q} \text{ for some } \mathfrak{q} \in \mathrm{Ass}(M)_{h_{r-i}},$$

which is a contradiction, as height $\mathfrak{p} \leq h_{r-i-1} <$ height $\mathfrak{q}$. It follows from Lemma A.2 that $\mathfrak{p} \in \mathrm{Ass}(M/M_i)$.

8. The argument is based on induction, and exactly the same as the one in part 4, using more information; from the inclusions

$$\mathrm{Ass}(M_i) \subseteq \mathrm{Ass}(M_{i+1}) \subseteq \mathrm{Ass}(M_i) \cup \mathrm{Ass}(M_{i+1}/M_i),$$

the induction hypothesis and part 6 we deduce that

$$\mathrm{Ass}(M)_{\geq h_{r-i+1}} \subseteq \mathrm{Ass}(M_{i+1}) \subseteq \mathrm{Ass}(M)_{\geq h_{r-i+1}} \cup \mathrm{Ass}(M)_{h_{r-i}},$$

which put together with part 4, along with Lemma A.2, produces the equality. □

Now suppose that as a submodule of M, $M_0 = 0$ has an irredundant primary decomposition of the form

$$M_0 = 0 = \bigcap_{1 \leq j \leq r} Q_1^{h_j} \cap \ldots \cap Q_{s_j}^{h_j} \tag{7}$$

where for a fixed $j \leq r$ and $e \leq s_j$, $Q_e^{h_j}$ is a primary submodule of M such that $\mathrm{Ass}(M/Q_e^{h_j}) = \{\mathfrak{p}_e^{h_j}\}$ and $\mathrm{Ass}(M)_{h_j} = \{\mathfrak{p}_1^{h_j}, \ldots, \mathfrak{p}_{s_j}^{h_j}\}$.

Theorem A.4 *Let M be a sequentially Cohen-Macaulay module with filtration as in Definition 4.9, and suppose that $M_0 = 0$ has a primary decomposition as in (7). Then for each $i = 0, \ldots, r-1$, M_i has the following primary decomposition:*

$$M_i = \bigcap_{1 \leq j \leq r-i} Q_1^{h_j} \cap \ldots \cap Q_{s_j}^{h_j}. \tag{8}$$

Proof: We prove this by induction on r (length of the filtration). The case $r = 1$ is clear, as the filtration is of the form $0 = M_0 \subset M$. Now consider M with filtration

$$0 = M_0 \subset M_1 \subset \ldots \subset M_r = M.$$

Since M_{r-1} is a sequentially Cohen-Macaulay module of length $r-1$, it satisfies the statement of the theorem. We first show that M_{r-1} has a primary decomposition as described in (8). From part 7 of Proposition A.3 it follows that

$$\text{Ass}(M/M_{r-1}) = \text{Ass}(M)_{h_1}$$

and so for some $\mathfrak{p}_e^{h_1}$-primary submodules $\mathcal{P}_e^{h_1}$ of M $(1 \leq e \leq s_j)$, we have

$$M_{r-1} = \mathcal{P}_1^{h_1} \cap \ldots \cap \mathcal{P}_{s_1}^{h_1}. \tag{9}$$

We would like to show that $Q_e^{h_1} = \mathcal{P}_e^{h_1}$ for $e = 1, \ldots, s_1$.
Fix $e = 1$ and assume $Q_1^{h_1} \not\subset \mathcal{P}_1^{h_1}$. From the inclusion $M_0 \subset \mathcal{P}_1^{h_1}$ and Lemma A.1 it follows that for some e and j (with $e \neq 1$ if $j = 1$), we have $\mathfrak{p}_e^{h_j} \subseteq \mathfrak{p}_1^{h_1}$. Because of the difference in heights of these ideals the only conclusion is $\mathfrak{p}_e^{h_j} = \mathfrak{p}_1^{h_1}$, which is not possible. With a similar argument we deduce that $Q_e^{h_1} \subset \mathcal{P}_e^{h_1}$, for $e = 1, \ldots, s_1$.

Now fix $j \in \{1, \ldots, r\}$ and $e \in \{1, \ldots, s_j\}$. If $M_{r-1} = Q_e^{h_j}$ we are done. Otherwise, note that for every j and $\mathfrak{p}_e^{h_j}$-primary submodule $Q_e^{h_j}$ of M,

$$Q_e^{h_j} \cap M_{r-1}$$

is a $\mathfrak{p}_e^{h_j}$-primary submodule of M_{r-1} (as we have $\emptyset \neq \text{Ass}(M_{r-1}/(Q_e^{h_j} \cap M_{r-1})) = \text{Ass}((M_{r-1} + Q_e^{h_j})/Q_e^{h_j}) \subseteq \text{Ass}(M/Q_e^{h_j}) = \{\mathfrak{p}_e^{h_j}\}$). So $M_0 = 0$ as a submodule of M_{r-1} has a primary decomposition

$$M_0 \cap M_{r-1} = 0 = \bigcap_{1 \leq j \leq r} (Q_1^{h_j} \cap M_{r-1}) \cap \ldots \cap (Q_{s_j}^{h_j} \cap M_{r-1}).$$

From Proposition A.3 part 8 it follows that

$$\mathrm{Ass}(M_{r-1}) = \mathrm{Ass}(M)_{\geq h_2}$$

so the components $Q_t^{h_1} \cap M_{r-1}$ are redundant for $t = 1, \ldots, s_1$, so for each such t we have

$$\bigcap_{Q_e^{h_j} \neq Q_t^{h_1}} (Q_1^{h_j} \cap M_{r-1}) \subseteq Q_t^{h_1} \cap M_{r-1}.$$

If $Q_e^{h_j} \cap M_{r-1} \not\subseteq Q_t^{h_1} \cap M_{r-1}$ for some e and j (with $Q_e^{h_j} \neq Q_t^{h_1}$), then using Lemma A.1 for some such e and j we have $\mathfrak{p}_e^{h_j} \subseteq \mathfrak{p}_t^{h_1}$, which is a contradiction (because of the difference of heights).

Therefore, for each t ($1 \leq t \leq s_1$), there exists indices e and j (with $Q_e^{h_j} \neq Q_t^{h_1}$) such that

$$Q_e^{h_j} \cap M_{r-1} \subseteq Q_t^{h_1} \cap M_{r-1}.$$

It follows now, from the primary decomposition of M_{r-1} in (9), that for a fixed t

$$\mathcal{P}_1^{h_1} \cap \ldots \cap \mathcal{P}_{s_1}^{h_1} \cap Q_e^{h_j} \subseteq Q_t^{h_1}.$$

Assume $\mathcal{P}_t^{h_1} \not\subseteq Q_t^{h_1}$. Applying Lemma A.1 again, we deduce that

$$\mathfrak{p}_e^{h_j} \subseteq \mathfrak{p}_t^{h_1}, \text{ or there is } t' \neq t \text{ such that } \mathfrak{p}_{t'}^{h_1} \subseteq \mathfrak{p}_t^{h_1}.$$

Neither of these is possible, so $\mathcal{P}_t^{h_1} \subseteq Q_t^{h_1}$ for all t.

We have therefore proved that

$$M_{r-1} = Q_1^{h_1} \cap \ldots \cap Q_{s_1}^{h_1}.$$

By the induction hypothesis, for each $i \leq r-2$, M_i has the following primary decomposition:

$$M_i = \bigcap_{2 \leq j \leq r-i} (Q_1^{h_j} \cap M_{r-1}) \cap \ldots \cap (Q_{s_j}^{h_j} \cap M_{r-1}) = \bigcap_{1 \leq j \leq r-i} Q_1^{h_j} \cap \ldots \cap Q_{s_j}^{h_j},$$

which proves the theorem.

□

References

[1] N. Bourbaki, *Commutative Algebra, Chapters 1–7,* Elements of Mathematics (Berlin), Springer-Verlag, Berlin, 1989.

[2] W. Bruns and J. Herzog, *Cohen-Macaulay Rings*, Cambridge Studies in Advanced Mathematics **39**, revised edition, 1998.

[3] M. Caboara, S. Faridi and P. Selinger, Simplicial tree computations, in preparation.

[4] J.A. Eagon and V. Reiner, Resolution of Stanley-Reisner rings and Alexander duality, *J. Pure Appl. Algebra* **130** (1998), 265–275.

[5] S. Faridi, The facet ideal of a simplicial complex, *Manuscripta Math.* **109** (2002), 159–174.

[6] S. Faridi, Cohen-Macaulay properties of square-free monomial ideals, preprint.

[7] S. Faridi, Simplicial trees are sequentially Cohen-Macaulay, *J. Pure Appl. Algebra* **190** (2004), 121–136.

[8] R. Fröberg, A study of graded extremal rings and of monomial rings, *Math. Scand.* **51** (1982), 22–34.

[9] J. Herzog and T. Hibi, Componentwise linear ideals, *Nagoya Math. J.* **153** (1999), 141–153.

[10] E. Miller, Alexander duality for monomial ideals and their resolutions, math.AG/9812095.

[11] G. Rota and J. Stein, Standard basis in supersymplectic algebras, *Proc. Nat. Acad. Sci. U.S.A.* **86** (1989), no. 8, 2521–2524.

[12] P. Schenzel, On the dimension filtration and Cohen-Macaulay filtered modules, *Commutative Algebra and Algebraic Geometry (Ferrara)*, 245–264, Lecture Notes in Pure and Appl. Math., **206**, Dekker, New York, 1999.

[13] P. Schwartau, Liaison addition and monomial ideals, Ph.D. Thesis, Brandeis University, 1982.

[14] R.P. Stanley, *Combinatorics and Commutative Algebra*, second edition, Progress in Mathematics **41**, Birkhäuser Boston, Inc., Boston, MA, 1996.

[15] J. Stückrad and W. Vogel, *Buchsbaum Rings and Applications*, Springer-Verlag, Berlin, 1986.

[16] R. Villarreal, *Monomial Algebras*, Monographs and Textbooks in Pure and Applied Mathematics **238**, Marcel Dekker, Inc., New York, 2001.

[17] J. Weyman, Ideals generated by monomials, Master's Thesis, N. Copernicus University, Torun, Poland in 1977.

When does the equality $I^2 = QI$ hold true in Buchsbaum rings?

SHIRO GOTO
Department of Mathematics, School of Science and Technology
Meiji University, 214-8571 Japan
E-mail: goto@math.meiji.ac.jp

HIDETO SAKURAI
Department of Mathematics, School of Science and Technology
Meiji University, 214-8571 Japan
E-mail: ee78052@math.meiji.ac.jp

1 Introduction

Let A be a Noetherian local ring with the maximal ideal $\mathfrak{m}$ and $d = \dim A$. Let Q be a parameter ideal in A and let $I = Q : \mathfrak{m}$. In this paper we shall study the problem of when the equality $I^2 = QI$ holds true. K. Yamagishi [**18**, **19**], the first author, and K. Nishida [**9**] have recently showed that the Rees algebras $\mathrm{R}(I) = \bigoplus_{n \geq 0} I^n$, the associated graded rings $\mathrm{G}(I) = \mathrm{R}(I)/I\mathrm{R}(I)$, and the fiber cones $\mathrm{F}(I) = \mathrm{R}(I)/\mathfrak{m}\mathrm{R}(I)$ are all Buchsbaum rings with very specific graded local cohomology modules, if $I^2 = QI$ and the base ring A is Buchsbaum. Our results will supply [**18**, **19**] and [**9**] with ample examples.

This paper is a continuation of [**10**, **11**], and our research dates back to the works of A. Corso, C. Huneke, C. Polini, and W. Vasconcelos [**1**, **2**, **3**]. We summarize their results into the following, in which the equivalence of assertions (2) and (3) is due to the first author [**7**, Theorem (3.1)].

Theorem 1.1 ([**1**, **2**, **3**, **7**]) *Let Q be a parameter ideal in a Cohen-Macaulay local ring A with $d = \dim A$ and let $I = Q : \mathfrak{m}$. Then the following three conditions are equivalent.*

(1) $I^2 \neq QI$.

(2) *Q is integrally closed in A.*

(3) *A is a regular local ring, which contains a regular system $a_1, a_2, \cdots, a_d$ of parameters such that $Q = (a_1, \cdots, a_{d-1}, a_d^q)$ for some $1 \leq q \in \mathbb{Z}$.*

Therefore, we have $I^2 = QI$ for every parameter ideal Q in A, if A is a Cohen-Macaulay local ring but not regular.

Our goal is to generalize Theorem 1.1 to local rings A which are not necessarily Cohen-Macaulay. In contrast to Theorem 1.1, almost nothing was known unless A is Cohen-Macaulay, and this marked contrast has hurried up the researches [**10**, **11**] in the Buchsbaum case. Here we would like to summarize a part of our previous results. To state them, let

$$\mathrm{r}(A) = \sup_{Q} \ \ell_A((Q : \mathfrak{m})/Q)$$

and call it the Cohen-Macaulay type of A, where Q runs through parameter ideals in A. Thanks to [**12**, Theorem (2.5)], we have the formula

$$\mathrm{r}(A) = \sum_{i=0}^{d-1} \binom{d}{i} h^i(A) + \mu_{\hat{A}}(\mathrm{K}_{\hat{A}})$$

for every Buchsbaum local ring A with $d = \dim A \geq 1$, where $h^i(A) = \ell_A(\mathrm{H}^i_{\mathfrak{m}}(A))$ denotes the length of the ith local cohomology module $\mathrm{H}^i_{\mathfrak{m}}(A)$ of A with respect to the maximal ideal $\mathfrak{m}$ and $\mu_{\hat{A}}(\mathrm{K}_{\hat{A}})$ denotes the number of generators for the canonical module $\mathrm{K}_{\hat{A}}$ of the $\mathfrak{m}$-adic completion $\widehat{A}$ of A ([**15**]). Let $\mathrm{e}(A) = \mathrm{e}^0_{\mathfrak{m}}(A)$ denote the multiplicity of A with respect to $\mathfrak{m}$. With this notation the main results of [**10**] are summarized into the following.

Theorem 1.2 ([**10**]) *Let A be a Buchsbaum local ring with $d = \dim A \geq 1$ and* $\mathrm{e}(A) > 1$. *The following assertions hold.*

(1) *The equality $I^2 = QI$ holds true for all parameter ideals Q in A of the form $Q = (a_1^n, a_2, \cdots, a_d)$, where $a_1, a_2, \cdots, a_d$ be any system of parameters in A, $n \geq 2$, and $I = Q : \mathfrak{m}$.*

(2) *Let Q be a parameter ideal in A and assume that $\ell_A((Q : \mathfrak{m})/Q) = \mathrm{r}(A)$. Then the equality $I^2 = QI$ holds true, where $I = Q : \mathfrak{m}$.*

(3) *There exists an integer $\ell > 0$ such that the equality $I^2 = QI$ holds true for every parameter ideal Q in A such that $Q \subseteq \mathfrak{m}^\ell$, where $I = Q : \mathfrak{m}$.*

Theorem 1.2 (2) is a natural generalization of the last statement of Theorem 1.1 and Theorem 1.2 (3) shows that inside Buchsbaum local rings, the parameter ideals Q for which the equality $I^2 = QI$ holds true are still the majority.

We now state our own results, explaining how this paper is organized. The present purpose consists of two parts. The first one is to give the following Theorem 1.3, which we shall prove in Section 2. It gives, generalizing Theorem 1.2 (1), a new sufficient condition for a parameter ideal Q in a Buchsbaum local ring A to satisfy the equality $I^2 = QI$. We shall discuss in Section 3 some applications.

Theorem 1.3 *Let A be a Buchsbaum local ring with $d = \dim A \geq 1$ and* $\mathrm{e}(A) > 1$. *Let $Q = (x_1, x_2, \cdots, x_d)$ be a parameter ideal in A and assume that $x_d \equiv ab \mod (0) : \mathfrak{m}$ for some $a, b \in \mathfrak{m}$. Then $I^2 = QI$ where $I = Q : \mathfrak{m}$.*

The following is a direct consequence of Theorem 1.3.

Corollary 1.4 *Let A be a Buchsbaum local ring with $d = \dim A \geq 2$ and* $\mathrm{e}(A) > 1$. *Let $a, b \in \mathfrak{m}$ such that* $\dim A/(a) = \dim A/(b) = d - 1$ *and let $w \in (0) : \mathfrak{m}$. Then the equality $I^2 = QI$ holds true for every parameter ideal Q in the Buchsbaum local ring $A/(ab + w)$.*

The second purpose of this paper is to explore homogeneous Buchsbaum rings of maximal embedding dimension. Let $R = \sum_{n \geq 0} R_n$ be a Buchsbaum graded ring with $k = R_0$ an infinite field. Assume that R is homogeneous, that is, $R = k[R_1]$, and let $M = R_+$. Then R is said to have maximal embedding dimension if

$$\dim_k R_1 = \mathrm{e}(R) + \dim R - 1 + \mathrm{I}(R)$$

where $\mathrm{e}(R)$ and $\mathrm{I}(R)$ denote, respectively, the multiplicity $\mathrm{e}_M^0(R)$ of R with respect to M and the Buchsbaum invariant of R. This condition is equivalent to saying that the equality $M^2 = QM$ holds true for some (hence *any*) linear parameter ideal Q in R ([**5**, (3.8)], [**6**]). In Section 4 we shall prove the following.

Theorem 1.5 *Let $R = k[R_1]$ be a homogeneous Buchsbaum ring over an infinite field k with $\mathrm{e}(R) > 1$. Assume that R has maximal embedding dimension. Then $I^2 = QI$ for every graded parameter ideal Q in R, where $I = Q : M$.*

The key in the proof of Theorem 1.5 is the fact that the canonical homomorphism

$$\mathrm{Ext}_R^d(R/M^2, R) \to \mathrm{H}_M^d(R) = \lim_{n \to \infty} \mathrm{Ext}_R^d(R/M^n, R)$$

is surjective on the socles. To see this, we need some analysis of the graded minimal injective resolution of R, which we will perform in Section 4.

Typical examples of homogeneous Buchsbaum rings R of maximal embedding dimension are Buchsbaum rings of multiplicity 2. For a Buchsbaum local ring A with $\mathrm{e}(A) = 2$ and $\operatorname{depth} A > 0$, it is known by [**11**] that the equality $I^2 = QI$ holds true for every parameter ideal Q in A. It might be worth noticing that [**11**, Theorem (1.1)] is not a direct consequence of Theorem 1.5. In fact, [**11**] deals with *all* the parameter ideals Q in such a ring A, whereas Theorem 1.5 deals exclusively with the *graded* parameter ideals Q in a homogeneous Buchsbaum ring R. The authors do not know whether the equality $I^2 = QI$ holds true for every parameter ideal Q in the Buchsbaum local ring $A = R_M$, where R is the homogeneous Buchsbaum ring studied in Theorem 1.5 .

Let $R = \sum_{n \geq 0} R_n$ be a Buchsbaum graded ring with $k = R_0$ a field. Let $M = R_+$. When $\dim R = 1$, there is a sufficiently effective condition for the equality $I^2 = QI$ stated in terms of the a-invariant

$$\mathrm{a}(R) = \max\{n \in \mathbb{Z} \mid [\mathrm{H}_M^1(R)]_n \neq (0)\}$$

of R (Theorem 5.1). When the hypothesis of Theorem 5.1 is not satisfied, the equality $I^2 = QI$ may fail to hold true. The first counterexamples appeared in [**10**] (see also [**11**, Section 4]) and the simplest one is the following. Let

$$R = k[X, Y, Z]/(Y^2 - XZ, XY, X^3) \tag{1.6}$$

where $k[X,Y,Z]$ is the polynomial ring over a field k. Then R is a homogeneous Buchsbaum ring with $\dim R = 1$ and $\mathrm{e}(R) = 3$, which does not satisfy the hypothesis of Theorem 5.1. Let z be the image of Z in R. Then $Q = (z)$ is a linear parameter ideal of R such that $I^3 = QI^2$ but $I^2 \neq QI$, where $I = Q : M$. The Buchsbaum ring R stated in (1.6) is constructed from the Cohen-Macaulay graded ring

$$k[X,Y,Z]/(Y^2 - XZ, XY, X^2)$$

in a canonical manner, and the method is applicable to a more general setting, which we discuss in Section 6.

We are now entering the details. Throughout, let $(A, \mathfrak{m})$ be a Noetherian local ring with $d = \dim A$. We denote by $\mathrm{e}(A) = \mathrm{e}^0_{\mathfrak{m}}(A)$ the multiplicity of A with respect to the maximal ideal $\mathfrak{m}$. Let $\mathrm{H}^i_{\mathfrak{m}}(*)$ $(i \in \mathbb{Z})$ denote the local cohomology functors with respect to $\mathfrak{m}$. Let $\ell_A(*)$ and $\mu_A(*)$ stand for the length and the number of generators, respectively. Let $Q = (x_1, x_2, \cdots, x_d)$ be a parameter ideal in A and, otherwise specified, we always denote by I the ideal $Q : \mathfrak{m}$. Let $\mathrm{Min}\, A$ be the set of minimal prime ideals in A.

For a given Noetherian graded ring $R = \sum_{n \geq 0} R_n$ with $k = R_0$ a field, we shall maintain a similar notation. Let $M = R_+$ and let $\mathrm{e}(R) = \mathrm{e}^0_M(R)$ denote the multiplicity of R with respect to M. Let $\mathrm{H}^i_M(*)$ $(i \in \mathbb{Z})$ be the graded local cohomology functors of R with respect to M. Otherwise specified, for each graded parameter ideal Q in R, we denote also by I the graded ideal $Q : M$ in R.

2 Proof of Theorem 1.3

The purpose of this section is to prove Theorem 1.3. Before starting the proof, let us give a brief review on Buchsbaum rings.

Let R be a commutative ring and $x_1, x_2, \cdots, x_s \in R$. Then $x_1, x_2, \cdots, x_s$ is called a d-sequence in R, if $(x_1, \cdots, x_{i-1}) : x_j = (x_1, \cdots, x_{i-1}) : x_i x_j$ whenever $1 \leq i \leq j \leq s$. See **[16]** for basic results on d-sequences. For example, if $x_1, x_2, \cdots, x_s$ is a d-sequence in R, then we have

$$\begin{aligned}(x_1, \cdots, x_{i-1}) : x_i^2 &= (x_1, \cdots, x_{i-1}) : x_i \\ &= (x_1, \cdots, x_{i-1}) : (x_1, x_2, \cdots, x_s)\end{aligned}$$

for all $1 \leq i \leq s$. Also, the equality

$$((x_1,\cdots,x_{i-1}):x_i)\cap(x_1,x_2,\cdots,x_s)^n = (x_1,\cdots,x_{i-1})\cdot(x_1,x_2,\cdots,x_s)^{n-1} \quad (2.1)$$

holds true for all integers $1 \leq i \leq s$ and $1 \leq n \in \mathbb{Z}$.

Let A be a Noetherian local ring with the maximal ideal $\mathfrak{m}$ and $d = \dim A$. Then A is said to be a Buchsbaum ring, if the difference

$$\mathrm{I}(A) = \ell_A(A/Q) - \mathrm{e}^0_Q(A)$$

is independent of the choice of a parameter ideal Q in A, where $\mathrm{e}^0_Q(A)$ denotes the multiplicity of A with respect to Q. This condition is equivalent to saying that every system $x_1, x_2, \cdots, x_d$ of parameters for A forms a weak A-sequence, that is, the equality

$$(x_1,\cdots,x_{i-1}):x_i = (x_1,\cdots,x_{i-1}):\mathfrak{m} \quad (2.2)$$

holds true for all $1 \leq i \leq d$. Hence every system of parameters for a Buchsbaum local ring forms a d-sequence. When $R = \sum_{n\geq 0} R_n$ is a Noetherian graded ring with $k = R_0$ a field, we simply say that R is a Buchsbaum ring, if $A = R_M$ is a Buchsbaum local ring, where $M = R_+$.

If A is a Buchsbaum local ring, then $\mathfrak{m}\cdot\mathrm{H}^i_\mathfrak{m}(A) = (0)$ $(i \neq d)$ and one has the equality

$$\mathrm{I}(A) = \sum_{i=0}^{d-1} \binom{d-1}{i} h^i(A),$$

where $h^i(A) = \ell_A(\mathrm{H}^i_\mathfrak{m}(A))$ ([**17**, Chap. I, (2.6)]). For given integers $d \geq 1$ and $h_i \geq 0$ $(0 \leq i \leq d-1)$, there exists a Buchsbaum local ring A such that $\dim A = d$ and $h^i(A) = h_i$ for all $0 \leq i \leq d-1$. One may choose the Buchsbaum ring A so that A is an integral domain (respectively a normal domain), if $h_0 = 0$ (respectively $d \geq 2$ and $h_0 = h_1 = 0$) ([**4**, Theorem (1.1)]). See the book [**17**] for more details on Buchsbaum rings.

We now start the proof of Theorem 1.3.

Lemma 2.3 *Let W, L, and M be ideals in a commutative ring R. Let $a, b \in M$ and assume that $aW = (0)$, $L : a^2 = L : a$, and $L : ab = L : b$. Then*

$$(L + (ab) + W) : M = [(L + W) : M] + [(L + (ab)) : M].$$

Proof: Let us begin with the following.

Claim 2.4 (1) $W \cap [L + (a)] \subseteq L$.
(2) $[L : a] \cap [L + (b)] = L$.

Proof of Claim 2.4: (1) Let $w \in W$ and assume that $w = \ell + ay$ for some $\ell \in L$ and $y \in R$. Then $a^2 y \in L$, since $aw = 0$. Hence $ay \in L$ because $L : a^2 = L : a$, so that we have $w = \ell + ay \in L$.

(2) Let $x \in L : a$ and assume that $x = \ell + by$ for some $\ell \in L$ and $y \in R$. Then $(ab)y \in L$, since $ax \in L$. Hence $by \in L$ because $L : ab = L : b$, and so $x = \ell + by \in L$. $\square$

Let $x \in (L + (ab) + W) : M$. We write $ax = \ell + (ab)y + w$ with $\ell \in L$, $y \in R$, and $w \in W$. Then $w \in L + (a)$ whence $w \in L$ by 2.4 (1), so that $x - by \in L : a$. Set $z = x - by$. Hence $x = by + z$ and $z \in L : a$. Let $\alpha \in M$ and write $\alpha x = \ell_1 + (ab)y_1 + w_1$ with $\ell_1 \in L$, $y_1 \in R$, and $w_1 \in W$. Then

$$\ell_1 + (ab)y_1 + w_1 = b(\alpha y) + \alpha z.$$

Hence, since $aW = (0)$, we have that $\alpha z - w_1 \in [L : a] \cap [L + (b)] = L$ by 2.4 (2). Thus $z \in (L + W) : M$ so that $by = x - z \in (L + (ab) + w) : M$ too. We will show that $by \in (L + (ab)) : M$. Choose $\alpha \in M$ and write $\alpha(by) = \ell_2 + (ab)y_2 + w_2$ with $\ell_2 \in L$, $y_2 \in R$, and $w_2 \in W$. Then $(ab)(\alpha y - ay_2) = a\ell_2 \in L$, because $aw_2 = 0$. Hence $b(\alpha y - ay_2) \in L$, because $L : ab = L : b$. Consequently $by \in [L + (ab)] : M$, and so we get $x = by + z \in [(L + W) : M] + [(L + (ab)) : M]$ as claimed. $\square$

Otherwise specified, for the rest of this section let A denote a Buchsbaum local ring. We assume that $d = \dim A \geq 1$ and $\mathrm{e}(A) > 1$. Let $W = \mathrm{H}^0_{\mathfrak{m}}(A) = (0) : \mathfrak{m}$.

Proposition 2.5 *Let $x_1, x_2, \cdots, x_d$ be a system of parameters in A and assume that $x_d = ab$ for some $a, b \in \mathfrak{m}$. Then $(Q + W) : \mathfrak{m} = Q : \mathfrak{m}$, where $Q = (x_1, x_2, \cdots, x_d)$.*

Proof: Let $L = (x_1, x_2, \cdots, x_{d-1})$. As both $\{x_1, \cdots, x_{d-1}, a\}$ and $\{x_1, \cdots, x_{d-1}, b\}$ are systems of parameters in the Buchsbaum local ring A, by (2.2) we get

$$L : a^2 = L : a = L : ab = L : b = L : \mathfrak{m}.$$

Hence $(Q+W):\mathfrak{m} = [(L+W):\mathfrak{m}] + [Q:\mathfrak{m}]$ by Lemma 2.3. We have $(L+W):\mathfrak{m} = L:\mathfrak{m}$, because $\mathfrak{m}^2 \cdot [(L+W):\mathfrak{m}] \subseteq L$ and $L:\mathfrak{m} = L:\mathfrak{m}^n$ for all $n \geq 1$. Hence $(Q+W):\mathfrak{m} = Q:\mathfrak{m}$. □

Lemma 2.6 *Let Q and Q' be parameter ideals in A. Then $Q:\mathfrak{m} \subseteq Q':\mathfrak{m}$ if $Q \subseteq Q'+W$.*

Proof: We get $\mathfrak{m}\cdot(Q:\mathfrak{m}) = \mathfrak{m}Q$ by [**10**, Lemma (2.2)], because $\mathrm{e}(A) > 1$. Hence $\mathfrak{m}\cdot(Q:\mathfrak{m}) = \mathfrak{m}Q \subseteq \mathfrak{m}\cdot(Q'+W) = \mathfrak{m}Q'$. Thus $Q:\mathfrak{m} \subseteq Q':\mathfrak{m}$. □

We are now ready to prove Theorem 1.3.

Proof of Theorem 1.3: Let $Q' = (x_1, \cdots, x_{d-1}, ab)$, $I = Q:\mathfrak{m}$, and $I' = Q':\mathfrak{m}$. Then Q' is a parameter ideal in A and $Q+W = Q'+W$. Hence $I = I'$ by Lemma 2.6. If $I'^2 = Q'I'$, then $I^2 = I'^2 = Q'I' = (Q'+W)I' = (Q+W)I = QI$. Thus we may assume that $x_d = ab$. Assume that $d = 1$ and put $\bar{A} = A/W$, $\mathfrak{n} = \mathfrak{m}/W$. Then $I\bar{A} = Q\bar{A}:\mathfrak{n}$ by Proposition 2.5. Hence $(I\bar{A})^2 = Q\bar{A}\cdot I\bar{A}$ by Theorem 1.1, because $\bar{A}$ is a Cohen-Macaulay ring with $\mathrm{e}(\bar{A}) = \mathrm{e}(A) > 1$. Thus $I^2 \subseteq QI + W$, whence $I^2 = QI$ because $I^2 \subseteq \mathfrak{m}I \subseteq Q$ and $Q \cap W = (0)$ by (2.1) and (2.2). Suppose that $d > 1$ and that our assertion holds true for $d-1$. Then we get, passing to the ring $A/(x_1)$, that $I^2 \subseteq (x_2, \cdots, x_d)I + (x_1)$, and the equality $(x_1) \cap I^2 = x_1 I$ now follows similarly as in the proof of [**10**, Claim (2.12)]. We note a brief proof for the sake of completeness. Let $y \in (x_1) \cap I^2$ and write $y = x_1 z$ with $z \in A$. Let $\alpha \in \mathfrak{m}$. Then $x_1(\alpha z) \in Q^2$, because $\mathfrak{m}I^2 \subseteq Q^2$ ([**10**, Lemma (2.2)]). Hence $x_1(\alpha z) = x_1 q$ for some $q \in Q$, because $(x_1) \cap Q^2 = x_1 Q$ by (2.1). Thus $\alpha z - q \in (0):x_1 = W$ so that $z \in (Q+W):\mathfrak{m} = I$ by Proposition 2.5. Hence $y = x_1 z \in x_1 I$ and so $(x_1) \cap I^2 = x_1 I$ as claimed. □

3 Applications of Theorem 1.3 to Buchsbaum local rings of dimension 1

The purpose of this section is to give some applications of Theorem 1.3. Let A be a Buchsbaum local ring. We assume that $\dim A = 1$ and $e = \mathrm{e}(A) > 1$. Let

$W = \mathrm{H}^0_{\mathfrak{m}}(A) = (0) : \mathfrak{m}$.

We begin with the following.

Proposition 3.1 *Let Q be a parameter ideal in A. Then $I^2 = QI$ if $Q \subseteq \mathfrak{m}^{e-1}$, where $I = Q : \mathfrak{m}$.*

Proof: Passing to the ring $A[T]_{\mathfrak{m}A[T]}$ (here T is an indeterminate over A), we may assume the field $A/\mathfrak{m}$ is infinite. Choose $a \in \mathfrak{m}$ so that $\mathfrak{m}^{n+1} = a\mathfrak{m}^n$ for some $n \geq 0$. Let $\bar{A} = A/W$ and $\mathfrak{n} = \mathfrak{m}/W$. Then $\bar{A}$ is a Cohen-Macaulay local ring of dimension 1 and $\mathrm{e}(\bar{A}) = \mathrm{e}(A) = e > 1$. Hence $\ell_{\bar{A}}(\bar{A}/a\bar{A}) = e$ so that $\mathfrak{n}^e \subseteq a\bar{A}$. Suppose $\bar{A}$ is a Gorenstein ring. Then we have either $I = Q + W$ or $I = (Q+W) : \mathfrak{m}$, because $Q + W \subseteq I \subseteq (Q+W) : \mathfrak{m}$, and in any case the equality $I^2 = QI$ readily follows; see Proof of Theorem 1.3. Suppose that $\bar{A}$ is not a Gorenstein ring. Hence $e = \mathrm{e}(\bar{A}) > 2$. We put $\bar{\mathfrak{n}} = \mathfrak{n}/a\bar{A}$. Then $\bar{\mathfrak{n}}^{e-1} = (0)$. In fact, if $\bar{\mathfrak{n}}^{e-1} \neq (0)$, then we have $\mu_{\bar{A}}(\bar{\mathfrak{n}}) = 1$ because $\bar{\mathfrak{n}}^e = (0)$ and $\ell_{\bar{A}}(\bar{A}/a\bar{A}) = e$. Hence the ring $\bar{A}/a\bar{A}$ is Gorenstein and so $\bar{A}$ is a Gorenstein ring too. This is impossible. Thus $\bar{\mathfrak{n}}^{e-1} = (0)$ whence $\mathfrak{m}^{e-1} \subseteq (a) + W$. Let $Q = (f)$ be a parameter ideal in A and assume that $Q \subseteq \mathfrak{m}^{e-1}$. We write $f = ab + w$ with $b \in A$ and $w \in W$. Let $\bar{\ }$ denote the image in $\bar{A} = A/W$. Then $\bar{a}\bar{b} = \bar{f} \in \mathfrak{n}^2$ since $e > 2$, so that we have $b \in \mathfrak{m}$ because $\bar{a} \notin \mathfrak{n}^2$. Thus $Q = (ab + w)$ for some $b \in \mathfrak{m}$ and $w \in W$, whence $I^2 = QI$ by Theorem 1.3. □

Theorem 3.2 *Suppose that* $\mathrm{Min}\, A = \{\mathfrak{p}\}$ *and that $A/\mathfrak{p}$ is a DVR. Choose an integer $\ell > 0$ so that $\mathfrak{p}^{\ell+1} = (0)$. Then $I^2 = QI$ for every parameter ideal Q in A such that $Q \subseteq \mathfrak{m}^\ell$, where $I = Q : \mathfrak{m}$.*

Proof: Choose $a \in \mathfrak{m}$ so that $\mathfrak{m} = (a) + \mathfrak{p}$. Then $\mathfrak{m}^\ell = a\mathfrak{m}^{\ell-1} + \mathfrak{p}^\ell$ and $\mathfrak{p} \cap \mathfrak{m}^\ell \subseteq a\mathfrak{p} + \mathfrak{p}^\ell$. Let $f \in \mathfrak{m}^\ell$ be a parameter in A and write $\bar{f} = \bar{a}^n \bar{\varepsilon}$ with $n \geq \ell$ and ε a unit of A, where $\bar{\ }$ denotes the image in $A/\mathfrak{p}$. Then $f = a^n \varepsilon + z$ for some $z \in \mathfrak{p}$. We may assume that $\varepsilon = 1$, because our purpose is to show $I^2 = QI$ where $Q = (f)$ and $I = Q : \mathfrak{m}$. Because $z = f - a^n \in \mathfrak{p} \cap \mathfrak{m}^\ell$, we have $z = ap + q$ with $p \in \mathfrak{p}$ and $q \in \mathfrak{p}^\ell$. Hence

$$f = a^n + ap + q.$$

Let $\mathrm{r}(A)$ be the Cohen-Macaulay type of A. Then $\mathrm{r}(A) = \ell_A(W) + \mathrm{r}(A/W)$ ([**12**, Theorem (2.5)]) and we have the following.

Claim 3.3 *Suppose that $n > 1$. Then $\ell_A(((f):\mathfrak{m})/(f)) = \mathrm{r}(A)$.*

Proof of Claim 3.3: Let $g = ab$ where $b = a^{n-1} + p \in \mathfrak{m}$. Then $g = f - q$ is a parameter in A and $f\mathfrak{p} = g\mathfrak{p}$, since $q\mathfrak{p} \subseteq \mathfrak{p}^{\ell+1} = (0)$. From the exact sequence $0 \to \mathfrak{p} \to A \to A/\mathfrak{p} \to 0$, we get

$$\begin{array}{ccccccccc} (1) & 0 & \longrightarrow & \mathfrak{p}/f\mathfrak{p} & \longrightarrow & A/(f) & \longrightarrow & A/[(f)+\mathfrak{p}] & \longrightarrow & 0 \quad \text{and} \\ & & & \| & & & & & & \\ (2) & 0 & \longrightarrow & \mathfrak{p}/g\mathfrak{p} & \stackrel{\varphi}{\longrightarrow} & A/(g) & \stackrel{\varepsilon}{\longrightarrow} & A/[(g)+\mathfrak{p}] & \longrightarrow & 0, \end{array}$$

where $(f)+\mathfrak{p} = (g)+\mathfrak{p} = (a^n)+\mathfrak{p}$. Notice that

$$\begin{aligned} \ell_A(((g):\mathfrak{m})/(g)) &= \ell_A([((g)+W):\mathfrak{m}]/[(g)+W]) + \ell_A([(g)+W]/(g)) \\ &= \ell_A([((g)+W):\mathfrak{m}]/[(g)+W]) + \ell_A(W), \end{aligned}$$

because $(g):\mathfrak{m} = ((g)+W):\mathfrak{m}$ by Proposition 2.5 and $(g)\cap W = (0)$ by (2.1) and (2.2). Hence $\ell_A(((g):\mathfrak{m})/(g)) = \mathrm{r}(A/W) + \ell_A(W) = \mathrm{r}(A)$. Assume now that $\ell_A(((f):\mathfrak{m})/(f)) < \mathrm{r}(A)$. Then, by exact sequences (1) and (2), we get

$$\begin{aligned} &\ell_A((0):_{\mathfrak{p}/g\mathfrak{p}}\mathfrak{m}) = \ell_A((0):_{\mathfrak{p}/f\mathfrak{p}}\mathfrak{m}) \\ &\qquad\qquad \leq \ell_A(((f):\mathfrak{m})/(f)) < \mathrm{r}(A) = \ell_A(((g):\mathfrak{m})/(g)). \end{aligned}$$

Therefore in exact sequence (2) the map φ cannot be bijective on the socles. Hence the map ε is surjective on the socles, because the ring $A/[(g)+\mathfrak{p}] = A/[(a^n)+\mathfrak{p}]$ is an Artinian Gorenstein local ring. Let $\bar{a}$ denote the reduction of a mod $(g)+\mathfrak{p} = (a^n)+\mathfrak{p}$. Then, since $(0):_{A/[(g)+\mathfrak{p}]}\mathfrak{m} = (\bar{a}^{n-1})$, we may choose $\xi \in (g):\mathfrak{m}$ so that $\bar{\xi} = \bar{a}^{n-1}$. Let us write

$$\xi = a^{n-1} + a^n x + y$$

with $x \in A$ and $y \in \mathfrak{p}$. Then $\xi = a^{n-1}{\cdot}(1+ax) + y$. Hence $a^{n-1} + z \in (g):\mathfrak{m}$ so that $a^n + az \in (g) = (ab)$, where $z = y{\cdot}(1+ax)^{-1} \in \mathfrak{p}$. We write

$$a^n + az = (ab)v \tag{3.4}$$

with $v \in A$. Then in the DVR $A/\mathfrak{p}$, $\bar{a}^n = (\bar{a}\bar{b})\bar{v}$. Hence $\bar{a}^n = (\bar{a}\bar{b})\bar{v} = (\bar{a}{\cdot}\bar{a}^{n-1})\bar{v} = \bar{a}^n\bar{v}$, because $b = a^{n-1} + p$. Hence $v \notin \mathfrak{m}$. On the other hand, since $a^{n-1} + z \in$

$(g):\mathfrak{m}$ and $bv-(a^{n-1}+z) \in (0):a = W \subseteq (g):\mathfrak{m}$ by (3.4), we get $bv \in (g):\mathfrak{m}$. Thus $b \in (g):\mathfrak{m}$ because $v \notin \mathfrak{m}$. Let $\alpha \in \mathfrak{m}$ and write $\alpha b = (ab)\beta$ with $\beta \in A$. Then $\alpha - a\beta \in (0):b = W$, whence $\mathfrak{m} = (a)+W$, so that A/W is a DVR. This is impossible, because $\mathrm{e}(A/W) = \mathrm{e}(A) > 1$. Hence $\ell_A(((f):\mathfrak{m})/(f)) = \mathrm{r}(A)$, if $n > 1$. □

By 3.3 and Theorem 1.2 (2) we have $I^2 = QI$ if $n > 1$. Assume that $n = 1$. Then $\ell = 1$ because $n \geq \ell$ and so $f \equiv a \bmod \mathfrak{p}$. Hence $\mathfrak{m} = (f)+\mathfrak{p}$. Because $\mathfrak{p}^2 = (0)$, we then have $\mathfrak{m}^2 = f\mathfrak{m}$. Thus $I = (f):\mathfrak{m} = \mathfrak{m}$ and the equality $I^2 = QI$ follows. □

Example 3.5 Let $k[\![X,Y,Z]\!]$ be the formal power series ring over a field k and let $A = k[\![X,Y,Z]\!]/(Y^2 - XZ, XY, X^3)$. Then A is a Buchsbaum local ring, $\dim A = 1$, and $\mathrm{e}(A) = 3$. Let x, y, and z denote the reductions of X, Y, and Z mod $(Y^2 - XZ, XY, X^3)$ and let $\mathfrak{p} = (x,y)$. Then $\operatorname{Min} A = \{\mathfrak{p}\}$, $A/\mathfrak{p}$ is a DVR, and $\mathfrak{p}^3 = (0)$. Hence by Proposition 3.1 and Theorem 3.2 we get the equality $I^2 = QI$ for every parameter ideal $Q \subseteq \mathfrak{m}^2$. We have $I^3 = QI^2$ but $I^2 \neq QI$ for the parameter ideal $Q = (z)$ in A ([**10**, Theorem (5.17)]).

Example 3.6 Let $S = k[\![X_1, X_2, \cdots, X_m, Y, Z]\!]$ $(m \geq 2)$ be the formal power series ring over a field k and let

$$\mathfrak{a} = (X_1, \cdots, X_{m-1})^2 + (X_m^2) + (X_iX_m \mid 1 \leq i \leq m) + (Z^2 - X_1Y).$$

We put $A = S/\mathfrak{a}$. Then A is a Buchsbaum local ring, $\dim A = 1$, and $\mathrm{e}(A) = 2m$. Let x_i, y, and z denote the reductions of X_i, Y, and Z mod $\mathfrak{a}$ and put $\mathfrak{p} = (x_1, x_2, \cdots, x_m, z)$. Then $\operatorname{Min} A = \{\mathfrak{p}\}$, $A/\mathfrak{p}$ is a DVR, and $\mathfrak{p}^3 = (0)$. Hence by Theorem 3.2, $I^2 = QI$ for every parameter ideal $Q \subseteq \mathfrak{m}^2$. We have $I^3 = QI^2$ but $I^2 \neq QI$ for the parameter ideal $Q = (y)$ in A ([**11**, Section 4]).

We explore one more example. Let $(R, \mathfrak{n})$ be a regular local ring with $\dim R \geq 2$ and write $\mathfrak{n} = (X_1, X_2, \cdots, X_n) + (Y)$ where $n = \dim R - 1$. We put $P = (X_1, X_2, \cdots, X_n)$. Let $\ell \geq 2$ be an integer and choose an ideal $\mathfrak{a}$ of R so that

$$\mathfrak{n}P^\ell \subseteq \mathfrak{a} \subseteq P^\ell.$$

We put $A = R/\mathfrak{a}$, $\mathfrak{m} = \mathfrak{n}/\mathfrak{a}$, and $\mathfrak{p} = P/\mathfrak{a}$. Then $\operatorname{Min} A = \{\mathfrak{p}\}$, $\mathfrak{p}^{\ell+1} = (0)$, and $A/\mathfrak{p}$ is a DVR.

Lemma 3.7 *A is a Buchsbaum local ring,* $\dim A = 1$, $\mathrm{H}^0_{\mathfrak{m}}(A) = \mathfrak{p}^\ell$, *and* $\mathrm{e}(A) = \binom{n+\ell-1}{n} > 1$.

Proof: Since $\mathfrak{n}{\cdot}(P^\ell/\mathfrak{a}) = (0)$, by the exact sequence $0 \to P^\ell/\mathfrak{a} \to A \to R/P^\ell \to 0$, we get $\mathrm{H}^0_{\mathfrak{m}}(A) = P^\ell/\mathfrak{a} = \mathfrak{p}^\ell$. Hence A is a Buchsbaum ring. We have $\mathrm{e}(A) = \ell_{A_\mathfrak{p}}(A_\mathfrak{p}) = \ell_{R_P}(R_P/P^\ell R_P) = \binom{n+\ell-1}{n} > 1$, because $\operatorname{Min} A = \{\mathfrak{p}\}$ and $A/\mathfrak{p}$ is a DVR. □

Hence by Theorem 3.2 we get $I^2 = QI$ for every parameter ideal Q in A such that $Q \subseteq \mathfrak{m}^\ell$. We actually furthermore have the following.

Proposition 3.8 $I^2 = QI$ *for every parameter ideal* Q *in* A, *where* $I = Q : \mathfrak{m}$.

Proof: Let $L = \mathfrak{p}^{\ell-1}/\mathfrak{p}^\ell$ and look at the exact sequence

$$0 \to L \to A/\mathfrak{p}^\ell \to A/\mathfrak{p}^{\ell-1} \to 0. \tag{3.9}$$

Since $\mathfrak{p}L = (0)$ and $A/\mathfrak{p}^\ell$ is a Cohen-Macaulay ring, the $A/\mathfrak{p}$-module L is finitely generated and free. Notice that

$$\begin{aligned}\operatorname{rank}_{A/\mathfrak{p}} L = \ell_{R_P}(P^{\ell-1}R_P/P^\ell R_P) &= \binom{n+\ell-2}{n-1} \\ &= \ell_R((0) :_{R/[P^\ell+(Y)]} \mathfrak{n}) = \mathrm{r}(A/\mathfrak{p}^\ell). \end{aligned} \tag{3.10}$$

Let $f \in \mathfrak{m}$ be a parameter of A. We may assume $\bar{f} = \bar{y}^m$ with $m > 0$, where $y = Y$ mod $\mathfrak{a}$ and $\bar{\ }$ denotes the image in $A/\mathfrak{p}$. We look at the exact sequence

$$0 \to L/fL \xrightarrow{\varphi} A/[\mathfrak{p}^\ell + (f)] \to A/[\mathfrak{p}^{\ell-1} + (f)] \to 0. \tag{3.11}$$

induced from exact sequence (3.9). We then have that the homomorphism φ in (3.11) is an isomorphism on the socles, because

$$\ell_A((0) :_{L/fL} \mathfrak{m}) = \operatorname{rank}_{A/\mathfrak{p}} L = \mathrm{r}(A/\mathfrak{p}^\ell) = \ell_A((0) :_{A/[\mathfrak{p}^\ell+(f)]} \mathfrak{m})$$

by (3.10). Hence $(0) :_{A/[\mathfrak{p}^\ell+(f)]} \mathfrak{m} = y^{m-1}{\cdot}\varphi(L/fL)$, so that we have

$$(Q+W) : \mathfrak{m} = (Q+W) + y^{m-1}\mathfrak{p}^{\ell-1}$$

where $Q = (f)$ and $W = \mathfrak{p}^\ell = \mathrm{H}^0_{\mathfrak{m}}(A)$. Let $a_1,\ a_2 \in I = Q : \mathfrak{m}$ and write $a_1 = q_1 + w_1 + y^{m-1}x_1$ and $a_2 = q_2 + w_2 + y^{m-1}x_2$ with $q_i \in Q$, $w_i \in W$, and $x_i \in \mathfrak{p}^{\ell-1}$. Then $y^{m-1}x_1, y^{m-1}x_2 \in I$ and

$$a_1a_2 = a_1(q_2 + w_2) + y^{m-1}a_1x_2 \text{ and } y^{m-1}a_1x_2 = (q_1 + w_1)y^{m-1}x_2 + y^{2m-2}x_1x_2.$$

Since $\mathfrak{m}W = (0)$ and $I \subseteq \mathfrak{m}$, we get $a_1(q_2 + w_2) = a_1 q_2 \in QI$ and $(q_1 + w_1)y^{m-1}x_2 = q_1 y^{m-1}x_2 \in QI$. In order to see $a_1 a_2 \in QI$, it suffices to show $y^{2m-2}x_1 x_2 = 0$. Since $y^{2m-2}x_1x_2 \in y^{2(m-1)}\mathfrak{p}^{2(\ell-1)}$ and $\mathfrak{p}^{\ell+1} = \mathfrak{m}\mathfrak{p}^{\ell} = (0)$, we may assume $\ell = 2$ and $m = 1$. Then $x_1, x_2 \in \mathfrak{p} \cap I$, so that $x_1x_2 \in Q \cap \mathfrak{p}^2 = Q \cap W = (0)$. Thus $y^{2m-2}x_1x_2 = 0$ in any case, whence $I^2 = QI$ as claimed. □

4 Proof of Theorem 1.5

The purpose of this section is to prove Theorem 1.5. Let $R = k[R_1]$ be a homogeneous Buchsbaum ring over an infinite field k. Let $d = \dim R$ and $M = R_+$. We denote by $\mathrm{H}^i_M(*)$ the ith graded local cohomology functor of R with respect to M. Assume that R has maximal embedding dimension, that is, the equality

$$\dim_k R_1 = \mathrm{e}(R) + \dim R - 1 + \mathrm{I}(R)$$

holds true, where $\mathrm{e}(R)$ (respectively $\mathrm{I}(R)$) denotes the multiplicity of R with respect to M (respectively the Buchsbaum invariant of R). This condition is equivalent to saying that $M^2 = QM$ for some (hence, *any*) linear parameter ideal Q in R ([**5**, (3.8)]). Hence for every $f \in R_1$ with $\dim R/(f) = d-1$, the Buchsbaum ring $R/(f)$ also has maximal embedding dimension. We put $L = (0) :_{\mathrm{H}^d_M(R)} M$ and let L_n $(n \in \mathbb{Z})$ denote the homogeneous components of L. Let $\mathrm{a}(R) = \max\{n \in \mathbb{Z} \mid [\mathrm{H}^d_M(R)]_n \neq (0)\}$. Hence

$$\mathrm{a}(R) = \max\{n \in \mathbb{Z} \mid L_n \neq (0)\}$$

and $\mathrm{a}(R) \leq 1 - d$ ([**6**, Corollary (3.3)]).

Let us begin with the following.

Lemma 4.1 $L = L_{-d} + L_{1-d}$.

Proof: If $d = 0$, then $R = k + R_1$ and there is nothing to prove. Suppose that $d > 0$ and that our assertion holds true for $d-1$. Choose $f \in R_1$ so that $\dim R/(f) = d-1$. Then $(0) : f = \mathrm{H}^0_M(R)$ and $f \cdot \mathrm{H}^{d-1}_M(R) = (0)$, because R is a Buchsbaum ring, and so we have the exact sequence

$$0 \to \mathrm{H}^{d-1}_M(R) \xrightarrow{\alpha} \mathrm{H}^{d-1}_M(R/(f)) \xrightarrow{\partial} [\mathrm{H}^d_M(R)](-1) \xrightarrow{f} \mathrm{H}^d_M(R) \to 0 \qquad (4.2)$$

of graded local cohomology modules, where $[\mathrm{H}^d_M(R)](-1)$ denotes the shift of grading. Let $0 \neq x \in L_n$ and assume that $n \neq -d, 1-d$. Then since $fx = 0$, in exact sequence (4.2) we have $x = \partial(y)$ for some $0 \neq y \in [\mathrm{H}^{d-1}_M(R/(f))]_{n+1}$. Since $n+1 \neq 1-d, 2-d$, by the hypothesis of induction on d we get $My \neq (0)$, while $\partial(My) = Mx = (0)$. Consequently, $(0) \neq My \subseteq \alpha(\left[\mathrm{H}^{d-1}_M(R)\right]_{n+2})$, so that $n+2 = 2-d$ because $\mathrm{H}^{d-1}_M(R) = [\mathrm{H}^{d-1}_M(R)]_{2-d}$ by [**6**, Corollary (3.3)]. This is impossible. □

The following is the key for the proof of Theorem 1.5.

Proposition 4.3 *The canonical homomorphism*

$$\mathrm{Ext}^d_R(R/M^2, R) \to \mathrm{H}^d_M(R) = \lim_{n\to\infty} \mathrm{Ext}^d_R(R/M^n, R)$$

is surjective on the socles.

Proof: We may assume $d > 0$. Let

$$E^\bullet : 0 \to R \to E^0 \overset{\partial^0}{\to} E^1 \overset{\partial^1}{\to} \cdots \overset{\partial^{d-1}}{\to} E^d \overset{\partial^d}{\to} \cdots$$

be the graded minimal injective resolution of R and let $I^i = \mathrm{H}^0_M(\mathrm{E}^i)$ $(i \in \mathbb{Z})$. (Here we put $E^i = (0)$ for $i < 0$.) Then we get the complex

$$I^\bullet : \cdots \to 0 \to I^0 \overset{\partial^0}{\to} I^1 \overset{\partial^1}{\to} \cdots \overset{\partial^{d-1}}{\to} I^d \overset{\partial^d}{\to} \cdots$$

of graded injective R-modules, whose cohomology modules $\mathrm{H}^i(I^\bullet)$ $(i \in \mathbb{Z})$ are exactly $\mathrm{H}^i_M(R)$. Recall that each I^i is a finite direct sum of copies of $E(-n)$ $(n \in \mathbb{Z})$ where

$$E = R^* := \underline{\mathrm{Hom}}_k(R, k)$$

denotes the graded k-dual of R ([**13**, (I.2.7)]). Let

$$I^i = \bigoplus_{n \in F} E(-n)^{\alpha_n^{(i)}}$$

where the integers $\alpha_n^{(i)} \geq 0$ denote the multiplicity of $E(-n)$ in the direct sum decomposition of I^i.

We then have the following. The next claim is known by [**6**, Proposition (3.1)] in the case where $i \leq d-1$. We give a brief proof, including the case $i \leq d-1$, for the sake of completeness.

Claim 4.4 *Let $n,\ i \in \mathbb{Z}$. Then $\alpha_n^{(i)} = 0$ if $n > 1-i$.*

Proof of Claim 4.4: Suppose that $\alpha_n^{(i)} > 0$ for some $n, i \in \mathbb{Z}$ with $n > 1-i$. Then $i \geq 0$ because $I^i \neq (0)$. We choose such an integer i as small as possible. Let $0 \neq x \in E_0 = [E(-n)]_n \subseteq [I^i]_n$. Then $\partial^i(x) = 0$, because $Mx = (0)$ and the resolution $E^\bullet$ is minimal. Hence the element x defines an element $\bar{x} \in [\mathrm{H}_M^i(R)]_n$ with $n > 1-i$. Notice that $[\mathrm{H}_M^i(R)]_m = (0)$ if $m > 1-i$ ([**6**, Corollary (3.3)]). We then have $x = \partial^{i-1}(y)$ for some $y \in [I^{i-1}]_n$. Hence $n = 2-i$, because $y \neq 0$ and $I^{i-1} = \bigoplus_{m \leq 2-i} E(-m)^{\alpha_m^{(i-1)}}$ by the choice of i. Consequently, the homogeneous element y of I^{i-1} with degree $2-i$ must be contained in the socle of I^{i-1}, and so the minimality of $E^\bullet$ forces $x = \partial^{i-1}(y) = 0$, which is impossible. Hence $\alpha_n^{(i)} = 0$ if $n > 1-i$. □

Now let $\{x_i\}_{1 \leq i \leq \ell}$ and $\{y_j\}_{1 \leq j \leq m}$ be a k-basis of L_{-d} and L_{1-d}, respectively. Let $\{c_t\}_{1 \leq t \leq v}$ be a k-basis of R_1. We choose for each $1 \leq i \leq \ell$ and $1 \leq j \leq m$, $f_i \in [I^d]_{-d}$ and $g_j \in [I^d]_{1-d}$ so that $\partial^d(f_i) = \partial^d(g_j) = 0$ and $x_i = \bar{f_i}$ and $y_j = \bar{g_j}$ in $\mathrm{H}_M^d(R)$, where $\bar{\ }$ denotes the image in the cohomology $\mathrm{H}^d(I^\bullet) = \mathrm{H}_M^d(R)$. Then because $ML = (0)$, we have $c_t f_i = \partial^{d-1}(\xi_{ti})$ and $c_t g_j = \partial^{d-1}(\eta_{tj})$ for some $\xi_{ti} \in [I^{d-1}]_{1-d}$ and $\eta_{tj} \in [I^{d-1}]_{2-d}$. Notice that by 4.4, $[I^{d-1}]_n = (0)$ if $n > 2-d$ and $[I^d]_n = (0)$ if $n > 1-d$. We then have that $M^2\xi_{ti} = M^2\eta_{tj} = (0)$ and $M^2 f_i = M^2 g_j = (0)$ for all $1 \leq i \leq \ell$, $1 \leq j \leq m$, and $1 \leq t \leq v$; that is, $f_i, g_j \in (0) :_{E^d} M^2$ and $\xi_{ti}, \eta_{tj} \in (0) :_{E^{d-1}} M^2$. Because

$$\partial^d(f_i) = \partial^d(g_j) = 0, \partial^{d-1}(\xi_{ti}) = c_t f_i, \text{ and} \partial^{d-1}(\eta_{tj}) = c_t g_j,$$

the elements f_i and g_j define socle elements

$$\bar{f_i},\ \bar{g_j} \in \mathrm{H}^d((0) :_{E^\bullet} M^2) = \mathrm{Ext}_R^d(R/M^2, R)$$

whose images under the canonical map

$$\varphi : \mathrm{Ext}_R^d(R/M^2, R) = \mathrm{H}^d((0) :_{E^\bullet} M^2) \to \mathrm{H}_M^d(R) = \mathrm{H}^d(I^\bullet)$$

are exactly x_i and y_j. Hence the canonical homomorphism φ is surjective on the socles. □

Let $W = \mathrm{H}_M^0(R)$. Then by [**10**, Proof of Theorem (3.9)] we get the following.

Corollary 4.5 *Suppose that* $d = \dim R > 0$ *and let* Q *be a graded parameter ideal in* R. *Assume that* $Q \subseteq M^2$. *Then* $(Q+W) : M = Q : M$. *If* $\mathrm{e}(R) > 1$, *then* $I^2 = QI$ *where* $I = Q : M$.

We are now ready to prove Theorem 1.5.

Proof of Theorem 1.5: We may assume that $d = \dim R > 0$ and that our assertion holds true for $d-1$. Let $Q = (f_1, f_2, \cdots, f_d)$ be a graded parameter ideal in R and let $n_i = \deg f_i$ $(1 \le i \le d)$. To prove the equality $I^2 = QI$, thanks to Corollary 4.5, we may assume that $Q \not\subseteq M^2$. Suppose that $n_i = 1$ for all $1 \le i \le d$. Then Q is a linear parameter ideal in R with $M^2 = QM$, because R has maximal embedding dimension (cf. [**5**, (3.8)]). Hence $I = Q : M = M$ so that $I^2 = QI$. Therefore, after renumbering of f_i's, we may assume that $n_i = 1$ if $1 \le i \le \ell$ while $n_i \ge 2$ if $\ell < i \le d$, where $1 \le \ell < d$. Then, passing to the ring $R/(f_1)$, we get $I^2 \subseteq (f_2, \cdots, f_d)I + (f_1)$. Hence similarly as in the proof of Theorem 1.3, we have the equality $I^2 = QI$ modulo the following.

Claim 4.6 $(Q+W) : M = Q : M$.

Proof of Claim 4.6: Let $\bar{R} = R/(f_1, f_2, \cdots, f_\ell)$ and $\bar{M} = \bar{R}_+$. Then because $Q\bar{R} \subseteq \bar{M}^2$, by Corollary 4.5 we get $(Q\bar{R} + \mathrm{H}^0_{\bar{M}}(\bar{R})) : \bar{M} = Q\bar{R} : \bar{M}$. Let $x \in (Q+W) : M$ and let $\bar{x}$ denote the reduction mod $(f_1, f_2, \cdots, f_\ell)$. Then, because $\bar{M}\bar{x} \subseteq Q\bar{R} + W\bar{R}$ and $W\bar{R} \subseteq (0) :_{\bar{R}} \bar{M} = \mathrm{H}^0_{\bar{M}}(\bar{R})$, we have $\bar{x} \in (Q\bar{R} + \mathrm{H}^0_{\bar{M}}(\bar{R})) : \bar{M} = Q\bar{R} : \bar{M}$, whence $x \in Q : M$ as claimed. □

Remark 4.7 Typical examples of Buchsbaum homogeneous rings $R = k[R_1]$ of maximal embedding dimension are as follows.

Let $S = k[X_1, X_2, \cdots, X_d, Y_1, Y_2, \cdots, Y_d]$ $(d > 0)$ be the polynomial ring over a field k and put

$$R = S/(X_1, X_2, \cdots, X_d) \cap (Y_1, Y_2, \cdots, Y_d).$$

Then R is a Buchsbaum ring, $\dim R = d$, $\operatorname{depth} R = 1$, $\mathrm{e}(R) = 2$, and R has maximal embedding dimension. Let $A = R_M$. Then the equality $I^2 = QI$ holds true for every parameter ideal Q in the Buchsbaum local ring A ([**11**, Theorem (1.1)]). This result is not a direct consequence of Theorem 1.5, and the authors do not know, in general, whether $I^2 = QI$ for every parameter ideal Q in the local ring $A = R_M$, where R is the homogeneous Buchsbaum ring studied in Theorem 1.5.

5 Graded Buchsbaum rings of dimension 1

Let $R = \sum_{n \geq 0} R_n$ be a Buchsbaum graded ring with $k = R_0$ a field. In this section we assume that $\dim R = 1$. Let $M = R_+$, $a = \mathrm{a}(R)$, and $L = (0) :_{\mathrm{H}_M^1(R)} M$. We put $W = \mathrm{H}_M^0(R)$. The purpose is to prove the following.

Theorem 5.1 *Let $b \in \mathbb{Z}$ with $b \leq a$ and assume that $L_i = (0)$ if $i \notin [b,a] = \{i \in \mathbb{Z} \mid b \leq i \leq a\}$. If $a < 2b$, then $I^2 = QI$ for every graded parameter ideal Q in R, where $I = Q : M$.*

Proof: Let $\bar{R} = R/W$. We have $0 < b \leq a$, since $2b > a \geq b$. Therefore $\bar{R}$ is not the polynomial ring over k, because $\mathrm{a}(\bar{R}) = \mathrm{a}(R) = a > 0$. In particular $\mathrm{e}(R) > 1$. Let $f \in R_n$ $(n > 0)$ be a homogeneous parameter in R and put $Q = (f)$. Then f is $\bar{R}$-regular and we have the exact sequence

$$0 \to \mathrm{H}_M^0(\bar{R}/f\bar{R}) \to \mathrm{H}_M^1(\bar{R})(-n) \xrightarrow{f} \mathrm{H}_M^1(\bar{R}) \to 0$$

of graded local cohomology modules. Because $\mathrm{H}_M^1(\bar{R}) \cong \mathrm{H}_M^1(R)$, we get an isomorphism

$$(0) :_{\bar{R}/f\bar{R}} M \cong L(-n)$$

whence the graded R-module $[(Q+W) : M]/(Q+W)$ is concentrated with degrees in the interval $[b+n, a+n]$. In particular $R_m \subseteq Q+W$ for all $m > a+n$. Let us choose homogeneous elements $\{\xi_i\}_{1 \leq i \leq \ell}$ of $I = Q : M$ so that their images $\{\bar{\xi}_i\}_{1 \leq i \leq \ell}$ in $I/(Q+W)$ form a k-basis of $I/(Q+W)$. Hence $I = (Q+W) + \sum_{i=1}^{\ell} k\xi_i$ and $\xi_i \notin Q+W$ for any $1 \leq i \leq \ell$. Let $\deg \xi_i = d_i$. Then $a+n \geq d_i \geq b+n$, since $\xi_i \in (Q+W) : M$ but $\xi_i \notin Q+W$. We shall show that $\xi_i\xi_j \in Q^2$ for all $1 \leq i, j \leq \ell$. Assume that $\xi_i\xi_j \notin Q^2$ for some $1 \leq i, j \leq \ell$ and write $\xi_i\xi_j = f\delta$ with $\delta \in R_{d_i+d_j-n}$. We then have $\delta \notin Q+W$ since $\xi_i\xi_j = f\delta \notin Q^2$, while $\delta \in Q+W$ because $d_i + d_j - n \geq (2b+2n) - n = 2b+n > a+n$, which is the required contradiction. Thus $\xi_i\xi_j \in Q^2$ for all $1 \leq i, j \leq \ell$ so that we have $I^2 = QI$. $\square$

The following is a direct consequence of Theorem 5.1. In Section 6 we will show that the assumption that $\mathrm{a}(R) > 0$ in Corollary (5.2) is crucial (cf. Theorem 6.1 (1)).

Corollary 5.2 *Let $a = \mathrm{a}(R) > 0$ and assume that $L = (0) :_{\mathrm{H}^1_M(R)} M$ is concentrated in degree a. Then $I^2 = QI$ for every graded parameter ideal Q in R, where $I = Q : M$.*

Here let us explore one example to illustrate Theorem 5.1. Let $S = k[X,Y,Z]$ be the polynomial ring over a field k and let $a,b,c > 0$ be integers. We consider S to be a graded ring with $\deg X = a, \deg Y = b$, and $\deg Z = c$. Let $T = k[t]$ be the polynomial ring and let $\varphi : S \to T$ be the k-algebra map defined by $\varphi(X) = t^a, \varphi(Y) = t^b$, and $\varphi(Z) = t^c$. We put $P = \operatorname{Ker}\varphi$ and assume that there exist integers $\alpha, \beta, \gamma, \alpha', \beta', \gamma' > 0$ such that

$$P = I_2 \begin{pmatrix} X^\alpha & Y^\beta & Z^\gamma \\ Y^{\beta'} & Z^{\gamma'} & X^{\alpha'} \end{pmatrix},$$

where $I_2 \begin{pmatrix} X^\alpha & Y^\beta & Z^\gamma \\ Y^{\beta'} & Z^{\gamma'} & X^{\alpha'} \end{pmatrix}$ denotes the ideal in S generated by the maximal minors of the matrix $\begin{pmatrix} X^\alpha & Y^\beta & Z^\gamma \\ Y^{\beta'} & Z^{\gamma'} & X^{\alpha'} \end{pmatrix}$. This assumption is satisfied under mild conditions on the integers a,b,c (cf. [**14**]). Let $\Delta_1 = X^{\alpha'}Y^\beta - Z^{\gamma+\gamma'}, \Delta_2 = Y^{\beta'}Z^\gamma - X^{\alpha+\alpha'}$, and $\Delta_3 = X^\alpha Z^{\gamma'} - Y^{\beta+\beta'}$. Then $P = (\Delta_1, \Delta_2, \Delta_3)$ is a perfect ideal of grade 2 and the semigroup ring $S/P = k[t^a, t^b, t^c]$ has the S- resolution of the form

$$0 \to \begin{matrix} S(-(a\alpha + c\gamma + c\gamma')) \\ \oplus \\ S(-(b\beta' + c\gamma + c\gamma')) \end{matrix} \xrightarrow{\begin{bmatrix} X^\alpha & Y^{\beta'} \\ Y^\beta & Z^{\gamma'} \\ Z^\gamma & X^{\alpha'} \end{bmatrix}} \begin{matrix} S(-(c\gamma + c\gamma')) \\ \oplus \\ S(-(a\alpha + a\alpha')) \\ \oplus \\ S(-(b\beta + b\beta')) \end{matrix} \tag{5.3}$$

$$\xrightarrow{[\Delta_1 \Delta_2 \Delta_3]} S \to S/P \to 0.$$

Let $\delta = a+b+c$ and $K_S = S(-\delta)$. Then, taking the K_S-dual of resolution (5.3), we get the following presentation

$$\begin{matrix} S(c\gamma+c\gamma'-\delta) & & \\ \oplus & & S(a\alpha+c\gamma+c\gamma'-\delta) \\ S(a\alpha+a\alpha'-\delta) & \xrightarrow{\begin{bmatrix} X^{\alpha} & Y^{\beta} & Z^{\gamma} \\ Y^{\beta'} & Z^{\gamma'} & X^{\alpha'} \end{bmatrix}} & \oplus & \to K_{S/P} \to 0, \\ \oplus & & S(b\beta'+c\gamma+c\gamma'-\delta) \\ S(b\beta+b\beta'-\delta) & & \end{matrix}$$

of the graded canonical module $K_{S/P}$ of S/P, which shows that the socle of $H^1_N(S/P)$, where $N = [S/P]_+$, is generated by two homogeneous elements ξ and η with

$$\deg\xi = a\alpha + c\gamma + c\gamma' - \delta \text{ and } \deg\eta = b\beta' + c\gamma + c\gamma' - \delta.$$

We have that $\deg\eta \geq \deg\xi$ and $2\cdot\deg\xi > \deg\eta$ if and only if

$$b\beta' \geq a\alpha \text{ and } 2(a\alpha + c\gamma + c\gamma') > \delta + b\beta'.$$

When this is the case, $\deg\eta = \mathrm{a}(S/P)$. Hence by Theorem 5.1 we get the following.

Example 5.4 Suppose that $b\beta' \geq a\alpha$ and $2(a\alpha + c\gamma + c\gamma') > \delta + b\beta'$. Let $R = \sum_{n\geq 0} R_n$ be a one-dimensional Buchsbaum graded ring with $R_0 = k$ and let $M = R_+$. Assume that $R/H^0_M(R) \cong k[t^a, t^b, t^c]$ as graded k-algebras. Then $I^2 = QI$ for every graded parameter ideal Q in R where $I = Q : M$.

For instance, let $a = 3, b = 7$, and $c = 8$. Then $P = I_2\begin{pmatrix} X^2 & Y & Z \\ Y & Z & X^3 \end{pmatrix}$ and $\deg\xi = 4, \deg\eta = 5$. Because $2\cdot\deg\xi > \deg\eta$, by Theorem 5.1 $I^2 = QI$ for every graded parameter ideal Q in R, where $R = \sum_{n\geq 0} R_n$ is any one-dimensional Buchsbaum graded ring with $R_0 = k$ such that $R/H^0_M(R) \cong k[t^3, t^7, t^8]$ as graded k-algebras. Choose $a = 3, b = 4$, and $c = 5$. Then $P = I_2\begin{pmatrix} X & Y & Z \\ Y & Z & X^2 \end{pmatrix}$ and $\deg\xi = 1, \deg\eta = 2$. For this choice we have $2\cdot\deg\xi = \deg\eta$ and, as is shown in [**10**, Section 4], there exists a one-dimensional Buchsbaum graded ring $R = \sum_{n\geq 0} R_n$ with $R_0 = k$ such that $R/H^0_M(R) \cong k[t^3, t^4, t^5]$ as graded k-algebras and R contains a graded parameter ideal Q with $I^3 = QI^2$ but $I^2 \neq QI$.

6 A method for constructing bad homogeneous Buchsbaum rings of dimension 1

Let k be a field and let $R = \sum_{n \geq 0} R_n$ be a Noetherian graded ring with $k = R_0$ a field. Let $M = R_+$ and let $\mathrm{e}(R)$ denote the multiplicity of R with respect to M. The purpose of this section is to give a general method for constructing a homogeneous Buchsbaum ring $R = k[R_1]$ with $\dim R = 1$ and $\mathrm{e}(R) = 3$ which fails to have the equality $I^2 = QI$ for some linear parameter ideal $Q = (f)$, where $I = Q : M$.

Let $S = k[X,Y,Z]$ be the polynomial ring over a field k and let $N = (X,Y,Z)$. We regard S to be a graded ring with $\deg X = \deg Y = \deg Z = 1$. For each $\xi = aX + bY + cZ \in S_1$ with $a, b, c \in k$, let

$$\mathfrak{a}(\xi) = \mathrm{I}_2 \begin{pmatrix} Y & Z & X \\ X & Y & \xi \end{pmatrix} = (Y^2 - XZ, XY - Z\xi, X^2 - Y\xi)$$

denote the ideal of S generated by the maximal minors of the matrix $\begin{pmatrix} Y & Z & X \\ X & Y & \xi \end{pmatrix}$. Then $\mathrm{ht}_S \mathfrak{a}(\xi) = 2$ and so $S/\mathfrak{a}(\xi)$ is a one-dimensional homogeneous Cohen-Macaulay ring. Because

$$\mathfrak{a}(\xi) + (Z) = (Y^2 - XZ, XY - Z\xi, X^2 - (aX + bY + cZ)Y) + (Z) = (X,Y)^2 + (Z),$$

we get that $\mathrm{e}(S/\mathfrak{a}(\xi)) = \dim_k S/[\mathfrak{a}(\xi) + (Z)] = 3$, $\mathrm{a}(S/\mathfrak{a}(\xi)) = \mathrm{a}(S/[\mathfrak{a}(\xi) + (Z)]) - 1 = 0$ by [**13**, 3.1.6], and the socle of $\mathrm{H}^1_N(S/\mathfrak{a}(\xi))$ is concentrated in degree 0. We furthermore have the following.

Theorem 6.1

(1) *For each* $\xi \in S_1$ *there exists a homogeneous Buchsbaum ring* $R = k[R_1]$ *with* $\dim R = 1$ *and* $\mathrm{e}(R) = 3$ *such that*

(i) $I^2 \neq QI$ *for some linear parameter ideal* Q *in* R,

(ii) $I^2 = QI$ *for every graded parameter ideal* Q *in* R *such that* $Q \subseteq M^2$ *where* $I = Q : M$, *and*

(iii) $R/\mathrm{H}^0_M(R) \cong S/\mathfrak{a}(\xi)$ *as graded k-algebras.*

(2) *Conversely, let* $R = k[R_1]$ *be a homogeneous Buchsbaum ring with* $\dim R = 1$ *and* $\mathrm{e}(R) = 3$. *Assume that* $I^2 \neq QI$ *for some linear parameter ideal* Q *in* R *where* $I = Q : M$. *Then for some* $\xi \in S_1$, *we have* $R/\mathrm{H}^0_M(R) \cong S/\mathfrak{a}(\xi)$ *as graded k-algebras.*

Proof of Theorem 6.1 (1): Let $\mathfrak{a} = \mathfrak{a}(\xi)$ and $\tilde{\mathfrak{a}} = (Y^2 - XZ, XY - Z\xi) + N\cdot(X^2 - Y\xi)$. We put $R = S/\tilde{\mathfrak{a}}$. Then $N\mathfrak{a} \subseteq \tilde{\mathfrak{a}} \subseteq \mathfrak{a}$ and $\tilde{\mathfrak{a}} = (Y^2 - XZ, XY - Z\xi, X^3 - XY\xi)$. Therefore, since $S/\mathfrak{a}$ is a Cohen-Macaulay ring with $\dim S/\mathfrak{a} = 1$ and $\mathrm{e}(S/\mathfrak{a}) = 3$, the homogeneous ring R is Buchsbaum, $\dim R = 1$, $\mathrm{e}(R) = 3$, and $\mathrm{H}^0_M(R) = \mathfrak{a}/\tilde{\mathfrak{a}} = (x^2 - yu)$, where x, y , z, and u denote the images of X, Y , Z, and ξ in R. We put $Q = (z)$ and $I = Q : M$. Then Q is a linear parameter ideal in R and $I = (z, y, x^2 - yu)$, because $\tilde{\mathfrak{a}} + (Z) = (Y^2 - XZ, XY - Z\xi, X^3 - (aX + bY + cZ)XY) + (Z) = (X^3, XY, Y^2, Z)$ where $\xi = aX + bY + cZ$ with $a, b, c \in k$. Hence $I^2 = zI + (y^2)$.

Claim 6.2 $y^2 \notin zI$.

Proof of Claim 6.2: Assume that $y^2 \in zI = (z^2, yz)$, that is, $Y^2 \in (Z^2, YZ) + (Y^2 - XZ, XY - Z\xi, X^3 - XY\xi)$. Then $Y^2 \in (Z^2, YZ, Y^2 - XZ, XY - Z\xi)$, since $Y^2 \in S_2$. Let $\xi = aX + bY + cZ$ with $a, b, c \in k$. Then $Y^2 \in (Z^2, YZ, Y^2 - XZ, XY - (aX + bY + cZ)Z) = (Z^2, YZ, Y^2 - XZ, XY - aXZ)$, which is impossible. In fact, assume that $Y^2 \in (Z^2, YZ, Y^2 - XZ, XY - aXZ)$ and write $Y^2 = \alpha Z^2 + \beta YZ + \gamma(Y^2 - XZ) + \delta(XY - aXZ)$ with $\alpha, \beta, \gamma, \delta \in k$. Then comparing the coefficients, we get $\alpha = 0$, $\beta = 0$, $\gamma = 1$, $\delta = 0$, and $\gamma + a\delta = 0$, which is absurd. □

Hence $y^2 \notin QI$ so that $I^2 \neq QI$, which proves Theorem 6.1 (1). See Proposition 3.1 for assertion (ii). □

Proof of Theorem 6.1 (2): Assume that $I^2 \neq QI$ for the linear parameter ideal $Q = (f)$. We put $\bar{R} = R/W$ and $\bar{M} = M/W$. Then $\bar{R}$ is a Cohen-Macaulay ring with $\dim \bar{R} = 1$ and $\mathrm{e}(\bar{R}) = 3$. Let $I = Q : M$. Then we have:

Claim 6.3 $\dim_k \bar{R}_1 = 3$, $\dim_k \bar{M}/f\bar{R} = 2$, *and* $Q + W \subsetneq I \subsetneq (Q + W) : M = M$.

Proof of Claim 6.3: If $I = Q + W$, then $I^2 = Q^2 = QI$, which is impossible. We get $I \neq (Q + W) : M$ similarly as in the proof of Theorem 1.3. Hence $Q + W \subsetneq I \subsetneq (Q + W) : M$ so that $\bar{R}$ is not a Gorenstein ring. Thus $\dim_k \bar{M}/f\bar{R} = 2$

and $\bar{M}^2 \subseteq f\bar{R}$, because $\dim_k \bar{R}/f\bar{R} = \mathrm{e}(\bar{R}) = 3$. Hence $\dim_k \bar{R}_1 = 3$ and $(Q+W) : M = M$. □

We write $M = Q+W+(g,h)$ with $g,h \in R_1$. Let $\bar{g},\bar{h}$ denote the images of g,h in $M/(Q+W)$. Then $\bar{g},\bar{h}$ form a k-basis of $M/(Q+W)$ and $\dim_k M/I = \dim_k I/(Q+W) = 1$ by 6.3. Since $I^2 \neq QI$, there exist homogeneous elements ϕ,ψ in I such that $\phi\psi \notin QI$. However, since $\dim_k I/(Q+W) = 1$ and $(Q+W)I = QI$, without loss of generality we may assume that $g \in I$ and $g^2 \notin QI$. Hence $I = Q+W+(g)$ and $gM \subseteq Q$. Let $g^2 = f\delta$ with $\delta \in R_1$ and write $\delta = af+bg+ch+z$ with $a,b,c \in k$ and $z \in W$. Then $\delta \equiv ch \bmod I$, whence $c \neq 0$ because $\delta \notin I$. Therefore $M/I = k\bar{h} = k\bar{\delta}$ where $^-$ denotes the image in M/I and so, replacing h by δ, we may assume that $g^2 = fh$. Since $gM \subseteq Q$ and $hM \subseteq Q+W$, we have $gh = fu$ and $h^2 = fv+w$ with $u,v \in R_1$ and $w \in W$. Then $f^2v = fh^2 = g^2h = fgu$, because $g^2 = fh$ and $fw = 0$. Hence $fv \equiv gu \bmod W$ so that $h^2 \equiv gu \bmod W$.

Now let $\Phi : S = k[X,Y,Z] \to \bar{R}$ denote the k-algebra map defined by $\Phi(X) = \bar{h}$, $\Phi(Y) = \bar{g}$, and $\Phi(Z) = \bar{f}$, where $^-$ denotes the image in $\bar{R}$. Then, since the k-space $\bar{R}_1$ is spanned by $\bar{f}$, $\bar{g}$, $\bar{h}$, the homomorphism Φ is surjective, so that $\bar{u} = \Phi(\xi)$ for some $\xi \in S_1$. Let $\mathfrak{a} = \operatorname{Ker}\Phi$. We then have the following, which completes the proof of Theorem 6.1 (2).

Claim 6.4 $\mathfrak{a} = \mathfrak{a}(\xi)$.

Proof Claim 6.4: Because $\bar{R}$ is a Cohen-Macaulay ring with $\dim_k \bar{R}/f\bar{R} = 3$, it suffices to check that $\mathfrak{a} \supseteq \mathfrak{a}(\xi)$ and $\dim_k S/[\mathfrak{a}(\xi)+(z)] = 3$. Since $\bar{g}^2 = \bar{f}\,\bar{h}$, $\bar{g}\bar{h} = \bar{f}\bar{u}$, and $\bar{h}^2 = \bar{g}\,\bar{u}$, we certainly have $\mathfrak{a} \supseteq \mathfrak{a}(\xi)$. Because $\mathfrak{a}(\xi)+(Z) = (X,Y)^2+(Z)$, we get $\dim_k S/[\mathfrak{a}(\xi)+(Z)] = 3$. Hence $\mathfrak{a} = \mathfrak{a}(\xi)$. □

Let $\mathfrak{a}$ be a graded ideal in $S = k[X,Y,Z]$ such that $\dim S/\mathfrak{a} = 1$, $\mathrm{e}(S/\mathfrak{a}) = 3$, and $S/\mathfrak{a}$ is not a Gorenstein ring. The next problem is to clarify when $S/\mathfrak{a} \cong S/\mathfrak{a}(\xi)$ as graded k-algebras for some $\xi \in S_1$. Because $\mathrm{e}(S/\mathfrak{a}) = 3$, the number of the associated prime ideals of $\mathfrak{a}$ is at most 3 and following is the classification table of such ideals $\mathfrak{a}$ according to $n = \#\operatorname{Ass}_S S/\mathfrak{a}$. The proof we have done to perform the classification is standard but needs a lot of very careful and boring computations, whose details we would like to leave to the forthcoming paper [**8**].

Classification Table 6.5

n	$\mathfrak{a}$	canonical form of the matrix	
3	(XY,YZ,ZX)	$\begin{pmatrix} Y & Z & X \\ X & Y & Y \end{pmatrix}$	if $p = \mathrm{ch}\, k \neq 2$;
		$\begin{pmatrix} Y & Z & X \\ X & Y & (a+1)X+aY \end{pmatrix}$	if $p = 2, k \neq \mathbb{Z}/2\mathbb{Z}$, and $a \in k \setminus \{0,1\}$
2	$(XZ,YZ,$ $X^2+aXY+bY^2)$	$\begin{pmatrix} Y & Z & X \\ X & Y & \frac{a}{b}X - \frac{1}{b}Y \end{pmatrix}$	where $a,b \in k$ with X^2+aX+b irreducible in $k[X]$
2	(XZ,XY,Y^2)	$\begin{pmatrix} Y & Z & X \\ X & Y & X \end{pmatrix}$	
1	$(X,Y)^2$	nothing	
1	$(Y^2-XZ,$ $XY,X^2)$	$\begin{pmatrix} Y & Z & X \\ X & Y & 0 \end{pmatrix}$	
1	$\mathrm{I}_2(\mathbb{M})$	$\mathbb{M} = \begin{pmatrix} Y & Z & X \\ X & Y & aX+bY+cZ \end{pmatrix}$	where $a,b,c \in k$ with $X^3 - aX^2 - bX - c$ irreducible in $k[X]$

Let us explain how to read the table. Let $\mathfrak{a}$ be a graded ideal in S such that $\dim S/\mathfrak{a} = 1$ and $\mathrm{e}(S/\mathfrak{a}) = 3$. Assume that $S/\mathfrak{a}$ is not a Gorenstein ring. Then as is well-known, the ideal $\mathfrak{a}$ is generated by the maximal minors of a matrix $\mathbb{M} = \begin{pmatrix} \xi_{11} & \xi_{12} & \xi_{13} \\ \xi_{21} & \xi_{22} & \xi_{23} \end{pmatrix}$ with entries $\xi_{ij} \in S_1$. The second column of Table 6.5 indicates that, e.g., when $n = 3$, after suitable elementary k-linear transformations of the indeterminates X,Y,Z, we may assume that $\mathbb{M} = \begin{pmatrix} X & Y & 0 \\ 0 & Y & Z \end{pmatrix}$, whence $\mathfrak{a} = (XY,YZ,ZX)$. The third column of the table says that, when $\mathrm{ch}\, k \neq 2$ (respectively $\mathrm{ch}\, k = 2$ but $k \neq \mathbb{Z}/2\mathbb{Z}$), after further elementary k-linear transformations of X,Y,Z, our ideal $\mathfrak{a}$ is exactly generated by the maximal minors of the matrix $\begin{pmatrix} Y & Z & X \\ X & Y & Y \end{pmatrix}$ (respectively $\begin{pmatrix} Y & Z & X \\ X & Y & (a+1)X+aY \end{pmatrix}$ with $a \in k \setminus \{0,1\}$). When $k = \mathbb{Z}/2\mathbb{Z}$, the matrix $\begin{pmatrix} X & Y & 0 \\ 0 & Y & Z \end{pmatrix}$ is never transformed into the matrix of the form $\begin{pmatrix} Y & Z & X \\ X & Y & \xi \end{pmatrix}$ with $\xi \in S_1$. The matrix $\begin{pmatrix} X & Y & 0 \\ 0 & X & Y \end{pmatrix}$ certainly cannot be transformed into the matrix of the form $\begin{pmatrix} Y & Z & X \\ X & Y & \xi \end{pmatrix}$ with $\xi \in S_1$. Our Table 6.5 claims that these two cases are the unique exceptions.

Except these two cases up to isomorphisms of graded k-algebras, for any graded ideal $\mathfrak{a}$ in the polynomial ring $S = k[X,Y,Z]$ such that $S/\mathfrak{a}$ is a non-Gorenstein Cohen-Macaulay ring with $\dim S/\mathfrak{a} = 1$ and $\mathrm{e}(S/\mathfrak{a}) = 3$, one can always construct at least one homogeneous Buchsbaum ring $R = k[R_1]$ with $\dim R = 1$ and $\mathrm{e}(R) = 3$ such that

(i) $I^2 \neq QI$ for some linear parameter ideal Q in R, but

(ii) $I^2 = QI$ for every graded parameter ideal Q in R such that $Q \subseteq M^2$, and

(iii) $R/\mathrm{H}^0_M(R) \cong S/\mathfrak{a}$ as graded k-algebras,

where $I = Q : M$.

References

[**1**] A. Corso, C. Huneke, and W.V. Vasconcelos, On the integral closure of ideals, *Manuscripta Math.* **95** (1998), 331–347.

[**2**] A. Corso and C. Polini, Links of prime ideals and their Rees algebras, *J. Algebra* **178** (1995), 224–238.

[**3**] A. Corso, C. Polini, and W.V. Vasconcelos, Links of prime ideals, *Math. Proc. Camb. Philos. Soc.* **115** (1994), 431–436.

[**4**] S. Goto, On Buchsbaum rings, *J. Algebra* **67** (1980), 272–279.

[**5**] S. Goto, Buchsbaum rings obtained by gluing, *Nagoya Math. J.* **83** (1981), 123–135.

[**6**] S. Goto, Buchsbaum rings of maximal embedding dimension, *J. Algebra* **76** (1982), 383–399.

[**7**] S. Goto, Integral closedness of complete-intersection ideals, *J. Algebra* **108** (1987), 151–160.

[**8**] S. Goto and N. Matsuoka, A method for constructing bad Buchsbaum graded rings and a classification of certain graded ideals of height two, in preparation.

[**9**] S. Goto and K. Nishida, Buchsbaumness of associated graded rings, *J. Pure Appl. Algebra* **181** (2003), 61–74.

[**10**] S. Goto and H. Sakurai, The equality $I^2 = QI$ in Buchsbaum rings, *Rend. Sem. Mat. Univ. Padova* **110** (2003), 25–56.

[**11**] S. Goto and H. Sakurai, The reduction exponent of socle ideals associated to parameter ideals in a Buchsbaum local ring of multiplicity two, *J. Math. Soc. Japan* **56** (2004), 1157–1168.

[**12**] S. Goto and N. Suzuki, Index of reducibility of parameter ideals in a local ring, *J. Algebra* **87** (1984), 53–88.

[**13**] S. Goto and K.-I. Watanabe, On graded rings, I, *J. Math. Soc. Japan* **30** (1978), 179–213.

[**14**] J. Herzog, Generators and relations of abelian semigroups and semigroup rings, *Manuscripta Math.* **3** (1970), 175–193.

[**15**] J. Herzog and E. Kunz, *Der kanonische Modul eines Cohen-Macaulay-Rings*, Lecture Notes in Math. **238**, Springer-Verlag, Berlin · Heidelberg · New York · Tokyo, 1971.

[**16**] C. Huneke, The theory of d-sequences and powers of ideals, *Adv. Math.* **46** (1982), 249–279.

[**17**] J. Stückrad and W. Vogel, *Buchsbaum Rings and Applications*, Springer-Verlag, Berlin · New York · Tokyo, 1986.

[**18**] K. Yamagishi, The associated graded modules of Buchsbaum modules with respect to $\mathfrak{m}$-primary ideals in the equi-$\mathbb{I}$-invariant case, *J. Algebra* **225** (2000), 1–27.

[**19**] K. Yamagishi, Buchsbaumness in Rees modules associated to ideals of minimal multiplicity in the equi-$\mathbb{I}$-invariant case, *J. Algebra* **251** (2002), 213–255.

Integral closures of ideals in completions of regular local domains

WILLIAM HEINZER
Department of Mathematics, Purdue University
West Lafayette, IN 47907, USA
E-mail: heinzer@math.purdue.edu

CHRISTEL ROTTHAUS
Department of Mathematics, Michigan State University
East Lansing, MI 48824, USA
E-mail: rotthaus@math.msu.edu

SYLVIA WIEGAND
Department of Mathematics and Statistics, University of Nebraska
Lincoln, NE 68588, USA
E-mail: swiegand@math.unl.edu

Abstract: In this paper we exhibit an example of a three-dimensional regular local domain $(A, \mathbf{n})$ having a height-two prime ideal P with the property that the extension $P\widehat{A}$ of P to the $\mathbf{n}$-adic completion $\widehat{A}$ of A is not integrally closed. We use a construction we have studied in earlier papers: For $R = k[x,y,z]$, where k is a field of characteristic zero and x,y,z are indeterminates over k, the example A is an intersection of the localization of the power series ring $k[y,z][[x]]$ at the maximal ideal (x,y,z) with the field $k(x,y,z,f,g)$, where f,g are elements of $(x,y,z)k[y,z][[x]]$ that are algebraically independent over $k(x,y,z)$. The elements f,g are chosen in such a way that, using results from our earlier papers, A is Noetherian and it is possible to describe A as a nested union of rings associated to A that are localized polynomial rings over k in five variables.

1 Introduction and background

We are interested in the general question: What can happen in the completion of a "nice" Noetherian ring? We are examining this question as part of a project of constructing Noetherian and non-Noetherian integral domains using power series rings. In this paper as a continuation of that project we display an example of a three-dimensional regular local domain $(A, \mathbf{n})$ having a height-two prime ideal P with the property that the extension $P\widehat{A}$ of P to the $\mathbf{n}$-adic completion $\widehat{A}$ of A is not integrally closed. The ring A in the example is a nested union of regular local domains of dimension five.

Let I be an ideal of a commutative ring R with identity. We recall that an element $r \in R$ is *integral over* I if there exists a monic polynomial $f(x) \in R[x]$, $f(x) = x^n + \sum_{i=1}^n a_i x^{n-i}$, where $a_i \in I^i$ for each $i, 1 \leq i \leq n$ and $f(r) = 0$. Thus $r \in R$ is integral over I if and only if $IJ^{n-1} = J^n$, where $J = (I, r)R$ and n is some positive integer. (Notice that $f(r) = 0$ implies $r^n = -\sum_{i=1}^n a_i r^{n-i} \in IJ^{n-1}$ and this implies $J^n \subseteq IJ^{n-1}$.) If $I \subseteq J$ are ideals and $IJ^{n-1} = J^n$, then I is said to be a *reduction* of J. The *integral closure* $\overline{I}$ of an ideal I is the set of elements of R integral over I. If $I = \overline{I}$, then I is said to be *integrally closed*. It is well known that $\overline{I}$ is an integrally closed ideal. An ideal is integrally closed if and only if it is not a reduction of a properly bigger ideal. A prime ideal is always integrally closed. An ideal is said to be *normal* if all the powers of the ideal are integrally closed.

We were motivated to construct the example given in this paper by a question asked by Craig Huneke as to whether there exists an analytically unramified Noetherian local ring $(A, \mathbf{n})$ having an integrally closed ideal I for which $I\widehat{A}$ is not integrally closed, where $\widehat{A}$ is the $\mathbf{n}$-adic completion of A. In Example 2.1, the ring A is a 3-dimensional regular local domain and $I = P = (f, g)A$ is a prime ideal of height two. Sam Huckaba asked if the ideal of our example is a normal ideal. The answer is "yes." Since f, g form a regular sequence and A is Cohen-Macaulay, the powers P^n of P have no embedded associated primes and therefore are P-primary [**8**, (16.F), p. 112], [**9**, Ex. 17.4, p. 139]. Since the powers of the maximal ideal of a regular local domain are integrally closed, the powers of P are integrally closed. Thus the Rees algebra $A[Pt] = A[ft, gt]$ is a normal domain while the Rees algebra $\widehat{A}[ft, gt]$ is not integrally closed.

A problem analogous to that considered here in the sense that it also deals with the behavior of ideals under extension to completion is addressed by Loepp and Rotthaus in [**7**]. They construct nonexcellent local Noetherian domains to demonstrate that tight closure need not commute with completion.

Remark 1.1 Without the assumption that A is analytically unramified, there exist examples even in dimension one where an integrally closed ideal of A fails to extend to an integrally closed ideal in $\widehat{A}$. If A is reduced but analytically ramified, then the zero ideal of A is integrally closed, but its extension to $\widehat{A}$ is not integrally closed. An example in characteristic zero of a one-dimensional Noetherian local domain that is analytically ramified is given by Akizuki in his 1935 paper [**1**]. An example in positive characteristic is given by F.K. Schmidt [**11**, pp. 445-447]. Another example due to Nagata is given in [**10**, Example 3, pp. 205-207]. (See also [**10**, (32.2), p. 114].)

Remark 1.2 Let R be a commutative ring and let R' be an R-algebra. We list cases where extensions to R' of integrally closed ideals of R are again integrally closed. The R-algebra R' is said to be *quasi-normal* if R' is flat over R and the following condition $(N_{R,R'})$ holds: If C is any R-algebra and D is a C-algebra in which C is integrally closed, then also $C \otimes_R R'$ is integrally closed in $D \otimes_R R'$.

1. By [**6**, Lemma 2.4], if R' is an R-algebra satisfying $(N_{R,R'})$ and I is an integrally closed ideal of R, then IR' is integrally closed in R'.

2. Let $(A, \mathbf{n})$ be a Noetherian local ring and let $\widehat{A}$ be the $\mathbf{n}$-adic completion of A. Since $A/\mathbf{q} \cong \widehat{A}/\mathbf{q}\widehat{A}$ for every $\mathbf{n}$-primary ideal $\mathbf{q}$ of A, the $\mathbf{n}$-primary ideals of A are in one-to-one inclusion preserving correspondence with the $\widehat{\mathbf{n}}$-primary ideals of $\widehat{A}$. It follows that an $\mathbf{n}$-primary ideal I of A is a reduction of a properly larger ideal of A if and only if $I\widehat{A}$ is a reduction of a properly larger ideal of $\widehat{A}$. Therefore an $\mathbf{n}$-primary ideal I of A is integrally closed if and only if $I\widehat{A}$ is integrally closed.

3. If A is excellent, then the map $A \to \widehat{A}$ is quasi-normal by [**2**, (7.4.6) and (6.14.5)], and in this case every integrally closed ideal of A extends to an integrally closed ideal of $\widehat{A}$.

4. If $(A, \mathbf{n})$ is a local domain and A^h is the Henselization of A, then every integrally closed ideal of A extends to an integrally closed ideal of A^h. This

follows because A^h is a filtered direct limit of étale A-algebras [**6**, (iii), (i), (vii) and (ix), pp. 800- 801].

5. In general, integral closedness of ideals is a local condition. Suppose R' is an R-algebra that is *locally normal* in the sense that for every prime ideal P' of R', the local ring $R'_{P'}$ is an integrally closed domain. Since principal ideals of an integrally closed domain are integrally closed, the extension to R' of every principal ideal of R is integrally closed. In particular, if $(A,\mathbf{n})$ is an analytically normal Noetherian local domain, then every principal ideal of A extends to an integrally closed ideal of $\widehat{A}$.

6. If R is an integrally closed domain, then for every ideal I and element x of R we have $\overline{xI} = x\overline{I}$. If $(A,\mathbf{n})$ is analytically normal and also a UFD, then every height-one prime ideal of A extends to an integrally closed ideal of $\widehat{A}$. In particular if A is a regular local domain, then $P\widehat{A}$ is integrally closed for every height-one prime P of A. If $(A,\mathbf{n})$ is a 2-dimensional regular local domain, then every nonprincipal integrally closed ideal of A has the form xI, where I is an $\mathbf{n}$-primary integrally closed ideal and $x \in A$. In view of item 2, every integrally closed ideal of A extends to an integrally closed ideal of $\widehat{A}$ in the case where A is a 2-dimensional regular local domain.

7. Suppose R and R' are Noetherian rings and assume that R' is a flat R-algebra. Let I be an integrally closed ideal of R. The flatness of R' over R implies that every $P' \in \text{Ass}\,(R'/IR')$ contracts in R to some $P \in \text{Ass}\,(R/I)$ [**9**, Theorem 23.2]. Since a regular map is quasi-normal, if the map $R \to R'_{P'}$ is regular for each $P' \in \text{Ass}\,(R'/IR')$, then IR' is integrally closed.

2 A non-integrally closed extension

In the construction of the following example we make use of results from [**3**]-[**5**].

Construction of Example 2.1 Let k be a field of characteristic zero and let x, y and z be indeterminates over k. Let $R := k[x,y,z]_{(x,y,z)}$ and let R^* be the (xR)-adic completion of R. Thus $R^* = k[y,z]_{(y,z)}[[x]]$, the formal power series ring in x over $k[y,z]_{(y,z)}$.

Let α and β be elements of $xk[[x]]$ which are algebraically independent over $k(x)$. Set

$$f = (y-\alpha)^2, \quad g = (z-\beta)^2, \quad \text{and } A = k(x,y,z,f,g) \cap R^*.$$

Then the (xA)-adic completion A^* of A is equal to R^* [**4**, Lemma 2.3.2, Prop. 2.4.4].

In order to better understand the structure of A, we recall some of the details of the construction of a nested union B of localized polynomial rings over k in 5 variables associated to A. (More details may be found in [**5**].)

Approximation Technique 2.2 With k,x,y,z,f,g,R and R^* as in (3.1), write

$$f = y^2 + \sum_{j=1}^{\infty} b_j x^j, \qquad g = z^2 + \sum_{j=1}^{\infty} c_j x^j,$$

for some $b_j, \in k[y]$ and $c_j \in k[z]$. There are natural sequences $\{f_r\}_{r=1}^{\infty}$, $\{g_r\}_{r=1}^{\infty}$ of elements in R^*, called the rth *endpieces* for f and g, respectively, which "approximate" f and g. These are defined for each $r \geq 1$ by:

$$f_r := \sum_{j=r}^{\infty} (b_j x^j)/x^r, \qquad g_r := \sum_{j=r}^{\infty} (c_j x^j)/x^r.$$

For each $r \geq 1$, define B_r to be $k[x,y,z,f_r,g_r]$ localized at the maximal ideal generated by $(x,y,z,f_r - b_r, g_r - c_r)$. Then define $B = \bigcup_{r=1}^{\infty} B_r$. The endpieces defined here are slightly different from the notation used in [**5**]. Also we are using here a localized polynomial ring for the base ring R. With minor adjustments, however, [**5**, Theorem 2.2] applies to our setup.

Theorem 2.3 *Let A be the ring constructed in* (3.1) *and let $P = (f,g)A$, where f and g are as in* (3.1) *and* (3.2). *Then*

1. *$A = B$ is a three-dimensional regular local domain that is a nested union of five-dimensional regular local domains.*

2. *P is a height-two prime ideal of A.*

3. *If A^* denotes the (xA)-adic completion of A, then $A^* = k[y,z]_{(y,z)}[[x]]$ and PA^* is not integrally closed.*

4. *If $\widehat{A}$ denotes the completion of A with respect to the powers of the maximal ideal of A, then $\widehat{A} = k[[x,y,z]]$ and $P\widehat{A}$ is not integrally closed.*

Proof: Notice that the polynomial ring $k[x,y,z,\alpha,\beta] = k[x,y,z,y-\alpha,z-\beta]$ is a free module of rank 4 over the polynomial subring $k[x,y,z,f,g]$ since $f = (y-\alpha)^2$ and $g = (z-\beta)^2$. Hence the extension

$$k[x,y,z,f,g] \to k[x,y,z,\alpha,\beta][1/x]$$

is flat. Thus item (1) follows from [**5**, Theorem 2.2].

For item (2), it suffices to observe that P has height two and that, for each positive integer r, $P_r := (f,g)B_r$ is a prime ideal of B_r. We have $f = xf_1 + y^2$ and $g = xg_1 + z^2$. It is clear that $(f,g)k[x,y,z,f,g]$ is a height-two prime ideal. Since B_1 is a localized polynomial ring over k in the variables x,y,z,f_1-b_1,g_1-c_1, we see that

$$P_1B_1[1/x] = (xf_1+y^2, xg_1+z^2)B_1[1/x]$$

is a height-two prime ideal of $B_1[1/x]$. Indeed, setting $f = g = 0$ is equivalent to setting $f_1 = -y^2/x$ and $g_1 = -z^2/x$. Therefore the residue class ring $(B_1/P_1)[1/x]$ is isomorphic to a localization of the integral domain $k[x,y,z][1/x]$. Since B_1 is Cohen-Macaulay and f,g form a regular sequence, and since $(x,f,g)B_1 = (x,y^2,z^2)B_1$ is an ideal of height three, we see that x is in no associated prime of $(f,g)B_1$ (see, for example [**9**, Theorem 17.6]). Therefore $P_1 = (f,g)B_1$ is a height-two prime ideal.

For $r > 1$, there exist elements $u_r \in k[x,y]$ and $v_r \in k[x,z]$ such that $f = x^r f_r + u_r x + y^2$ and $g = x^r g_r + v_r x + z^2$. An argument similar to that given above shows that $P_r = (f,g)B_r$ is a height-two prime of B_r. Therefore $(f,g)B$ is a height-two prime of $B = A$.

For items 3 and 4, $R^* = B^* = A^*$ by Construction 2.1 and it follows that $\widehat{A} = k[[x,y,z]]$. To see that $PA^* = (f,g)A^*$ and $P\widehat{A} = (f,g)\widehat{A}$ are not integrally closed, observe that $\xi := (y-\alpha)(z-\beta)$ is integral over PA^* and $P\widehat{A}$ since $\xi^2 = fg \in P^2$. On the other hand, $y-\alpha$ and $z-\beta$ are nonassociate prime elements in the local unique factorization domains A^* and $\widehat{A}$. An easy computation shows that $\xi \notin P\widehat{A}$. Since $PA^* \subseteq P\widehat{A}$, this completes the proof. □

Remark 2.4 In a similar manner it is possible to construct for each integer $d \geq 3$ an example of a d-dimensional regular local domain $(A, \mathbf{n})$ having a prime ideal P of height $h := d - 1$ such that $P\widehat{A}$ is not integrally closed. Indeed, let k be a field of characteristic zero and let $x, y_1, \ldots, y_h$ be indeterminates over k. Let $\alpha_1, \ldots, \alpha_h \in xk[\![x]\!]$ be algebraically independent over $k(x)$. For each i with $1 \leq i \leq h$, define $f_i = (y_i - \alpha_i)^h$. Proceeding in a manner similar to what is done in (3.1) we obtain a d-dimensional regular local domain A and a prime ideal $P = (f_1, \ldots, f_h)A$ of height h such that $y_i - \alpha_i \in \widehat{A}$. Let $\xi = \prod_{i=1}^{h}(y_i - \alpha_i)$. Then $\xi^h = f_1 \cdots f_h \in P^h$ implies ξ is integral over $P\widehat{A}$, but using that $y_1 - \alpha_1, \ldots, y_h - \alpha_h$ is a regular sequence in $\widehat{A}$, we see that $\xi \notin P\widehat{A}$.

3 Comments and questions

In connection with Theorem 2.3 it is natural to ask the following question.

Question 3.1 For P and A as in Theorem 2.3, is P the only prime of A that does not extend to an integrally closed ideal of $\widehat{A}$?

Comments 3.2 In relation to the example given in Theorem 2.3 and to Question 3.1, we have the following commutative diagram, where all the maps shown are the natural inclusions:

$$\begin{array}{ccccc} B = A & \xrightarrow{\gamma_1} & A' := k(x,y,z,\alpha,\beta) \cap R^* & \xrightarrow{\gamma_2} & R^* = A^* \\ \delta_1 \uparrow & & \delta_2 \uparrow & & \psi \uparrow \\ S := k[x,y,z,f,g] & \xrightarrow{\varphi} & T := k[x,y,z,\alpha,\beta] & = & T \end{array} \tag{1}$$

Let $\gamma = \gamma_2 \cdot \gamma_1$. Referring to the diagram above, we observe the following:

1. The discussion in [**4**, bottom p. 668 to top p. 669] implies that [**4**, Thm. 3.2] applies to the setting of Theorem 3.3. By [**4**, Prop. 4.1 and Thm. 3.2], $A'[1/x]$ is a localization of T. By Theorem 3.3 and [**4**, Thm 3.2], $A[1/x]$ is a localization of S. Furthermore, by [**4**, Prop. 4.1] A' is excellent. (Notice, however, that A is not excellent since there exists a prime ideal P of A

such that $P\widehat{A}$ is not integrally closed.) The excellence of A' implies that if $Q^* \in \text{Spec } A^*$ and $x \notin Q^*$, then the map $\psi_{Q^*} : T \to A^*_{Q^*}$ is regular [**2**, (7.8.3 v)].

2. Let $Q^* \in \text{Spec } A^*$ be such that $x \notin Q^*$ and let $\mathbf{q}' = Q^* \cap T$. By [**9**, Theorem 32.1] and Item 1 above, if $\varphi_{\mathbf{q}'} : S \to T_{\mathbf{q}'}$ is regular, then $\gamma_{Q^*} : A \to A^*_{Q^*}$ is regular.

3. Let I be an ideal of A. Since A' and A^* are excellent and both have completion $\widehat{A}$, Remark 1.2.3 shows that the ideals IA', IA^* and $I\widehat{A}$ are either all integrally closed or all fail to be integrally closed.

4. The Jacobian ideal of the extension $\varphi : S = k[x,y,z,f,g] \to T = k[x,y,z,\alpha,\beta]$ is the ideal of T generated by the determinant of the matrix

$$\mathcal{J} := \begin{pmatrix} \frac{\partial f}{\partial \alpha} & \frac{\partial g}{\partial \alpha} \\ \frac{\partial f}{\partial \beta} & \frac{\partial g}{\partial \beta} \end{pmatrix}.$$

Since the characteristic of the field k is zero, this ideal is $(y-\alpha)(z-\beta)T$.

In Proposition 3.3, we relate the behavior of integrally closed ideals in the extension $\varphi : S \to T$ to the behavior of integrally closed ideals in the extension $\gamma : A \to A^*$.

Proposition 3.3 *With the setting of* Theorem 2.3 *and* Comment 3.2.2, *let I be an integrally closed ideal of A such that $x \notin Q$ for each $Q \in Ass\ (A/I)$. Let $J = I \cap S$. If JT is integrally closed (resp. a radical ideal) then IA^* is integrally closed (resp. a radical ideal).*

Proof: Since the map $A \to A^*$ is flat, x is not in any associated prime of IA^*. Therefore IA^* is contracted from $A^*[1/x]$ and it suffices to show $IA^*[1/x]$ is integrally closed (resp. a radical ideal). Our hypothesis implies $I = IA[1/x] \cap A$. By Comment 3.2.1, $A[1/x]$ is a localization of S. Thus every ideal of $A[1/x]$ is the extension of its contraction to S. It follows that $IA[1/x] = JA[1/x]$. Thus $IA^*[1/x] = JA^*[1/x]$.

Also by Comment 3.2.1, the map $T \to A^*[1/x]$ is regular. If JT is integrally closed, then Remark 1.2.7 implies that $JA^*[1/x]$ is integrally closed. If JT is a radical ideal, then the regularity of the map $T \to A^*[1/x]$ implies that $JA^*[1/x]$ is a radical ideal. □

Proposition 3.4 *With the setting of* Theorem 2.3 *and* Comment 3.2, *let* $Q \in$ *Spec* A *be such that* $Q\widehat{A}$ *(or equivalently* QA^**) is not integrally closed. Then*

1. Q *has height two and* $x \notin Q$.
2. *There exists a minimal prime* Q^* *of* QA^* *such that with* $\mathbf{q}' = Q^* \cap T$, *the map* $\varphi_{\mathbf{q}'} : S \to T_{\mathbf{q}'}$ *is not regular.*
3. Q *contains* $f = (y-\alpha)^2$ *or* $g = (z-\beta)^2$.
4. Q *contains no element that is a regular parameter of* A.

Proof: By Remark 1.2.6, the height of Q is two. Since $A^*/xA^* = A/xA = R/xR$, we see that $x \notin Q$. This proves item 1.

By Remark 1.2.7, there exists a minimal prime Q^* of QA^* such that $\gamma_{Q^*} : A \to A^*_{Q^*}$ is not regular. Thus item 2 follows from Comment 3.2.2.

For item 3, let Q^* and $\mathbf{q}'$ be as in item 2. Since γ_{Q^*} is not regular it is not essentially smooth [**2**, 6.8.1]. By [**5**, (2.7)], $(y-\alpha)(z-\beta) \in \mathbf{q}'$. Hence $f = (y-\alpha)^2$ or $g = (z-\beta)^2$ is in $\mathbf{q}'$ and thus in Q. This proves item 3.

Suppose $w \in Q$ is a regular parameter for A. Then A/wA and A^*/wA^* are two-dimensional regular local domains. By Remark 1.2.6, QA^*/wA^* is integrally closed, but this implies that QA^* is integrally closed, which contradicts our hypothesis that QA^* is not integrally closed. This proves item 4. □

Question 3.5 In the setting of Theorem 2.3 and Comment 3.2, let $Q \in$ Spec A with $x \notin Q$ and let $\mathbf{q} = Q \cap S$. If QA^* is integrally closed, does it follow that $\mathbf{q}T$ is integrally closed?

Question 3.6 In the setting of Theorem 2.3 and Comment 3.2, if a prime ideal Q of A contains f or g, but not both, and does not contain a regular parameter of A, does it follow that QA^* is integrally closed ?

In Example 2.1, the three-dimensional regular local domain A contains height-one prime ideals P such that $\widehat{A}/P\widehat{A}$ is not reduced. This motivates us to ask:

Question 3.7 Let $(A, \mathbf{n})$ be a three-dimensional regular local domain and let $\widehat{A}$ denote the $\mathbf{n}$-adic completion of A. If for each height-one prime P of A, the extension $P\widehat{A}$ is a radical ideal, i.e., the ring $\widehat{A}/P\widehat{A}$ is reduced, does it follow that $P\widehat{A}$ is integrally closed for each $P \in \operatorname{Spec} A$?

Acknowledgments

The authors are grateful for the hospitality and cooperation of Michigan State, Nebraska and Purdue, where several work sessions on this research were conducted. The authors also thank the referee for helpful suggestions and comments about the paper.

References

[**1**] Y. Akizuki, Einige Bemerkungen über primäre Integritätsbereiche mit Teilerkettensatz, *Proc. Phys.-Math. Soc. Japan* **17** (1935), 327–336.

[**2**] A. Grothendieck, *Éléments de Géométrie Algébrique IV, Publ. Math. Inst. Hautes Étud. Sci.* **24**, 1965.

[**3**] W. Heinzer, C. Rotthaus and S. Wiegand, Noetherian rings between a semilocal domain and its completion, *J. Algebra* **198** (1997), 627–655.

[**4**] W. Heinzer, C. Rotthaus and S. Wiegand, Noetherian domains inside a homomorphic image of a completion, *J. Algebra* **215** (1999), 666–681.

[**5**] W. Heinzer, C. Rotthaus and S. Wiegand, Examples of integral domains inside power series rings, in *Commutative Ring Theory and Applications*, Fontana, Kabbaj and Wiegand, Eds., 233–254, Marcel Dekker, Inc., 2002.

[**6**] J. Lipman, Relative Lipshitz saturation, *Amer. J. Math.* **97** (1975), 791–813.

[**7**] S. Loepp and C. Rotthaus, Some results on tight closure and completion, *J. Algebra* **246** (2001), 859–880.

[**8**] H. Matsumura, *Commutative Algebra, Second Edition*, Benjamin, 1980.

[**9**] H. Matsumura, *Commutative Ring Theory*, Cambridge University Press, 1989.

[**10**] M. Nagata, *Local Rings*, John Wiley, 1962.

[**11**] F.K. Schmidt, Über die Erhaltung der Kettensätze der Idealtheorie bei beliebigen endlichen Körpererweiterungen, *Math. Z.* **41** (1936), 443–450.

The components of a variety of matrices with square zero and submaximal rank

DIKRAN KARAGUEUZIAN
Mathematics Department, SUNY Binghamton, Binghamton, NY 13902, USA
E-mail: dikran@math.binghamton.edu

BOB OLIVER[1]
LAGA, Université Paris 13
99, Avenue Jean-Baptiste Clément, 93430 Villetaneuse, France
E-mail: bob@math.univ-paris13.fr

JOANA VENTURA[2]
Departamento de Matemática, Instituto Superior Técnico
Av. Rovisco Pais, 1049-001 Lisboa, Portugal
E-mail: jventura@math.ist.utl.pt

Abstract: The structure of the variety of upper-triangular square-zero matrices was investigated by Rothbach, who introduced techniques enabling him to determine its irreducible components. In this paper, we fix a particular irreducible component of this variety and study the structure of the subvariety of matrices of submaximal rank in this component. We use Rothbach's techniques to determine the components of this variety. We also show that this subvariety contains the support variety for a certain universal homology module.

[1]Bob Oliver was partly supported by UMR 7539 of the CNRS.
[2]Joana Ventura was fully supported by FCT grant SFRH/BPD/8034/2002.

1 Introduction

A longstanding conjecture in algebraic topology describes the free rank of symmetry of a product of spheres. This conjecture states that if the elementary abelian group $(\mathbb{Z}/p)^r$ acts freely on $S^{n_1} \times \cdots \times S^{n_s}$, then $r \leq s$. A more ambitious generalization of this conjecture states that if $(\mathbb{Z}/p)^r$ acts freely on a manifold M, then $\sum_i \dim_{\mathbb{F}_p} H_i(M, \mathbb{F}_p) \geq 2^r$. For a survey of conjectures of this type, and partial results, the reader might consult Section 2 of [**1**].

In [**2**] G. Carlsson produced a functorial translation of the second conjecture (for $p = 2$) into the language of commutative algebra. Carlsson was able to obtain partial results [**2**, **3**] on his version of the conjecture using techniques of commutative algebra; this yielded new results on the topological side as well.

The key point in the proofs in [**3**] is to show that an upper-triangular square-zero matrix over a polynomial ring can, through some specialization of variables, be forced to have submaximal rank. The matrix arises as the differential of a free differential graded module over a polynomial ring, and in the case of interest the module has even rank. The generic $2n \times 2n$ square-zero matrix has rank n; by a "matrix of submaximal rank" we mean a matrix of rank less than n. The structure of the variety of upper-triangular square-zero matrices was later investigated by Rothbach [**5**], who introduced techniques enabling him to determine its irreducible components.

Throughout this paper, we fix one particular irreducible component Z of this variety, and study the structure of the subvariety of matrices of submaximal rank in Z. We use Rothbach's techniques to determine the components of this variety. Also, following a suggestion of Carlsson, we show that this subvariety contains the support variety for a certain universal homology module. The hope is that this universal homology module and the component result will be useful for the commutative algebra version of the conjecture, but we have not yet made progress in that direction.

The structure of the rest of the paper is as follows: In Section 2 we introduce notation and terminology for the objects of study, restricting attention to an irreducible component Z of the variety of upper-triangular $2n \times 2n$ square-zero matrices. In Section 3 we relate our submaximal rank subvariety $Y \subset Z$ to the support variety for the universal homology module. In Section 4 we review Rothbach's techniques, and in Section 5 we determine the irreducible components of Y.

The work for this paper was done while the first and third authors were visiting the University of Paris 13; it is a pleasure to thank the faculty and staff there for their hospitality.

2 Definitions and notation

In this paper, we work over an algebraically closed field k, which will be the ground field for all polynomial rings. We also regard all varieties as being defined over k. In view of the motivation mentioned in the introduction, the reader may wish to take $k = \overline{\mathbb{F}}_2$, but this restriction is not necessary for the results.

The following notation will be used throughout the rest of the paper.

- U_{2n} is the variety of strictly upper-triangular $2n \times 2n$ matrices over k.
- V_{2n} is the variety of square-zero matrices in U_{2n}.
- Z denotes a particular irreducible component of V_{2n}.
- R is the coordinate ring of Z.
- Y is the subvariety of matrices of rank less than n in Z.
- I denotes the ideal of R corresponding to Y.

The coordinate ring of U_{2n} is $R(U_{2n}) = k[x_{ij} \mid i < j]$. There are surjections of coordinate rings $R(U_{2n}) \to R(V_{2n}) \to R$ corresponding to the inclusions $Z \hookrightarrow V_{2n} \hookrightarrow U_{2n}$. Using these surjections, we can regard the images of the x_{ij} as elements of R. Let $M \in M_{2n}(R)$ be the $2n \times 2n$ upper triangular matrix whose (i, j)-entry is the image in R of x_{ij}. In particular, $M^2 = 0$. We regard M (the *universal matrix*) as a differential on the R-module R^{2n}.

Note that I is the radical of the ideal generated by all $n \times n$ minors of the universal matrix M.

Definition 2.1 The *universal homology* of Z, written $H(M)$, is defined to be the R-module $\mathrm{Ker}(M)/\mathrm{Im}(M)$.

3 The support variety for universal homology

In this section, we show that Y contains the support variety for the universal homology module $H(M)$ of Z. We recall that by definition, the support variety of a module N is the variety corresponding to the annihiliator ideal of N. The two statements in the following proposition are thus equivalent.

Proposition 3.1 *$Y \supseteq \mathrm{supp}\,H(M)$ and $I \subseteq \sqrt{\mathrm{Ann}\,H(M)}$.*

Proof: Let J be the ideal generated by the $n \times n$ minors of M. We show that $J \subseteq \mathrm{Ann}\,H(M)$. Since $I = \sqrt{J}$, this then implies that $I \subseteq \sqrt{\mathrm{Ann}\,H(M)}$, and that $Y \supseteq \mathrm{supp}\,H(M)$.

We must show, for each $x \in J$, that $x \cdot H(M) = 0$. We can restate this last condition as "for each $x \in J$ and for each $v \in \mathrm{Ker}M$, there is a $u \in R^{2n}$ such that $Mu = xv$." Of course, it is enough to show this for the $n \times n$ minors which generate J. We will show this one minor at a time, by constructing an explicit linear map $N\colon R^{2n} \to R^{2n}$, so that $M(Nv) = xv$ for all $v \in \mathrm{Ker}(M)$.

Let X be an $(n \times n)$ submatrix of M, and set $x = \det X$. We can assume $x \neq 0$; otherwise there is nothing to prove. Set $U = R^{2n}$, and write $U = V \oplus V' = W \oplus W'$, where the decompositions correspond to the choices of rows and columns used to define the submatrix X. Thus $M : U \longrightarrow U$ has components

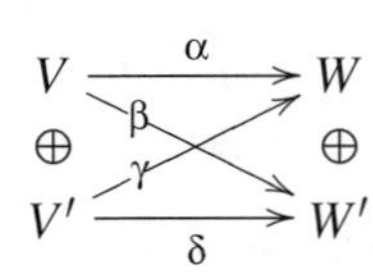

where α has matrix X and determinant x. These two decompositions of U thus correspond to "pushing X" to the upper left corner of the matrix M. Define $N : U \longrightarrow U$ to be the map

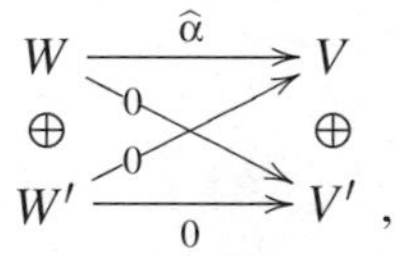

where $\widehat{\alpha}$ is the linear map whose matrix is the cofactor matrix of X (thus $\alpha\widehat{\alpha} = x \cdot \mathrm{Id}_W$).

Set $\bar{R} = R[1/x]$. Since R is an integral domain, and $x \neq 0$ by assumption, R is a subring of the localized ring $\bar{R}$. Since x is invertible in $\bar{R}$, the submatrix X of M chosen above (and corresponding to the minor x) is invertible over $\bar{R}$.

Set $\bar{U} = \bar{R} \otimes_R U$, $\bar{V} = \bar{R} \otimes_R V$, and similarly for $\bar{V}'$, $\bar{W}$, and $\bar{W}'$. Since $V \xrightarrow{\alpha} W$ is the map with matrix X, which is invertible over $\bar{R}$, α becomes invertible as a map from $\bar{V}$ to $\bar{W}$. Set $\mathcal{K} = \mathrm{Ker}(M) \subseteq \bar{U}$, $I = \mathrm{Im}(M) \subseteq \bar{U}$ and

$$\mathcal{K}' = \{(\alpha^{-1}\gamma(v'), -v') \mid v' \in \bar{V}'\},$$
$$I' = \{(w, \beta\alpha^{-1}(w)) \mid w \in \bar{W}\}.$$

Thus $\mathcal{K}'$ is the space of vectors in $\bar{V} \oplus \bar{V}'$ whose image under M lies in $\bar{W}'$ (has zero $\bar{W}$-component), while I' is the image of $\bar{V}$. Obviously $I' \subseteq I \subseteq \mathcal{K} \subseteq \mathcal{K}'$.

By definition, $(\bar{W}', I')$ and $(\bar{V}, \mathcal{K}')$ are both pairs of complementary subspaces of $\bar{U}$, where $I' \subseteq \mathcal{K}'$. Hence

$$\bar{R}^n \cong \bar{W}' \cong \bar{U}/I' \cong (\bar{V} \oplus \mathcal{K}')/I' \cong \bar{V} \oplus (\mathcal{K}'/I') \cong \bar{R}^n \oplus (\mathcal{K}'/I').$$

Since $\bar{R}$ is noetherian, this implies that $\mathcal{K}'/I' = 0$, and hence that $\mathcal{K}' = I'$. Thus all four of the submodules $I, \mathcal{K}, I', \mathcal{K}'$ are equal.

From the definitions, it follows easily that $MN(u) = x \cdot u$ for all $u \in I'$. Since $I' = \mathcal{K} = \mathrm{Ker}(M)$ and R is a subring of $\bar{R}$, this completes the proof of Proposition 3.1. □

We conclude this section with the natural

Conjecture 3.2 $I = \sqrt{\mathrm{Ann}\, H(M)}$, *or equivalently,* $Y = \mathrm{supp}\, H(M)$.

4 The structure of V_{2n}

In this section, we review Rothbach's work on the structure of V_{2n} and its irreducible components; our decomposition of Y into irreducible components is obtained by similar methods. The reader familiar with **[5]** can safely skip this section. It should be noted that because of our motivation, and to minimize technical difficulties, we have opted only to consider components of V_{2n}. However, the work in **[5]** applies to $(n \times n)$-matrices for odd n as well.

Rothbach's work is based on the decomposition of V_{2n} into Borel orbits.

Definition 4.1 The *Borel orbits* in V_{2n} are the orbits of the conjugation action of the Borel group of all invertible upper-triangular matrices on V_{2n}.

Each Borel orbit contains a unique matrix of the type described in the following definition.

Definition 4.2 A *partial permutation matrix* X is a matrix of 0's and 1's in which each row and each column contains at most one 1.

To an upper-triangular partial permutation matrix we can associate a sequence of non-negative integers $(a_1, a_2, \ldots, a_{2n})$ by setting

$$a_i = \begin{cases} j & \text{if } Xe_i = e_j, \\ 0 & \text{if } Xe_i = 0. \end{cases}$$

Definition 4.3 A *valid X^2 word* is a sequence of non-negative integers $(a_1, \ldots, a_{2n})$ associated to a partial permutation matrix X with $X^2 = 0$. The integers a_i are the *letters* of the word. If v is a valid X^2 word, we write rank(v) for the number of nonzero integers a_i in v, i.e., the rank of the partial permutation matrix associated to v.

There is a one-to-one correspondence between the Borel orbits and valid X^2 words. Rothbach describes the ordering induced on valid X^2 words via certain *moves*, where $w < w'$ if and only if there is a sequence of moves which transforms w' into w.

Remark 4.4 It follows from the definition of partial permutation matrix that the nonzero letters in a valid X^2 word are distinct.

We can now describe the correspondence between Borel orbits and valid X^2 words. We will show that each Borel orbit contains a unique partial permutation matrix, and thus to each Borel orbit is associated a unique valid X^2 word. The closure of a Borel orbit is the closure of an image of the Borel group, which is an irreducible variety, so these closures are themselves irreducible varieties (cf. [**4**, Proposition I.8.1]). Clearly, the closure of a Borel orbit is itself a union of Borel orbits. Thus, V_{2n} is a finite union of irreducible varieties (closures of all Borel orbits), which are partially ordered by inclusion, and the components of V_{2n} are therefore the maximal elements of this poset. In this way, the problem of determining the components of V_{2n} is reduced to the combinatorics of the poset of valid X^2 words.

In order to determine which Borel orbits are contained in the closure of a given Borel orbit, in terms of the corresponding valid X^2 words, Rothbach defined certain "moves" which give an order relation on the valid X^2 words. To explain this, we introduce the following terminology. Let $(a_1,\ldots,a_{2n})$ be a valid X^2 word.

Definition 4.5 We say that the i-th letter a_i is *bound* if $a_i = 0$ and there exists a j such that $a_j = i$. If the letter a_i is not bound, then it is *free*.

It is helpful to regard valid X^2 words as "partial permutations" of the set $\{1,\ldots, 2n\}$. The word $(a_1,\ldots,a_{2n})$ is thought of as the partial permutation with domain $\{i\,|\,a_i \neq 0\}$, which sends i to a_i. The $X^2 = 0$ condition translates to saying that the domain and range of the permutation are disjoint. These can be illustrated by diagrams with arrows. For example, the words 002041 and 010003 correspond to the diagrams

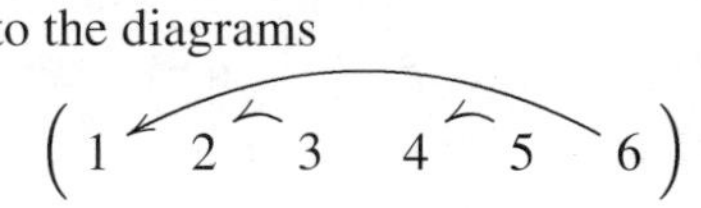

and

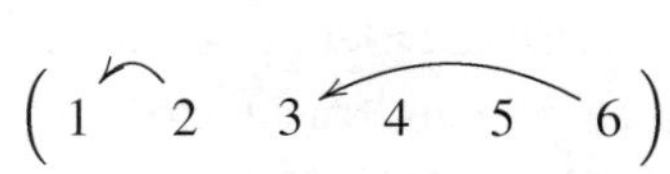

In the following descriptions, whenever we show a "subdiagram" of a partial permutation by restricting to some subset of indices $I \subseteq \{1,\ldots,2n\}$, it is understood that no index $i \notin I$ is sent to any index $j \in I$, and no index $i \in I$ is sent to any nonzero index $j \notin I$.

The three moves are the following:

- A *move of type 1* takes a nonzero letter a_k and replaces it with a_k^*, the largest integer less than a_k such that the replacement yields a new valid X^2 word. (Note that a_k^* always exists since replacement with 0 always yields a valid X^2 word.) In other words, if we set $j = a_k$ and $i = a_k^*$ (so $i < j < k$), then this move sends

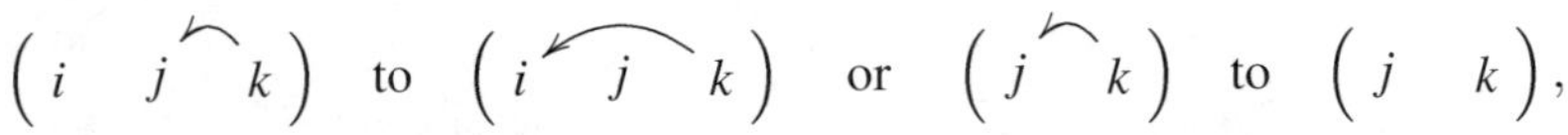

  when $i \neq 0$ or $i = 0$, respectively.

- A *move of type 2* takes two free letters a_k, a_l such that $k < l$ and $a_k > a_l$, and swaps their locations. In other words, it either sends

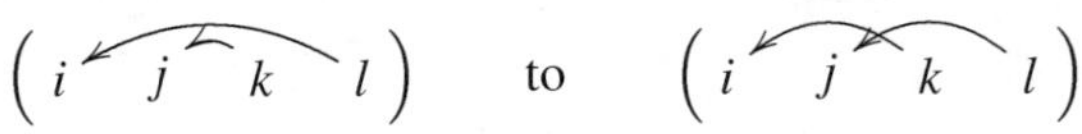

  if $a_k = j$ and $a_l = i \neq 0$ (and thus $i < j < k < l$), or else it sends

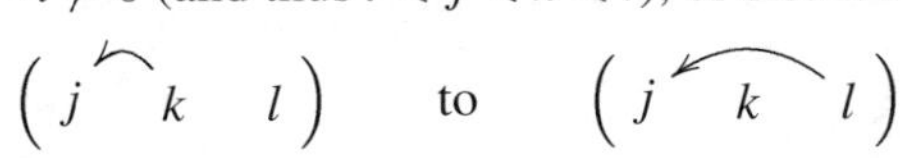

  if $a_k = j$ and $a_l = 0$ (thus $j < k < l$).

- A *move of type 3* is defined whenever there are indices $i < j < k < l$ such that $i = a_j$ and $k = a_l$ (hence $a_i = a_k = 0$), and replaces a_l by j, a_k by i, and a_j by 0. Pictorially, it sends

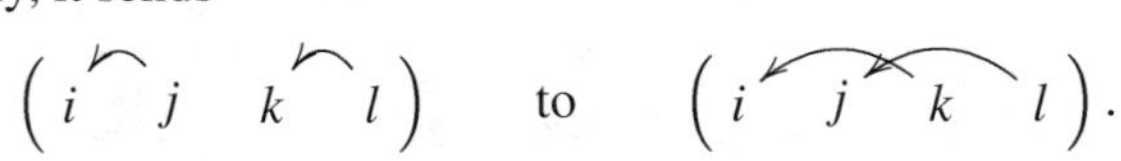

Observe that a move of type 2 or 3 preserves the rank of words. In fact, the only way of getting a word of smaller rank is to replace a letter by zero. This corresponds to applying move 1 one or more times. A sequence of moves of type 1 which results in a letter being replaced by zero will be called a move of type $1'$.

The partial ordering on valid X^2 words is defined by letting $w \geq w'$ if and only if w can be transformed into w' by a (possibly empty) finite sequence of moves. The maximal valid X^2 words are thus those which are not the result of any of the three types of moves.

Example 4.6 The word $(0,1,0,3)$ is transformed to $(0,0,1,2)$ by a move of type 3, so in the ordering defined above, $(0,0,1,2) < (0,1,0,3)$.

Finally, the maximal valid X^2 words are also called *bracket words* because there is a one-to-one correspondence between maximal valid X^2 words and sequences of left and right parentheses of length $2n$ which form valid LISP expressions. A bracket word corresponds to the valid X^2 word $(a_1, \ldots, a_{2n})$ where $a_i = 0$ if the i-th parenthesis in the bracket word is a left parenthesis, and $a_i = j$ if the i-th parenthesis is a right parenthesis which closes the j-th parenthesis.

Remark 4.7 For a bracket word w of length $2n$, we have $\text{rank}(w) = n$.

The key theorem of Rothbach's paper is:

Theorem 4.8 (Rothbach) *For any pair of valid X^2 words v, w, the Borel orbit O_v associated to v is contained in the closure of the Borel orbit O_w associated to w if and only if $v \leq w$. The irreducible components of V_{2n} are thus the closures of the Borel orbits associated to the maximal valid X^2 words; and the irreducible component of V_{2n} associated to a maximal valid X^2 word w is the union of the Borel orbits associated to the valid X^2 words which are less than or equal to w.*

Since Rothbach's paper is not generally available, we give a very brief sketch here of his techniques. For each $i \leq 2n$, let $k^i \subseteq k^{2n}$ be the subspace of elements $(x_1, \ldots, x_i, 0, \ldots, 0)$ for $x_1, \ldots, x_i \in k$. These are the subspaces of k^{2n} which are invariant under the action of all elements in the Borel group. For any $X \in V_{2n}$ and any $0 \leq j < i$, define $r(i, j, X) = \dim_k(X(k^i) + k^j)$. One easily sees that $r(i, j, X) = r(i, j, Y)$ if X and Y are in the same Borel orbit. For any valid X^2 word v, associated to a partial permutation matrix X, set $v_{ij} = r(i, j, X)$. Rothbach then shows:

- Two matrices $X, Y \in V_{2n}$ are in the same Borel orbit if and only if $r(i, j, X) = r(i, j, Y)$ for all i, j. The Borel orbit associated to v is therefore the set $\{X \in V_{2n} \,|\, r(i, j, X) = v_{ij} \ \forall i, j\}$.
- For any two valid X^2 words v, w, $v \leq w$ (as defined above via moves) if and only if $v_{ij} \leq w_{ij}$ for all i, j.
- If v is obtained from w by a move of one of the above types, then the Borel orbit O_v is in the closure of the Borel orbit O_w.

For any given valid X^2 word w, the union of the Borel orbits associated to words $v \leq w$ is just the set

$$\{X \in V_{2n} \,|\, r(i,j,X) \leq w_{ij} \ \forall i,j\}.$$

This is an algebraic set (hence closed), since it is defined by requiring determinants of certain submatrices to vanish. So together with the three points above, this proves that it is the closure of the Borel orbit associated to w.

Rothbach's theorem says that the irreducible components of V_{2n} are determined by the poset of all valid X^2 words. In the next section, we will study the subposet of words associated to orbits contained in Y, and thus determine the irreducible components of Y.

5 The irreducible components of Y

In this section, we will identify the irreducible components of Y. More specifically, if Z is an irreducible component of V_{2n} corresponding to a valid X^2 word w, we will describe the components of the subvariety $Y \subseteq Z$ in terms of the structure of w.

Definition 5.1 We say that a bracket word is *irreducible* if it cannot be expressed as the concatenation of bracket words of smaller length.

Example 5.2 The bracket word $(()())$ is irreducible; the bracket word $()(())$ is expressible as a concatenation of the irreducible bracket words $()$ and $(())$.

Let w be a bracket word of length $2n$. Observe that w can be expressed as a concatenation of irreducible bracket words $w = w_1 \cdots w_m$. (If w is irreducible then $m = 1$.) We shall use w, w_i to denote not only the irreducible bracket words in this factorization, but also the corresponding valid X^2 words. For a bracket word w factored in this way we make the following:

Definition 5.3 For each $i \in \{1, \ldots, m\}$, let $w^{(i)}$ be the valid X^2 word obtained from w by replacing the last letter of w_i by a zero.

Notice that the words $w^{(i)}$ defined in 5.3 are all obtained from w by a move of type $1'$, and so all have rank $n-1$. Notice also that, in general, there are other words obtained by a move $1'$. For example, if $w = 002041$ then 002001 is one such word.

Now we can describe the components of Y:

Theorem 5.4 *The irreducible components of Y are the closures of the Borel orbits corresponding to the words $w^{(i)}$. Alternatively, the irreducible component of Y corresponding to $w^{(i)}$ is the union of the Borel orbits corresponding to the valid X^2 words which are less than or equal to $w^{(i)}$.*

Example 5.5 Let $n = 3$ and let Z be the component corresponding to the bracket word $()(())$. Then the corresponding valid X^2 word is $w = 010043$. In this case, we express w as the concatenation of $()$ and $(())$. Thus, in the notation of the paragraph after Example 5.2, $m = 2$, $w_1 = ()$, and $w_2 = (())$. Writing this decomposition in terms of valid X^2 words, we have $w_1 = 01$, $w_2 = 0043$ and so $w^{(1)} = 000043$ and $w^{(2)} = 010040$. Thus the subvariety $Y \subseteq Z$ has two components.

In this case, we can describe these varieties in a simple way in terms of matrices. The component Z consists of all 6×6 matrices of the form

$$\begin{bmatrix} 0 & a & b & c & d & e \\ 0 & 0 & 0 & 0 & f & g \\ 0 & 0 & 0 & 0 & h & i \\ 0 & 0 & 0 & 0 & j & k \\ 0 & 0 & 0 & 0 & 0 & 0 \\ 0 & 0 & 0 & 0 & 0 & 0 \end{bmatrix} \quad \text{where} \quad \begin{bmatrix} a & b & c \end{bmatrix} \cdot \begin{bmatrix} f & g \\ h & i \\ j & k \end{bmatrix} = \begin{bmatrix} 0 & 0 \end{bmatrix}.$$

The subvariety Y has two components: one where

$$\operatorname{rank} \begin{bmatrix} f & g \\ h & i \\ j & k \end{bmatrix} \leq 1, \quad \text{and another where} \quad \begin{bmatrix} a & b & c \end{bmatrix} = \begin{bmatrix} 0 & 0 & 0 \end{bmatrix}.$$

Unfortunately it is not always possible to describe the components in this fashion.

Proof: [Proof of Theorem 5.4] The closure of each Borel orbit in Y is the closure of a continuous image of the Borel group of upper triangular matrices, and hence is irreducible (cf. [**4**, Proposition I.8.1]). Clearly, the closure of any Borel orbit is a union of Borel orbits, and hence the components of Y are just the maximal closures of Borel orbits. So by Rothbach's theorem (Theorem 4.8), the components of Y are the closures of the orbits associated to valid X^2 words which are maximal among those in Y.

It thus suffices to prove that if w is a maximal valid X^2 word, then any v with $\mathrm{rank}(v) < n$ and $v \leq w$ satisfies $v \leq w^{(i)}$ for some i. For any such v, there is a sequence of moves that we can apply which transforms w into v, and one of these moves must be of type $1'$ (since that is the only type of move which decreases the rank). We must show two things: that we can always make a move of type $1'$ first, and that the words $w^{(i)}$ are maximal among those obtained from w by a move of type $1'$. The first statement is proved in Lemma 5.6, and the second in Lemma 5.7. □

Lemma 5.6 *Let w be a bracket word of length $2n$ and let v be a valid X^2 word such that $\mathrm{rank}(v) < n$ and $v \leq w$. Then there is a valid X^2 word u obtained from w by a move of type $1'$ such that $v \leq u \leq w$.*

Proof: By induction on the number of moves applied on w to get v, it is enough to show that a move of any type followed by a move of type $1'$ is the same as a move of type $1'$ followed by some other move. Note, however, that the letters involved in each of the moves $1'$ may not be the same and that the type of the other move may change as we "commute" it past the move of type $1'$. We prove this in cases, according to the type of move which is being composed with the move of type $1'$. In what follows we will assume that the move of type $1'$ is applied to one of the letters involved in the other move; if this is not the case the two moves commute and the conclusion of the lemma follows immediately.

Case I: move 1 followed by move $1'$. A move of type 1 followed by a move of type $1'$ applied to the same index is equal to the move of type $1'$ applied to that index. This is illustrated by the following diagram:

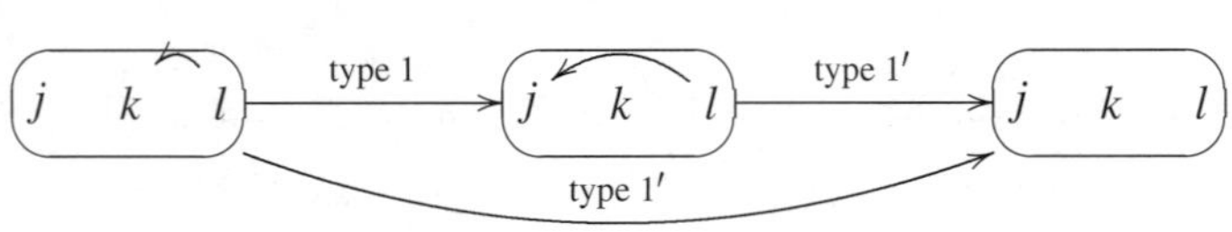

Case II: move 2 followed by move $1'$. Suppose we have a valid X^2 word $a_1 \cdots a_{2n}$, and indices $i < j < k < l$ such that $a_l = i$ and $a_k = j$ (and thus $a_i = a_j = 0$). So we can apply move 2 to the pair a_k, a_l and then apply move $1'$ to either of these letters. The following two commutative squares of diagrams of moves show that the composite of these two moves is always a type $1'$ move followed by a move of type 2 or 1.

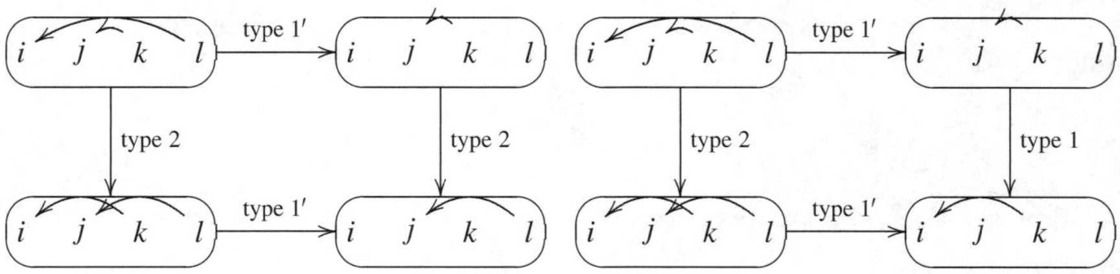

It remains to consider the possibility of a type 2 move which switches two letters of which one is zero. The composite of such a move followed by a type $1'$ move is itself a type $1'$ move, as illustrated by the following diagram.

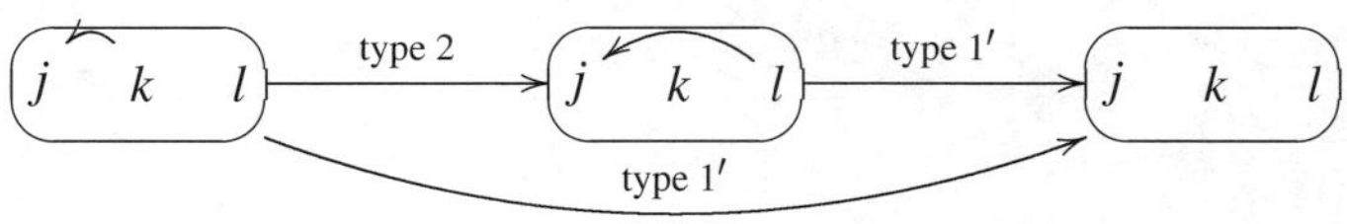

Case III: move 3 followed by move $1'$. Suppose we are given a valid X^2 word $a_1 \cdots a_{2n}$ to which we can apply a move of type 3. This means that there are indices $i < j < k < l$ such that $a_j = i$ and $a_l = k$ (and hence $a_k = a_i = 0$). After applying move 3 to this word, we can then apply move $1'$ to the letter in the l-th or k-th position, as illustrated in the bottom side of the following two squares. The first square illustrates the subcase where we apply the move of type $1'$ to the index k, and the second the subcase where we apply the move of type $1'$ to the index l.

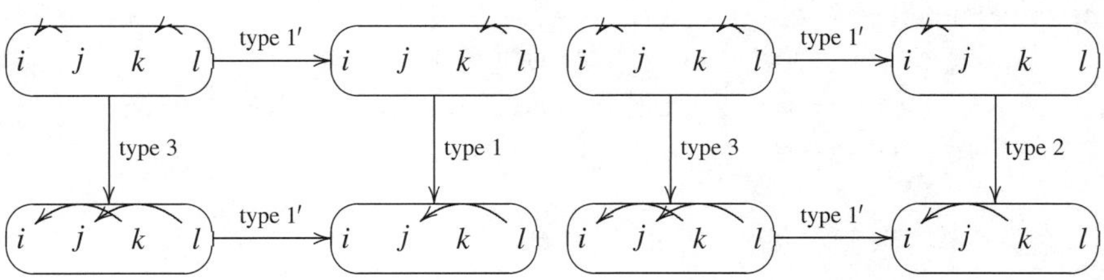

Thus both composites of moves can also be described as a type $1'$ move followed by a move of type 1 or 2. □

Lemma 5.7 *If u is a valid X^2 word with rank $n-1$ obtained from a bracket word w by a move of type $1'$, then $u \leq w^{(s)}$ for some s.*

Proof: Write $w = (a_1, \ldots, a_{2n})$. Let $j < k$ be indices such that $a_k = j$, and such that u is obtained from w by replacing a_k by 0. If u is not equal to $w^{(s)}$ for any s, then there are indices $i < j < k < l$ such that $a_l = i$ (hence $a_i = a_j = 0$), and some $w^{(s)}$ obtained from w by a move of type $1'$ where a_l is replaced by 0. In other words,

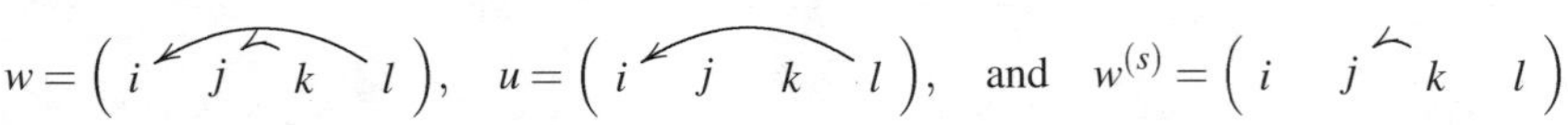

We now see that u is obtained from $w^{(s)}$ by a type 2 move followed by a type 1 move:

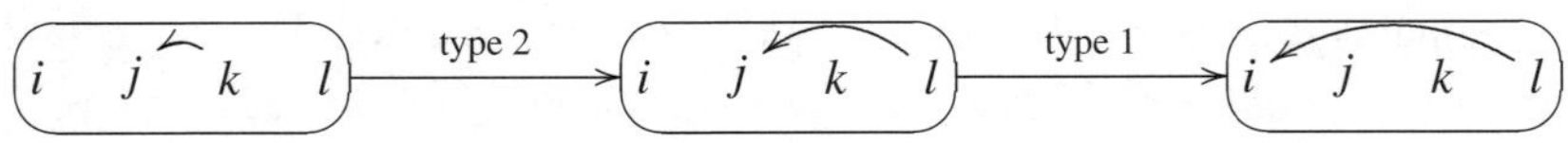

and thus $u \leq w^{(s)}$. □

References

[1] A. Adem and J. Davis, Topics in transformation groups, *Handbook of Geometric Topology*, North-Holland, 2002, 1–54.

[2] G. Carlsson, On the homology of finite free $(Z/2)^n$-complexes, *Invent. Math.* **74** (1983), 139–147.

[3] G. Carlsson, Free $(Z/2)^3$-actions on finite complexes, *Algebraic Topology and Algebraic K-Theory*, *Ann. of Math. Studies* **113**, Princeton Univ. Press, 1987, 332–344.

[4] D. Mumford, *The Red Book of Varieties and Schemes, Second Expanded Edition*, Lecture notes in mathematics 1358, Springer-Verlag, 1999.

[5] B. Rothbach, $X^2 = 0$ and strictly upper triangular matrices, preprint, 2002 (`http://math.berkeley.edu/~rothbach`).

The support of top graded local cohomology modules

MORDECHAI KATZMAN
Department of Pure Mathematics, University of Sheffield, United Kingdom
E-mail: M.Katzman@sheffield.ac.uk

1 Introduction

Let R_0 be any domain, let $R = R_0[U_1, \dots, U_s]/I$, where $U_1, \dots, U_s$ are indeterminates of positive degrees $d_1, \dots, d_s$, and $I \subset R_0[U_1, \dots, U_s]$ is a homogeneous ideal. The main theorem in this paper is Theorem 2.6, a generalization of Theorem 1.5 in [**KS**], which states that all the associated primes of $H := H^s_{R_+}(R)$ contain a certain non-zero ideal $\mathrm{c}(I)$ of R_0 called the "content" of I (see Definition 2.4.) It follows that the support of H is simply $\mathrm{V}(\mathrm{c}(I)R + R_+)$ (Corollary 1.8) and, in particular, H vanishes if and only if $\mathrm{c}(I)$ is the unit ideal. These results raise the question of whether local cohomology modules have finitely many minimal associated primes — this paper provides further evidence in favor of such a result (Theorem 2.10 and Remark 2.12.) Finally, we give a very short proof of a weak version of the monomial conjecture based on Theorem 2.6.

2 The vanishing of top local cohomology modules

Throughout this section R_0 will denote an arbitrary commutative Noetherian domain. We set $S = R_0[U_1, \dots, U_s]$ where $U_1, \dots, U_s$ are indeterminates of degrees $d_1, \dots, d_s$, and $R = S/I$ where $I \subset R_0[U_1, \dots, U_s]$ is an homogeneous ideal. We define $\Delta = d_1 + \cdots + d_s$ and denote with $\mathcal{D}$ the sub-semi-group of $\mathbb{N}$ generated by $d_1, \dots, d_s$. For $t \in \mathbb{Z}$, we shall denote by $(\,\bullet\,)(t)$ the t-th shift functor (on the category of graded R-modules and homogeneous homomorphisms). For any multi-index $\lambda = (\lambda^{(1)}, \dots, \lambda^{(s)}) \in \mathbb{Z}^s$ we shall write U^λ for $U_1^{\lambda^{(1)}} \dots U_s^{\lambda^{(s)}}$ and we shall set $|\lambda| = \lambda^{(1)} + \cdots + \lambda^{(s)}$.

Lemma 2.1 *Let I be generated by homogeneous elements $f_1, \dots, f_r \in S$. Then there is an exact sequence of graded S-modules and homogeneous homomorphisms*

$$\bigoplus_{i=1}^{r} H^s_{S_+}(S)(-\deg f_i) \xrightarrow{(f_1,\dots,f_r)} H^s_{S_+}(S) \longrightarrow H^s_{R_+}(R) \longrightarrow 0.$$

Proof. The functor $H^s_{S_+}(\bullet)$ is right exact and the natural equivalence between $H^s_{S_+}(\bullet)$ and $(\,\bullet\,) \otimes_S H^s_{S_+}(S)$ (see [**BS**, 6.1.8 & 6.1.9]) actually yields a homogeneous S-isomorphism

$$H^s_{S_+}(S)/(f_1, \dots, f_r)H^s_{S_+}(S) \cong H^s_{S_+}(R).$$

To complete the proof, just note that there is an isomorphism of graded S-modules $H^s_{S_+}(R) \cong H^s_{R_+}(R)$, by the Graded Independence Theorem [**BS**, 13.1.6]. □

We can realize $H^s_{S_+}(S)$ as the module $R_0[U_1^-, \dots, U_s^-]$ of inverse polynomials described in [**BS**, 12.4.1]: this graded R-module vanishes beyond degree $-\Delta$. More generally $R_0[U_1^-, \dots, U_s^-]_{-d} \neq 0$ if and only if $d \in \mathcal{D}$. For each $d \in \mathcal{D}$, we have that $R_0[U_1^-, \dots, U_s^-]_{-d}$ is a free R_0-module with base $\mathcal{B}(d) := \left(U^\lambda\right)_{-\lambda \in \mathbb{N}^s, |\lambda| = -d}$. We combine this realization with the previous lemma to find a presentation of each homogeneous component of $H^s_{R_+}(R)$ as the cokernel of a matrix with entries in R_0. Assume first that I is generated by one homogeneous element f of degree δ. For any $d \in \mathcal{D}$ we have, in view of Lemma 2.1, a graded exact sequence

$$R_0[U_1^-, \dots, U_s^-]_{-d-\delta} \xrightarrow{\phi_d} R_0[U_1^-, \dots, U_s^-]_{-d} \longrightarrow H^s_{R_+}(R)_{-d} \longrightarrow 0.$$

The map of free R_0-modules ϕ_d is given by multiplication on the left by a $\#\mathcal{B}(d) \times \#\mathcal{B}(d+\delta)$ matrix which we shall denote later by $M(f;d)$. In the general situation, where I is generated by homogeneous elements $f_1,\ldots,f_r \in S$, it follows from Lemma 2.1 that the R_0-module $H^s_{R_+}(R)_{-d}$ is the cokernel of a matrix $M(f_1,\ldots,f_r;d)$ whose columns consist of all the columns of $M(f_1,d),\ldots,$ $M(f_r,d)$. Consider a homogeneous $f \in S$ of degree δ. We shall now describe the matrix $M(f;d)$ in more detail and to do so we start by ordering the bases of the source and target of ϕ_d as follows. For any $\lambda,\mu \in \mathbb{Z}^s$ with negative entries we declare that $U^\lambda < U^\mu$ if and only if $U^{-\lambda} <_{\text{Lex}} U^{-\mu}$ where "$<_{\text{Lex}}$" is the lexicographical term ordering in S with $U_1 > \cdots > U_s$. We order the bases $\mathcal{B}(d)$, and by doing so also the columns and rows of $M(f;d)$, in ascending order. We notice that the entry in $M(f;d)$ in the U^α row and U^β column is now the coefficient of U^α in fU^β.

Lemma 2.2 *Let $\nu \in \mathbb{Z}^s$ have negative entries and let $\lambda_1,\lambda_2 \in \mathbb{N}^s$. If $U^{\lambda_1} <_{\text{Lex}} U^{\lambda_2}$ and $U^\nu U^{\lambda_1}$, $U^\nu U^{\lambda_2} \in R_0[U_1^-,\ldots,U_s^-]$ do not vanish then $U^\nu U^{\lambda_1} > U^\nu U^{\lambda_2}$.*

Proof. Let j be the first coordinate in which λ_1 and λ_2 differ. Then $\lambda_1^{(j)} < \lambda_2^{(j)}$ and so also $-\nu^{(j)} - \lambda_1^{(j)} > -\nu^{(j)} - \lambda_2^{(j)}$; this implies that $U^{-\nu-\lambda_1} >_{\text{Lex}} U^{-\nu-\lambda_2}$ and $U^{\nu+\lambda_1} > U^{\nu+\lambda_2}$. □

Lemma 2.3 *Let $f \neq 0$ be a homogeneous element in S. Then, for all $d \in \mathcal{D}$, the matrix $M(f;d)$ has maximal rank.*

Proof. We prove the lemma by producing a non-zero maximal minor of $M(f;d)$. Write $f = \sum_{\lambda\in\Lambda} a_\lambda U^\lambda$ where $a_\lambda \in R_0 \setminus \{0\}$ for all $\lambda \in \Lambda$ and let λ_0 be such that U^{λ_0} is the minimal member of $\{U^\lambda : \lambda \in \Lambda\}$ with respect to the lexicographical term order in S. Let δ be the degree of f. Each column of $M(f;d)$ corresponds to a monomial $U^\lambda \in \mathcal{B}(d+\delta)$; its ρ-th entry is the coefficient of U^ρ in $fU^\lambda \in R_0[U_1^-,\ldots,U_s^-]_{-d}$. Fix any $U^\nu \in \mathcal{B}(d)$ and consider the column c_ν corresponding to $U^{\nu-\lambda_0} \in \mathcal{B}(d+\delta)$. The ν-th entry of c_ν is obviously a_{λ_0}. By the previous lemma all entries in c_ν below the νth row vanish. Consider the square submatrix of $M(f;d)$ whose columns are the c_ν ($\nu \in \mathcal{B}(d)$); its determinant is clearly a power of a_{λ_0} and hence is non-zero. □

Definition 2.4 *For any $f \in R_0[U_1,\dots,U_s]$ write $f = \sum_{\lambda\in\Lambda} a_\lambda U^\lambda$ where $a_\lambda \in R_0$ for all $\lambda \in \Lambda$. For such an $f \in R_0[U_1,\dots,U_s]$ we define the* content $\mathrm{c}(f)$ *of f to be the ideal $\langle a_\lambda : \lambda \in \Lambda\rangle$ of R_0 generated by all the coefficients of f. If $J \subset R_0[U_1,\dots,U_s]$ is an ideal, we define its* content $\mathrm{c}(J)$ *to be the ideal of R_0 generated by the contents of all the elements of J. It is easy to see that if J is generated by $f_1,\dots,f_r$, then $\mathrm{c}(J) = \mathrm{c}(f_1)+\cdots+\mathrm{c}(f_r)$.*

Lemma 2.5 *Suppose that I is generated by homogeneous elements $f_1,\dots,f_r \in S$. Fix any $d \in \mathcal{D}$. Let $t := \operatorname{rank} M(f_1,\dots,f_r;d)$ and let I_d be the ideal generated by all $t\times t$ minors of $M(f_1,\dots,f_r;d)$. Then $\mathrm{c}(I) \subseteq \sqrt{I_d}$.*

Proof. It is enough to prove the lemma when $r = 1$; let $f = f_1$. Write $f = \sum_{\lambda\in\Lambda} a_\lambda U^\lambda$ where $a_\lambda \in R_0 \setminus \{0\}$ for all $\lambda \in \Lambda$. Assume that $\mathrm{c}(I) \not\subseteq \sqrt{I_d}$ and pick λ_0 so that U^{λ_0} is the minimal element in $\{U^\lambda : \lambda \in \Lambda\}$ (with respect to the lexicographical term order in S) for which $a_\lambda \notin \sqrt{I_d}$. Notice that the proof of Lemma 2.3 shows that U^{λ_0} cannot be the minimal element of $\{U^\lambda : \lambda\in\Lambda\}$. Fix any $U^\nu \in \mathcal{B}(d)$ and consider the column c_ν corresponding to $U^{\nu-\lambda_0} \in \mathcal{B}(d+\delta)$. The ν-th entry of c_ν is obviously a_{λ_0}. Lemma 2.2 shows that, for any other $\lambda_1 \in \Lambda$ with $U^{\lambda_1} >_{\mathrm{Lex}} U^{\lambda_0}$, either $\nu-\lambda_0+\lambda_1$ has a non-negative entry, in which case the corresponding term of $fU^{\nu-\lambda_0} \in R_0[U_1^-,\dots,U_s^-]_{-d}$ is zero, or $U^\nu > U^{\nu-\lambda_0+\lambda_1}$. Similarly, for any other $\lambda_1\in\Lambda$ with $U^{\lambda_1} <_{\mathrm{Lex}} U^{\lambda_0}$, either $\nu-\lambda_0+\lambda_1$ has a non-negative entry, in which case the corresponding term of $fU^{\nu-\lambda_0} \in R_0[U_1^-,\dots,U_s^-]_{-d}$ is zero, or $U^\nu < U^{\nu-\lambda_0+\lambda_1}$. We have shown that all the entries below the ν-th row of c_ν are in $\sqrt{I_d}$. Consider the matrix M whose columns are c_ν ($\nu \in \mathcal{B}(d)$) and let $\overline{} : R_0 \to R_0/\sqrt{I_d}$ denote the quotient map. We have

$$0 = \overline{\det(M)} = \det(\overline{M}) = \overline{a_{\lambda_0}}^{\binom{d-1}{s-1}}$$

and, therefore, $a_{\lambda_0} \in \sqrt{I_d}$, a contradiction. □

Theorem 2.6 *Suppose that I is generated by homogeneous elements $f_1,\dots,f_r \in S$. Fix any $d \in \mathcal{D}$. Then each associated prime of $H^s_{R_+}(R)_{-d}$ contains $\mathrm{c}(I)$. In particular $H^s_{R_+}(R)_{-d} = 0$ if and only if $\mathrm{c}(I) = R_0$.*

Proof. Recall that for any $p,q \in \mathbb{N}$ with $p \le q$ and any $p\times q$ matrix M of maximal rank with entries in any domain, $\operatorname{Coker} M = 0$ if and only if the ideal generated by the maximal minors of M is the unit ideal. Let $M = M(f_1, \ldots, f_r; d)$, so that $H^s_{R_+}(R)_{-d} \cong \operatorname{Coker} M$. In view of Lemmas 2.3 and 2.5, the ideal $\mathrm{c}(I)$ is contained in the radical of the ideal generated by the maximal minors of M; therefore, for each $x \in \mathrm{c}(I)$, the localization of $\operatorname{Coker} M$ at x is zero; we deduce that $\mathrm{c}(I)$ is contained in all associated primes of $\operatorname{Coker} M$. To prove the second statement, assume first that $\mathrm{c}(I)$ is not the unit ideal. Since all minors of M are contained in $\mathrm{c}(I)$, these cannot generate the unit ideal and $\operatorname{Coker} M \neq 0$. If, on the other hand, $\mathrm{c}(I) = R_0$ then $\operatorname{Coker} M$ has no associated prime and $\operatorname{Coker} M = 0$. □

Corollary 2.7 *Let the situation be as in* 2.6. *The following statements are equivalent:*

1. $\mathrm{c}(I) = R_0$;
2. $H^s_{R_+}(R)_{-d} = 0$ *for some* $d \in \mathcal{D}$;
3. $H^s_{R_+}(R)_{-d} = 0$ *for all* $d \in \mathcal{D}$.

Consequently, $H^s_{R_+}(R)$ *is asymptotically gap-free in the sense of* [**BH**, (4.1)].

Corollary 2.8 *The R-module* $H^s_{R_+}(R)$ *has finitely many minimal associated primes, and these are just the minimal primes of the ideal* $\mathrm{c}(I)R + R_+$.

Proof. Let $r \in \mathrm{c}(I)$. By Theorem 2.6, the localization of $H^s_{R_+}(R)$ at r is zero. Hence each associated prime of $H^s_{R_+}(R)$ contains $\mathrm{c}(I)R$. Such an associated prime must contain R_+, since $H^s_{R_+}(R)$ is R_+-torsion. On the other hand, $H^s_{R_+}(R)_{-\Delta} \cong R_0/\mathrm{c}(I)$ and $H^s_{R_+}(R)_i = 0$ for all $i > -\Delta$; therefore there is an element of the $(-\Delta)$-th component of $H^s_{R_+}(R)$ that has annihilator (over R) equal to $\mathrm{c}(I)R + R_+$. All the claims now follow from these observations. □

Remark 2.9 In [**Hu**, Conjecture 5.1], Craig Huneke conjectured that every local cohomology module (with respect to any ideal) of a finitely generated module over a local Noetherian ring has only finitely many associated primes. This conjecture was shown to be false (cf. [**K**, Corollary 1.3]) but Corollary 2.8 provides some evidence in support of the weaker conjecture that every local cohomology module (with respect to any ideal) of a finitely generated module over a local Noetherian ring has only finitely many *minimal* associated primes. The following theorem due to Gennady Lyubeznik ([**L**]) gives further support for this conjecture:

Theorem 2.10 *Let R be any Noetherian ring of prime characteristic p and let $I \subset R$ be any ideal generated by $f_1, \dots, f_s \in R$. The support of $H_I^s(R)$ is Zariski closed.*

Proof. We first notice that the localization of $H_I^s(R)$ at a prime $P \subset R$ vanishes if and only if there exist positive integers α and β such that

$$(f_1 \cdot \dots \cdot f_s)^{\alpha} \in \langle f_1^{\alpha+\beta}, \dots, f_s^{\alpha+\beta} \rangle$$

in the localization R_P. This is because if we can find such α and β we can then take $q := p^e$ powers and obtain

$$(f_1 \cdot \dots \cdot f_s)^{q\alpha} \in \langle f_1^{q\alpha+q\beta}, \dots, f_s^{q\alpha+q\beta} \rangle$$

for all such q. This shows that all elements in the direct limit sequence

$$R/\langle f_1, \dots f_s \rangle \xrightarrow{f_1 \cdot \dots \cdot f_s} R/\langle f_1^2, \dots f_s^2 \rangle \xrightarrow{f_1 \cdot \dots \cdot f_s} \dots$$

map to 0 in the direct limit and hence $H_I^s(R) = 0$. But if

$$(f_1 \cdot \dots \cdot f_s)^{\alpha} \in \langle f_1^{\alpha+\beta}, \dots, f_s^{\alpha+\beta} \rangle$$

in R_P, we may clear denominators and deduce that this occurs on a Zariski open subset containing P. Thus the complement of the support is a Zariski open subset. □

It may be reasonable to expect that non-top local cohomology modules might also have finitely many minimal associated primes; the only examples known to me of non-top local cohomology modules with infinitely many associated primes are the following: Let k be any field, let $R_0 = k[x, y, s, t]$ and let S be the localization of $R_0[u, v, a_1, \dots, a_n]$ at the maximal ideal $\mathfrak{m}$ generated by $x, y, s, t, u, v, a_1, \dots, a_n$. Let $f = sx^2v^2 - (t+s)xyuv + ty^2u^2 \in S$ and let $R = S/fS$. Denote by I the ideal of S generated by u, v and by A the ideal of S generated by $a_1, \dots, a_n$.

Theorem 2.11 *Assume that $n \geq 2$. The local cohomology module $\mathrm{H}^2_{I \cap A}(R)$ has infinitely many associated primes and $\mathrm{H}^{n+1}_{I \cap A}(R) \neq 0$.*

Proof. Consider the following segment of the Mayer-Vietoris sequence

$$\cdots \to \mathrm{H}^2_{I+A}(R) \to \mathrm{H}^2_I(R) \oplus \mathrm{H}^2_A(R) \to \mathrm{H}^2_{I\cap A}(R) \to \ldots.$$

Note that $a_1, \ldots, a_n, u$ form a regular sequence on R so $\operatorname{depth}_{I+A} R \geq n+1 \geq 3$ and the leftmost module vanishes. Thus $\mathrm{H}^2_I(R)$ injects into $\mathrm{H}^2_{I\cap A}(R)$ and Corollary 1.3 in [**K**] shows that $\mathrm{H}^2_{I\cap A}(R)$ has infinitely many associated primes. Consider now the following segment of the Mayer-Vietoris sequence

$$\cdots \to \mathrm{H}^{n+1}_{I\cap A}(R) \to \mathrm{H}^{n+2}_{I+A}(R) \to \mathrm{H}^{n+2}_I(R) \oplus \mathrm{H}^{n+2}_A(R) \to \ldots.$$

The direct summands in the rightmost module vanish since both I and A can be generated by less than $n+2$ elements, so $\mathrm{H}^{n+1}_{I\cap A}(R)$ surjects onto $\mathrm{H}^{n+2}_{I+A}(R)$. Now $\mathrm{c}(f)$ is the ideal of R_0 generated by $sx^2, -(t+s)xy$ and ty^2 so $\mathrm{c}(f) \subset \langle x, y\rangle \neq R_0$. Corollary 2.7 now shows that $\mathrm{H}^{n+2}_{I+A}(R)$ does not vanish and, therefore, nor does $\mathrm{H}^{n+1}_{I\cap A}(R)$. □

Remark 2.12 When $n \geq 3$, one has that $\mathrm{H}^3_{I+A}(R) = 0$ and the argument above shows that $\mathrm{H}^2_I(R) \oplus \mathrm{H}^2_A(R) \cong \mathrm{H}^2_{I\cap A}(R)$. Corollary 2.8 implies that $\mathrm{H}^2_I(R)$ has finitely many minimal primes and since the only associated prime of $\mathrm{H}^2_A(R)$ is A, $\mathrm{H}^2_{I\cap A}(R)$ has finitely many minimal primes. When $n = 2$ we obtain a short exact sequence

$$0 \to \mathrm{H}^2_I(R) \oplus \mathrm{H}^2_A(R) \to \mathrm{H}^2_{I\cap A}(R) \to \mathrm{H}^3_{I+A}(R) \to 0.$$

The short exact sequence

$$0 \to S \xrightarrow{f} S \to R \to 0$$

implies that $\mathrm{H}^3_{I+A}(R)$ injects into the local cohomology module $\mathrm{H}^4_{I+A}(S)$ whose only associated prime is $I+A$, so again we see that $\mathrm{H}^2_{I\cap A}(R)$ has finitely many minimal associated primes.

3 An application: a weak form of the Monomial Conjecture.

In [**Ho**] Mel Hochster suggested reducing the Monomial Conjecture to the problem of showing the vanishing of certain local cohomology modules which we now describe. Let C be either $\mathbb{Z}$ or a field of characteristic $p > 0$, let $R_0 = C[A_1, \dots, A_s]$ where $A_1, \dots, A_s$ are indeterminates, $S = R_0[U_s, \dots, U_s]$ where $U_1, \dots, U_s$ are indeterminates and $R = S/F_{s,t}S$ where

$$F_{s,t} = (U_1 \cdot \dots \cdot U_s)^t - \sum_{i=1}^{s} A_i U_i^{t+1}.$$

Suppose that

$$H_{s,t} := H^s_{\langle U_1, \dots, U_s \rangle}(R)$$

vanishes with $C = \mathbb{Z}$. If for some local ring T we can find a system of parameters $x_1, \dots, x_s$ so that $(x_1 \cdot \dots \cdot x_s)^t \in \langle x_1^{t+1}, \dots, x_s^{t+1} \rangle$, i.e., if there exist $a_1, \dots, a_s \in T$ so that $(x_1 \cdot \dots \cdot x_s)^t = \sum_{i=1}^{t} a_i x_i^{t+1}$, we can define an homomorphism $R \to T$ by mapping A_i to a_i and U_i to x_i. We can view T as an R-module and we have an isomorphism of T-modules

$$H^s_{\langle x_1, \dots, x_s \rangle}(T) \cong H^s_{\langle U_1, \dots, U_s \rangle}(R) \otimes_R T$$

and we deduce that

$$H^s_{\langle x_1, \dots, x_s \rangle}(T) = 0$$

but this cannot happen since $x_1, \dots, x_s$ form a system of parameters in T. We have just shown that the vanishing of $H_{s,t}$ for all $t \geq 1$ implies the Monomial Conjecture in dimension s. In [**Ho**] Mel Hochster proved that these local cohomology modules vanish when $s = 2$ or when C has characteristic $p > 0$, but in [**R**] Paul Roberts showed that, when $C = \mathbb{Z}$, $H_{3,2} \neq 0$, showing that Hochster's approach cannot be used for proving the Monomial Conjecture in dimension 3. This can be generalized further:

Proposition 3.1 *When $C = \mathbb{Z}$, $H_{s,2} \neq 0$ for all $s \geq 3$.*

Proof. We proceed by induction on s; the case $s = 3$ is proved in [**R**]. Assume that for some $s \geq 1$, $\alpha \geq 0$ and $\delta > \alpha$ the monomial $x_1^{\alpha} \dots x_{s+1}^{\alpha}$ is in the ideal of $C[x_1, \dots, x_{s+1}, a_1, \dots, a_{s+1}]$ generated by $x_1^{\alpha+\beta}, \dots, x_{s+1}^{\alpha+\beta}$ and $F_{s+1,t}$. Define $G_{s+1,2}$ to be the result of substituting $a_{s+1} = 0$ in $F_{s+1,2}$, i.e.,

$$G_{s+1,2} = (x_1 \dots x_{s+1})^2 - \sum_{i=1}^{s} a_i x_i^3.$$

If

$$x_1^{\alpha} \dots x_{s+1}^{\alpha} \in \langle x_1^{\alpha+\beta}, \dots, x_{s+1}^{\alpha+\beta}, F_{s+1,2} \rangle \tag{1}$$

then by setting $a_{s+1} = 0$ we see that

$$x_1^{\alpha} \dots x_{s+1}^{\alpha} \in \langle x_1^{\alpha+\beta}, \dots, x_{s+1}^{\alpha+\beta}, G_{s+1,2} \rangle.$$

If we assign degree 0 to $x_1, \dots, x_s$, degree 1 to x_{s+1} and degree 2 to $a_1, \dots, a_s$ then the ideal $\langle x_1^{\alpha+\beta}, \dots, x_{s+1}^{\alpha+\beta}, G_{s+1,2} \rangle$ is homogeneous and we must have

$$x_1^{\alpha} \dots x_{s+1}^{\alpha} \in \langle x_1^{\alpha+\beta}, \dots, x_s^{\alpha+\beta}, G_{s+1,2} \rangle.$$

If we now set $x_{s+1} = 1$ we obtain

$$x_1^{\alpha} \dots x_s^{\alpha} \in \langle x_1^{\alpha+\beta}, \dots, x_s^{\alpha+\beta}, F_{s,2} \rangle. \tag{2}$$

Now $H_{s+1,2} = 0$ if and only if for each $\beta \geq 1$ we can find an $\alpha \geq 0$ so that equation (1) holds, and this implies that for each $\beta \geq 1$ we can find an $\alpha \geq 0$ so that equation (2) holds, which is equivalent to $H_{s,2} = 0$. The induction hypothesis implies that $H_{s,2} \neq 0$ and so $H_{s+1,2} \neq 0$. □

The local cohomology modules $H_{s,t}$ are a good illustration for the failure of the methods of the previous section in the non-graded case. For example, one cannot decide whether $H_{s,t}$ is zero just by looking at $F_{s,t}$: the vanishing of $H_{s,t}$ depends on the characteristic of C! Compare this situation to the following graded problem.

Theorem 3.2 (A Weaker Monomial Conjecture) *Let T be a local ring with system of parameters $x_1, \dots, x_s$. For all $t \geq 0$ we have*

$$(x_1 \cdot \dots \cdot x_s)^t \notin \langle x_1^{st}, \dots, x_s^{st} \rangle.$$

Proof. Let $S = \mathbb{Z}[A_1, \dots, A_s][X_1, \dots, X_s]$ where $\deg A_i = 0$ and $\deg X_i = 1$ for all $1 \leq i \leq s$. Following Hochster's argument we reduce to the problem of showing that

$$H^s_{\langle X_1, \dots, X_s \rangle}(S/fS) = 0$$

where

$$f = (X_1 \cdot \dots \cdot X_s)^t - \sum_{i=1}^{s} A_i X_i^{st}.$$

Since f is homogeneous and $\mathrm{c}(f)$ is the unit ideal, the result follows from Theorem 2.6. □

References

[**BH**] M.P. Brodmann and M. Hellus, Cohomological patterns of coherent sheaves over projective schemes, *J. Pure Appl. Algebra* **172** (2002), 165–182.

[**BS**] M.P. Brodmann and R.Y. Sharp, *Local Cohomology: An Algebraic Introduction with Geometric Applications,* Cambridge University Press, 1998.

[**Ho**] M. Hochster, Canonical elements in local cohomology modules and the direct summand conjecture, *J. Algebra* **84** (1983), 503–553.

[**Hu**] C. Huneke, Problems on local cohomology, in *Free Resolutions in Commutative Algebra and Algebraic Geometry, Sundance 90*, D. Eisenbud and C. Huneke, Eds., 93–108, Research Notes in Mathematics **2**, Jones and Bartlett Publishers, Boston, 1992.

[**K**] M. Katzman, An example of an infinite set of associated primes of a local cohomology module, *J. Algebra* **252** (2002), 161–166.

[**KS**] M. Katzman and R.Y. Sharp, Some properties of top graded local cohomology modules, *J. Algebra* **259** (2003) , 599–612.

[**L**] G. Lyubeznik, private communication.

[**R**] P. Roberts, A computation of local cohomology, in *Commutative algebra: syzygies, multiplicities, and birational algebra (South Hadley, MA, 1992),* 351–356, *Contemp. Math.*, **159**, Amer. Math. Soc., Providence, RI, 1994.

An approach to the uniform Artin-Rees theorems from the notion of relation type

FRANCESC PLANAS-VILANOVA
Departament de Matemàtica Aplicada 1
Universitat Politècnica de Catalunya, Barcelona, Spain
E-mail: francesc.planas@upc.edu

1 Introduction

Our purpose in this work is to emphasize the tight connection between the Artin-Rees lemma and the notion of the relation type of an ideal with respect to a module. We present new proofs of some already known uniform Artin-Rees theorems as well as a new approach to the subject.

We begin by recalling the well-known Artin-Rees lemma, which plays an essential role in the completion process of rings and modules (see [**2**, **3**, **17**, **18**, **29**, **40**] for proofs, consequences and historical notes). The standard proof will provide us with perhaps the first evidence of the above cited strong connection between the Artin-Rees lemma and the notion of relation type, since it involves what is nowadays called the Rees module of an ideal with respect to a module.

Lemma 1.1 (The Artin-Rees lemma) *Let A be a noetherian ring, I an ideal of A and let $N \subseteq M$ be two finitely generated A-modules. Then there exists an integer*

$s \geq 1$, depending on N, M and I, such that for all $n \geq s$,

$$I^n M \cap N = I^{n-s}(I^s M \cap N).$$

Proof: Let $\mathcal{R}(I) = \bigoplus_{n\geq 0} I^n$ be the Rees ring of the ideal I, which is noetherian since A is noetherian. Let $\mathcal{R}(I;M) = \bigoplus_{n\geq 0} I^n M$ be the Rees module of I with respect to the module M, which is a graded finitely generated $\mathcal{R}(I)$-module since M is finitely generated. Consider the graded submodule $L = \bigoplus_{n\geq 0} I^n M \cap N$, which is finitely generated since $\mathcal{R}(I;M)$ is a noetherian module. Take s as the largest degree of a set of generators. Then $L_n \subset I^{n-s} L_s$ for all $n \geq s$, which implies the assertion. □

If we denote by $E(N,M;\{I\})_n = I^n M \cap N / I(I^{n-1} M \cap N)$, then the lemma is equivalent to stating that there exists an integer $s \geq 1$, depending on N, M and I, such that $E(N,M;\{I\})_n = 0$ for all $n \geq s+1$. A weaker result, which follows immediately from the lemma, and which will play an important role on the sequel, is that there exists an integer $s \geq 1$, depending on N, M and I, such that $I^n M \cap N \subseteq I^{n-s} N$.

2 A problem posed by Eisenbud and Hochster

Eisenbud and Hochster in [**7**], originally motivated by a question of Wehrfritz about the existence of a uniform index of nilpotency (Corollary 2 in [**7**]), gave a new proof of Zariski's main lemma on holomorphic functions on an algebraic variety ([**43**]) by establishing what they called a "Nullstellensatz with nilpotents." They also remarked that a different approach to their theorem could be obtained by answering the following question of uniform Artin-Rees type:

Question 2.1 *Let A be an affine ring and let $N \subseteq M$ be two finitely generated A-modules. Does there exist an integer $s \geq 1$ such that, for all integers $n \geq s$ and for all maximal ideals $\mathfrak{m}$ of A,*

$$\mathfrak{m}^n M \cap N = \mathfrak{m}^{n-s}(\mathfrak{m}^s M \cap N)?$$

In the same paper they already gave a negative answer to this problem (in the non-affine case) by building a 2-dimensional regular ring A whose maximal ideals form a countable set, say $\{\mathfrak{m}_1, \mathfrak{m}_2, \ldots\}$, such that $\cap_i \mathfrak{m}_i$ is a nonzero principal prime ideal generated by an element f, where $f \in \mathfrak{m}_i^i$ for every $i \geq 1$. In particular and since A is a domain, $\mathfrak{m}_i \left(\mathfrak{m}_i^{i-1} \cap (f)\right) = \mathfrak{m}_i(f) \subsetneq (f)$ whereas $\mathfrak{m}_i^i \cap (f) = (f)$.

An answer to a slightly weaker version of this question was given by O'Carroll in [**19**] under the hypothesis of A being an excellent ring.

Theorem 2.2 (see [19]) *Let A be an excellent ring and let $N \subseteq M$ be two finitely generated A-modules. Then there exists an integer $s \geq 1$ such that, for all integers $n \geq s$ and for all maximal ideals $\mathfrak{m}$ of A,*

$$\mathfrak{m}^n M \cap N \subset \mathfrak{m}^{n-s} N.$$

And the full answer to the question of Eisenbud and Hochster was given two years later by Duncan and O'Carroll.

Theorem 2.3 (see [6]) *Let A be an excellent ring and let $N \subseteq M$ be two finitely generated A-modules. Then there exists an integer $s \geq 1$ such that, for all integers $n \geq s$ and for all maximal ideals $\mathfrak{m}$ of A,*

$$\mathfrak{m}^n M \cap N = \mathfrak{m}^{n-s}(\mathfrak{m}^s M \cap N).$$

Later on, O'Carroll showed the uniformity of the Artin-Rees property for the set of principal ideals of a noetherian ring.

Theorem 2.4 (see [20]) *Let A be a noetherian ring and let $N \subseteq M$ be two finitely generated A-modules. Then there exists an integer $s \geq 1$ such that, for all integers $n \geq s$ and for all elements x of A,*

$$x^n M \cap N = x^{n-s}(x^s M \cap N).$$

And soon after, he noted that Theorem 2.4 allows for a quick proof of Theorem 2.2 in the case of an affine ground ring over a perfect field (see [**21**]).

3 Uniform Artin-Rees on arbitrary sets of ideals

The results of O'Carroll ([**19**, **20**]) and Duncan and O'Carroll ([**6**]) gave a complete answer to the question posed by Eisenbud and Hochster and, in the meantime, suggested new and interesting problems undertaken by Huneke in [**11**]. In order to facilitate the reading of the next results, we introduce a notation very close to the one proposed by Huneke in [**11**]. Given a noetherian ring A and two finitely generated A-modules $N \subseteq M$, it is said that the pair (N,M) has the strong (weak) uniform Artin-Rees property with respect to a set of ideals $\mathcal{W}$ of A and with strong (weak) uniform Artin-Rees number s, the integer s depending on N, M and $\mathcal{W}$, if, for all integer $n \geq s$ and for every ideal I of $\mathcal{W}$, $I^nM \cap N = I^{n-s}(I^sM \cap N)$ ($I^nM \cap N \subseteq I^{n-s}N$). Clearly, if s is a uniform Artin-Rees number and t is an integer such that $t \geq s$, then t is also a uniform Artin-Rees number. The minimum of the strong (weak) uniform Artin-Rees numbers is denoted by $s = s(N,M;\mathcal{W})$ ($w = w(N,M;\mathcal{W})$). If every pair $N \subset M$ of finitely generated A-modules has the strong (weak) uniform Artin-Rees property with respect to $\mathcal{W}$ it is said that A has the strong (weak) uniform Artin-Rees property with respect to $\mathcal{W}$. If $\mathcal{W}$ is the set of all ideals of A, then the phrase "with respect to $\mathcal{W}$" is deleted in these definitions and one writes $s = s(N,M)$ ($w = w(N,M)$).

In [**11**] (Theorems 4.12 and 5.11), and with the aid of the tight closure theory, Huneke strengthened the study of the uniform Artin-Rees properties to the whole set of ideals of A by weakening the thesis from the strong uniform Artin-Rees property to the weak uniform Artin-Rees property.

Theorem 3.1 (see [11]) *Let A be a noetherian ring and let $N \subset M$ be two finitely generated A-modules. If A satisfies at least one of the conditions below, then there exists an integer $s \geq 1$ such that, for all integers $n \geq s$ and for all ideals I of A,*

$$I^nM \cap N \subseteq I^{n-s}N.$$

(i) *A is essentially of finite type over a noetherian local ring.*

(ii) *A is of characteristic p and A is a finite module over A^p.*

(iii) *A is essentially of finite type over $\mathbb{Z}$.*

(iv) *A is an excellent 3-dimensional domain.*

And in the same paper he conjectured that excellent noetherian rings of finite Krull dimension have the weak uniform Artin-Rees property (see [**11**], Conjecture 1.3). Recently, Huneke showed a new case in which the conjecture holds.

Theorem 3.2 (see [13]) *Let A be an excellent noetherian ring which is the homomorphic image of a regular ring of finite Krull dimension. Assume that for all primes $\mathfrak{p}$ of A, the integral closure of $A/\mathfrak{p}$ has a resolution of singularities obtained by blowing up an ideal. Then A has the weak uniform Artin-Rees property.*

4 The Relation Type Conjecture

But the search for uniform properties in noetherian rings has been far from being circumscribed to the Artin-Rees properties. Uniform bounds in the theory of test elements ([**12**]), uniform Briançon-Skoda theorems ([**11**, **12**]), uniform bounds for annihilators of local cohomology ([**27**]) or uniform bounds for the relation type of parameter ideals ([**12**, **15**, **42**]) are some examples of the great interest awakened by the subject. It is the last one which we want to survey now.

Before that, let us revise the definition of relation type of an ideal (see, for instance, [**14**, **15**, **24**, **28**, **32**, **35**], for definitions and properties). Let A be a noetherian ring and let I be an ideal of A generated by elements $x_1,\dots,x_n$ of A. Let $\varphi : A[T_1,\dots,T_n] \to \mathcal{R}(I) = A[It]$ be the natural homogenous map which sends T_i to $x_i t$. The relation type of I, $\mathrm{rt}(I)$, is the largest degree of any minimal system of generators of the kernel of φ. The ideal I is said to be an ideal of linear type if $\mathrm{rt}(I) = 1$ (see [**8**] and the references there). An already classical result concerning ideals of linear type, proved simultaneously by Huneke [**9**] and Valla [**38**] (see [**34**] for the case of modules), is that ideals generated by d-sequences are of linear type. Regular sequences are examples of d-sequences. It is well-known that in a Cohen-Macaulay local ring any system of parameters forms a regular sequence, in particular, a d-sequence. In fact, Buchsbaum local rings are the rings in which every system of parameters is a d-sequence ([**10**]). Therefore, in Buchsbaum rings every ideal generated by a system of parameters has relation type 1. Moreover, the relation type is controlled through faithfully flat extensions and the quotient by modules of finite length (see, e.g., [**24**, **25**] and [**42**]). From this point of view it is natural to ask the following question:

Question 4.1 (Relation Type Conjecture, [12]) *Let A be a complete local equidimensional noetherian ring. Does there exist an integer $s \geq 1$ such that, for every ideal I of A generated by a system of parameters, the relation type of I is bounded* *above by s?*

Wang in [**41**] first proved this conjecture for 2-dimensional noetherian local rings. Concretely:

Theorem 4.2 (see [41]) *Let A be a 2-dimensional noetherian local ring. Then there exists an integer $s \geq 1$ such that, for every ideal I of A generated by a system of parameters, the relation type of I is bounded above by s.*

And afterwards, he proved the conjecture for rings with finite local cohomology.

Theorem 4.3 (see [42]) *Let $(A, \mathfrak{m})$ be a d-dimensional noetherian local ring such that $H^i_{\mathfrak{m}}(A)$ is finitely generated for $i = 0, \ldots, d-1$. Then there exists an integer $s \geq 1$ such that, for every ideal I of A generated by a system of parameters, the relation type of I is bounded above by s.*

For rings of dimension 3 or more, see the very recent work of Aberbach, Ghezzi and Hà [**1**].

5 A long-time known connection

The clear connection between the Artin-Rees lemma and the notion of relation type, at least for the case of ideals of linear type, appeared explicitly in the following two results. The first one is due to Huneke ([**9**]) and states that an ideal I such that its extension $I + J/J$ on the quotient ring A/J is generated by a d-sequence verifies a certain Artin-Rees property.

Proposition 5.1 (see [9]) *Let A be a commutative ring. If J is an ideal of A and I is an ideal of A generated by a d-sequence modulo J, then for all integers $n \geq 1$,*

$$I^n \cap J \subset I^{n-1}J.$$

The second one is due to Herzog, Simis and Vasconcelos ([**8**]) and is what they called the "Artin-Rees lemma on the nose." There, the hypothesis was on both ideals I and I/J.

Proposition 5.2 (see [8]) *Let A be a commutative ring. If $J \subset I$ are two ideals of A such that I and I/J are ideals of linear type, then for all integers $n \geq 1$,*

$$I^n \cap J = I^{n-1}J.$$

Much later, and already in the context of uniform properties, Lai proved in [**15**] that the Relation Type Conjecture for the quotient ring of a Buchsbaum local ring A by an ideal J is equivalent to the strong uniform Artin-Rees property for the pair (J,A) and with respect to the set of ideals of A whose image in A/J is a parameter ideal. Concretely:

Theorem 5.3 (see [15]) *Let A be a Buchsbaum local ring and let J be an ideal of A. Then the following are equivalent:*

(i) *The pair (J,A) has the strong uniform Artin-Rees property with respect to the set of ideals of A whose image in A/J is a parameter ideal of A/J and with strong uniform Artin-Rees number s.*

(ii) *The relation type of parameter ideals of A/J is uniformly bounded by s.*

A similar result to this was proved by Wang in [**41**] where a bound for the relation type of an ideal is deduced from a strong uniform Artin-Rees number. Concretely:

Proposition 5.4 (see [41]) *Let A be a noetherian local ring and let I be an ideal of A generated by a regular sequence. If J is an ideal of A such that, for all integers $n \geq s$, $I^n \cap J = I^{n-s}(I^s \cap J)$, then $\mathrm{rt}(I+J/J) \leq s$.*

And using this fact, he was able to prove that the strong uniform Artin-Rees property cannot hold for the class of all ideals, even in a 3-dimensional regular local ring, by displaying a family of ideals with unbounded relation type in a 2-dimensional regular local ring ([**41**]).

Example 5.5 (see [41]) Let k be a field and let $A = k[\![X,Y,Z]\!]$ be the power series ring in three variables. Let $J = (Z)$ be the ideal of A generated by Z. For each integer $r \geq 1$, consider the ideal $I_r = (X^r, Y^r, X^{r-1}Y + Z^r)$ of A, which is generated by a regular sequence. Clearly $\mathrm{rt}(I_r + J/J) \geq r$. In particular, there does not exist an integer $s \geq 1$ such that, for all integers $n \geq s$ and for all integers $r \geq 1$,

$$I_r^n \cap J = I_r^{n-s}(I_r^s \cap J).$$

6 The exact sequence linking AR with RT

So far we have seen that the strong uniform Artin-Rees property holds for the set of maximal ideals and the set of principal ideals; that the weak Artin-Rees property holds for the whole set of ideals under some conditions on the base ring; and that Artin-Rees theory and Relation Type are two linked subjects so that questions on uniform Artin-Rees may be translated into questions on relation type and vice versa, for instance, the Relation Type Conjecture or that the strong uniform Artin-Rees property does not hold for the whole class of ideals of a ring. Our next purpose now is to deepen on this link. In order to do that, we recall some notations and properties beginning with the (module of effective) relations of the Rees module of an ideal I with respect to a module M (see [**25**]).

If A is a commutative ring, we mean by a standard A-algebra a commutative graded algebra $U = \oplus_{n\geq 0}U_n$, with $U_0 = A$ and U generated by the elements of degree 1. Two examples of standard algebras are the Rees algebra $\mathcal{R}(I)$ of an ideal I of A, which is a standard A-algebra, and the associated graded ring $\mathcal{G}(I)$ of I, which is a standard A/I-algebra. Let $U_+ = \oplus_{n>0}U_n$ be the irrelevant ideal of U. If $E = \oplus_{n\geq 0}E_n$ is a graded U-module, let $s(E) = \min\{r \geq 1 \mid E_n = 0 \text{ for all } n \geq r+1\}$, which may possibly be infinite.

A standard U-module will be a graded U-module $F = \oplus_{n\geq 0}F_n$ such that $F_n = U_nF_0$ for all $n \geq 0$. The Rees module $\mathcal{R}(I;M) = \oplus_{n\geq 0}I^nM$ of an ideal I of A with respect to an A-module M is a standard $\mathcal{R}(I)$-module and the associated graded module $\mathcal{G}(I;M) = \oplus_{n\geq 0}I^nM/I^{n+1}M$ of I with respect to an A-module M is a standard $\mathcal{G}(I)$-module.

Given F and G, two standard U-modules, and $\varphi : G \to F$, a surjective graded morphism of U-modules, let $E(\varphi)$ denote the graded A-module

$$E(\varphi) = \ker\varphi/U_+\ker\varphi = \ker\varphi_0 \oplus (\oplus_{n\geq 1}\ker\varphi_n/U_1\ker\varphi_{n-1}) = \oplus_{n\geq 0}E(\varphi)_n.$$

The next remark is a simple but very useful property, which will be decisive in what follows: if $\varphi : G \to F$, $\psi : H \to G$ are two surjective graded morphisms of graded U-modules, then there exists a graded exact sequence of A-modules

$$E(\psi) \to E(\varphi\circ\psi) \overset{\psi}{\to} E(\varphi) \to 0. \tag{1}$$

In particular, $s(E(\varphi)) \leq s(E(\varphi\circ\psi)) \leq \max(s(E(\varphi)), s(E(\psi)))$ (see [**25**, Lemma

2.3]).

For a standard U-module F and $\gamma : \mathbf{S}(U_1) \otimes F_0 \stackrel{\alpha \otimes 1}{\to} U \otimes F_0 \to F$, where $\alpha : \mathbf{S}(U_1) \to U$ is the canonical symmetric presentation of U and $U \otimes F_0 \to F$ is the structural morphism, the module of effective n-relations of F is defined to be $E(F)_n = E(\gamma)_n = \ker\gamma_n / U_1 \ker\gamma_{n-1}$ (where, for $n = 0$, $E(F)_n = 0$). Put $E(F) = \oplus_{n\geq 1} E(F)_n = \oplus_{n\geq 1} E(\gamma)_n = E(\gamma) = \ker\gamma / \mathbf{S}_+(U_1)\ker\gamma$. The relation type of F is defined to be $\mathrm{rt}(F) = s(E(F))$; that is, $\mathrm{rt}(F)$ is the minimum integer $r \geq 1$ such that the effective n-relations are zero for all $n \geq r+1$. Using the exact sequence (1), it can be shown that the module of effective n-relations and the relation type do not depend on the chosen symmetric presentation ([**25**]).

For an ideal I of A and an A-module M, the module of effective n-relations and the relation type of I with respect to M are defined to be $E(I;M)_n = E(\mathcal{R}(I;M))_n$ and $\mathrm{rt}(I;M) = \mathrm{rt}(\mathcal{R}(I;M))$ (see also [**8**], page 106, and [**34**], page 41). Analogously to what happens with the Castelnuovo regularity ([**22**]), with the symmetric canonical presentations ([**38**]) and with further work on a-invariants ([**36**]), the relation type of I with respect to M coincides with the relation type of $\mathcal{G}(I;M)$ ([**25**]).

Now, given an ideal I of A and two A-modules $N \subset M$, take the standard $\mathbf{S}(I)$-modules $F = \mathcal{R}(I;M/N)$, $G = \mathcal{R}(I;M)$ and $H = \mathbf{S}(I) \otimes M$. Consider the surjective graded morphisms of $\mathbf{S}(I)$-modules $\varphi : G \to F$ defined by $\varphi_n : G_n = I^n M \to I^n M / I^n M \cap N = I^n M + N / N = F_n$ and $\psi : H \to G$ induced by the natural graded morphism $\alpha : \mathbf{S}(I) \to \mathcal{R}(I)$. Since ψ is a symmetric presentation of $G = \mathcal{R}(I;M)$, $E(\psi) = E(I;M)$ and $s(E(\psi)) = \mathrm{rt}(I;M)$. Since $\varphi \circ \psi$ is a symmetric presentation of $F = \mathcal{R}(I;M/N)$, $E(\varphi \circ \psi) = E(I;M/N)$ and $s(E(\varphi \circ \psi)) = \mathrm{rt}(I;M/N)$. Finally, it is clear that $E(\varphi)_n = I^n M \cap N / I(I^{n-1} M \cap N)$, which coincides with the module $E(N,M;\{I\})_n$ introduced in Section 1. In particular, $s(E(\varphi)) = s(N,M;\{I\})$. Using the exact sequence (1), we arrive at the desired exact sequence of A-modules which links Artin-Rees theory with relation type (see [**25**]):

$$E(I;M)_n \to E(I;M/N)_n \to E(N,M;\{I\})_n \to 0. \tag{2}$$

Evaluating the "function" s on (2), we obtain the inequalities:

$$s(N,M;\{I\}) \leq \mathrm{rt}(I;M/N) \leq \max(\mathrm{rt}(I;M), s(N,M;\{I\})). \tag{3}$$

7 Uniform Relation Type

The purpose of this section is to prove the results of Section 5 using the inequalities (3). Let us begin with the results of Huneke and of Herzog, Simis and Vasconcelos.

Proof of Proposition 5.1 and Proposition 5.2: If I is of linear type modulo J (for instance, if it is generated by a d-sequence modulo J), then $\mathrm{rt}(I+J/J) = 1$. From the first inequality in (3), $s(J,A;\{I\}) \leq \mathrm{rt}(I;A/J) = 1$. In other words $I^n \cap J = I^{n-1}(I \cap J)$ for all $n \geq 1$. In particular, $I^n \cap J \subset I^{n-1}J$ and, if $J \subset I$, $I^n \cap J = I^{n-1}J$. □

Remark that the hypothesis I being generated by a d-sequence modulo J has been weakened to I being of linear type modulo J and that the hypothesis I being of linear type is not needed. With respect to the result of Wang, we have the following:

Proof of Proposition 5.4: If I is of linear type (for instance, if I is generated by a regular sequence), $\mathrm{rt}(I;A) = 1$. If $I^n \cap J = I^{n-s}(I^s \cap J)$ for all $n \geq s$, then $s(J,A;\{I\}) \leq s$. From the second inequality in (3), we obtain that $\mathrm{rt}(I;A/J) \leq \max(\mathrm{rt}(I;A), s(J,A;\{I\})) \leq s$. So $\mathrm{rt}(I+J/J) \leq s$. □

Before giving a proof of the result of Lai, let us introduce the following useful notation which is in some sense implicit in everything we have been doing before.

Given a commutative ring A and an A-module M, the uniform relation type of M with respect to a set of ideals $\mathcal{W}$ of A is defined to be $\mathrm{urt}(M;\mathcal{W}) = \sup\{\mathrm{rt}(I;M) \mid I \in \mathcal{W}\}$. If $\mathcal{W}$ is the set of all ideals of A, then the phrase "with respect to $\mathcal{W}$" is deleted and we simply write $\mathrm{urt}(M)$. Taking suprema in the key inequalities (3), one has

$$s(N,M;\mathcal{W}) \leq \mathrm{urt}(M/N;\mathcal{W}) \leq \max(\mathrm{urt}(M;\mathcal{W}), s(N,M;\mathcal{W})). \tag{4}$$

In other words, one way to prove a strong uniform Artin-Rees property for a pair (N,M) with respect to a set of ideals $\mathcal{W}$ consists in bounding above the uniform relation type of the quotient M/N with respect to the same set of ideals.

Now, we prove the result of Lai:

Proof of Theorem 5.3: Let $\mathcal{V}$ be the set of parameter ideals I of A such that $I+J/J$ is a parameter ideal of A/J and let $\mathcal{W}$ be the set of ideals I of A such that $I+J/J$ is a parameter ideal of A/J. Clearly $\mathcal{V} \subset \mathcal{W}$ and $s(J,A;\mathcal{V}) \leq s(J,A;\mathcal{W})$. If $I \in \mathcal{W}$, then, by prime avoidance, one can find an ideal K of A such that K is generated by a system of parameters of A and $K+J/J = I+J/J$. In particular $\text{rt}(I+J/J) = \text{rt}(K+J/J)$ and $\text{urt}(A/J;\mathcal{W}) = \text{urt}(A/J;\mathcal{V})$. On the other hand, since A is a Buchsbaum local ring, wherein parameter ideals are generated by d-sequences, $\text{urt}(A;\mathcal{V}) = 1$. Using the inequalities (4), we obtain:

$$s(J,A;\mathcal{V}) \leq s(J,A;\mathcal{W}) \leq \text{urt}(A/J;\mathcal{W}) = \text{urt}(A/J;\mathcal{V}) \text{ and}$$
$$\text{urt}(A/J;\mathcal{V}) \leq \max(\text{urt}(A;\mathcal{V}), s(J,A;\mathcal{V})) = s(J,A;\mathcal{V}).$$

Thus $s(J,A;\mathcal{W}) = \text{urt}(A/J;\mathcal{W})$, proving the equivalence between the strong uniform Artin-Rees property for the pair (J,A) with respect to the set of parameter ideals modulo J and the relation type conjecture for the ring A/J. □

From now on we will focus our attention on bounding the Uniform Relation Type.

8 Uniform Relation Type in dimension 1

Since the strong uniform Artin-Rees property cannot hold for the class of all ideals, there will not be in general a uniform bound for the relation type of all ideals of a ring. So the first question is to characterize which are the rings A where $\text{urt}(A)$ is finite. A particular case was solved by Costa in [**5**] where he proved that for a commutative (not necessarily noetherian local) domain A, $\text{urt}(A) = 1$ is equivalent to A being a Prüfer domain. And a little more generally, commutative rings A with $\text{urt}(A) = 1$ are known to be the rings of weak dimension one or less ([**23**]). In particular, for a notherian local ring A, $\text{urt}(A) = 1$ if and only if A is a discrete valuation ring or a field. But if the intention is to characterize just the finiteness of $\text{urt}(A)$, then the works of Cohen [**4**] and Sally [**30**], wherein noetherian local rings with a uniform bound on the number of generators of all ideals are characterized as the rings of Krull dimension at most one, might be a source of inspiration. And indeed, the following result goes in the same direction.

Theorem 8.1 (see [25]) *Let A be an excellent ring. The following conditions are equivalent:*

(i) $\operatorname{urt}(M) < \infty$ *for all finitely generated A-modules M.*

(ii) $\operatorname{urt}(A) < \infty$.

(iii) *There exists an integer $s \geq 1$ such that, for every three-generated ideal I of A, the relation type of I is bounded by s.*

(iv) $\dim A \leq 1$.

In particular, and using the inequalities (4), we have the strong uniform Artin-Rees property in codimension one.

Corollary 8.2 (see [25]) *Let A be an excellent ring and let $N \subseteq M$ be two finitely generated A-modules such that* $\dim(M/N) \leq 1$. *Then there exists an integer $s \geq 1$ such that, for all integers $n \geq s$ and for all ideals I of A,*

$$I^n M \cap N = I^{n-s}(I^s M \cap N)\,.$$

With respect to the proof of Theorem 8.1, it is clear that $(i) \Rightarrow (ii) \Rightarrow (iii)$. The proof of $(iii) \Rightarrow (iv)$ takes into account that in a 2-dimensional (or greater) noetherian local ring $(A, \mathfrak{m})$, there exist two $\mathfrak{m}$-independent elements $x, y \in A$ (see, for instance, [**37**]). With these two elements, one can consider the family of ideals of A defined by $I_r = (x^r, y^r, x^{r-1}y)$ (remark that, modulo J, it is the same family considered by Wang in Example 5.5). Then, it can easily be checked that the module of effective r-relations of I_r is generated by the class of the polynomial $T_1^{r-1}T_2 - T_3^r$ and that the relation type of I_r is exactly r. The proof of $(iv) \Rightarrow (i)$ consists in reducing the problem to a 1-dimensional maximal Cohen-Macaulay module. In this case, if I is an $\mathfrak{m}$-primary ideal, then the relation type of I with respect to M is bounded above by $e(A)$, the multiplicity of A (see [**25**] for the details). For instance, if A is a 1-dimensional noetherian local reduced ring, one proves that $\operatorname{urt}(A) \leq e(A) + 1$. And if A is a domain, then $\operatorname{urt}(A) \leq e(A)$ (see, for instance, [**25**] and [**31**] for more examples).

9 Uniform Relation Type with respect to maximal ideals

Having in mind the strong uniform Artin-Rees property with respect to some sets of ideals, it was shown in [**26**] that the uniform relation type with respect to the set of all maximal ideals of an excellent ring is finite (see also [**33**] for another proof).

Theorem 9.1 (see [33, 26]) *Let A be an excellent ring and let M be a finitely generated A-module. Then there exists an integer $s \geq 1$ such that, for all maximal ideals $\mathfrak{m}$ of A, the relation type of $\mathfrak{m}$ with respect to M satisfies*

$$\mathrm{rt}(\mathfrak{m};M) \leq s.$$

In particular, and using again the inequalities (4), one can recover Theorem 2.3 due to Duncan and O'Carroll.

Corollary 9.2 (see [6]) *Let A be an excellent ring and let $N \subseteq M$ be two finitely generated A-modules. Then there exists an integer $s \geq 1$ such that, for all integers $n \geq s$ and for all maximal ideals $\mathfrak{m}$ of A,*

$$\mathfrak{m}^n M \cap N = \mathfrak{m}^{n-s}(\mathfrak{m}^s M \cap N).$$

The proof of Theorem 9.1 has three main steps. The first is a characterization of when the natural surjective morphism between the tensor product of two associated graded rings with the associated graded ring of the sum is an isomorphism, in terms of normal transversality (see also [**39**], pages 130 and following). Concretely, if I and J are two ideals of a noetherian ring A, it is shown that $\mathcal{G}(I) \otimes \mathcal{G}(J) \simeq \mathcal{G}(I+J)$ if and only if $\mathrm{Tor}_i(A/I^p, A/J^q) = 0$, for all integers $p, q \geq 1$ and for the integers $i = 1, 2$. In the second step, one proves that the relation type of the tensor product of two standard modules is bounded above by the maximun of the relation type of each one, that is, $\mathrm{rt}(F \otimes G) \leq \max(\mathrm{rt}(F), \mathrm{rt}(G))$, being F and G two standard modules. Then the third step is obtained from the first two. In fact one deduces that if $(A, \mathfrak{m})$ is a noetherian local ring, if M is a finitely generated A-module and if $\mathfrak{p}$ is a prime ideal such that $A/\mathfrak{p}$ is regular local and $\mathcal{G}(\mathfrak{p})$ and $\mathcal{G}(\mathfrak{p};M)$ are free $A/\mathfrak{p}$-modules, then

$$\mathrm{rt}(\mathfrak{m}, M) \leq \mathrm{rt}(\mathfrak{p};M).$$

To finish the whole proof, one constructs, by using an argument involving generic flatness (see Theorem 22.A [**16**]), a cover of the spectrum of the ring A by finitely many locally closed sets where the hypotheses of the third step are fulfilled.

10 Uniform Relation Type with respect to principal ideals

Another set of ideals in which there is a known bound for the uniform relation type is the set of principal ideals of a noetherian ring.

Theorem 10.1 (see [25]) *Let A be an excellent ring and let M be a finitely generated A-module. Then there exists an integer $s \geq 1$ such that, for all element $x \in A$, the relation type of (x) with respect to M satisfies*

$$\mathrm{rt}((x);M) \leq s.$$

And from inequalities (4), one can deduce Theorem 2.4 due to O'Carroll.

Corollary 10.2 (see [20]) *Let A be a noetherian ring and let $N \subseteq M$ be two finitely generated A-modules. Then there exists an integer $s \geq 1$ such that, for all integers $n \geq s$ and for all elements x of A,*

$$x^n M \cap N = x^{n-s}(x^s M \cap N).$$

The proof of Theorem 10.1 is essentially the same as the one by O'Carroll in his work ([**20**]). It consists in taking a minimal primary decomposition of 0 in M, $Q_1 \cap \ldots \cap Q_r = 0$, and then realizing that $\mathrm{rt}((x);M) \leq s$, for an integer $s \geq 1$ such that every Q_i contains $r_M(Q_i)^s M$.

Acknowledgments

I would like to thank the organizers of the Lisbon Conference on Commutative Algebra for giving me the opportunity to participate and enjoy the conference. Special thanks are due to Ana Luisa Correia, Maria Vaz Pinto and Santiago Zarzuela for their willingness to help and provide a very comfortable atmosphere for work in this wonderful and friendly city. To all of them: muito obrigado. I am also grateful to Philippe Gimenez, Cristodor Ionescu, José M. Giral and Wolmer V. Vasconcelos for their helpful comments on the subject of this talk. Finally, I would also like to thank the referee for the careful reading of this paper. This work has been partially supported by the BFM2001-3584 grant.

References

[1] I.M. Aberbach, L. Ghezzi and H.T. Hà, Homology multipliers and the relation type of parameter ideals, preprint.

[2] M.F. Atiyah and I.G. Macdonald, *Introduction to Commutative Algebra*, Addison-Wesley Publishing Co., Reading, 1969.

[3] N. Bourbaki, *Commutative Algebra: Chapters 1-7* (translated from the French), Elements of Mathematics, Springer-Verlag, Berlin, 1989.

[4] I.S. Cohen, Commutative rings with restricted minimum condition, *Duke Math. J.* **17** (1950), 27–42.

[5] D. Costa, On the torsion-freeness of the symmetric powers of an ideal, *J. Algebra* **80** (1983), 152–158.

[6] A.J. Duncan and L. O'Carroll, A full uniform Artin-Rees theorem, *J. Reine Angew. Math.* **394** (1989), 203–207.

[7] D. Eisenbud and M. Hochster, A Nullstellensatz with nilpotents and Zariski's main lemma on holomorphic functions, *J. Algebra* **58** (1979), 157–161.

[8] J. Herzog, A. Simis and W.V. Vasconcelos, Koszul homology and blowing-up rings, *Lecture Notes in Pure and Appl. Math.* **84** (1982), 79–169, Marcel Dekker, New York.

[9] C. Huneke, On the symmetric and Rees algebra of an ideal generated by a d-sequence, *J. Algebra* **62** (1980), 268–275.

[10] C. Huneke, The theory of d-sequences and powers of ideals, *Adv. Math.* **46** (1982), 249–279.

[11] C. Huneke, Uniform bounds in noetherian rings, *Invent. Math.* **107** (1992), 203-223.

[12] C. Huneke, Tight closure and its applications, *CBMS Lecture Notes* **88** (1996), American Mathematical Society, Providence.

[13] C. Huneke, Desingularizations and the uniform Artin-Rees theorem, *J. London Math. Soc.* **62** (2000), 740–756.

[14] B. Johnson and D. Katz, On the relation type of large powers of an ideal, *Mathematika* **41** (1994), 209–214.

[15] Y.H. Lai, On the relation type of systems of parameters, *J. Algebra* **175** (1995), 339–358.

[16] H. Matsumura, *Commutative Algebra*, Second Edition, Mathematics Lecture Note Series, Reading, Massachusetts 1980.

[17] H. Matsumura, *Commutative Ring Theory*, Cambridge University Press, 1986.

[18] M. Nagata, *Local Rings*, Interscience, 1962.

[19] L. O'Carroll, A uniform Artin-Rees theorem and Zariski's main lemma on holomorphic functions, *Invent. Math.* **90** (1987), 674–682.

[20] L. O'Carroll, A note on Artin-Rees numbers, *Bull. London Math. Soc.* **23** (1991), 209–212.

[21] L. O'Carroll, Addendum to: A note on Artin-Rees numbers, *Bull. London Math. Soc.* **23** (1991), 555–556.

[22] A. Ooishi, Genera and arithmetic genera of commutative rings, *Hiroshima Math. J.* **17** (1987), 47–66.

[23] F. Planas-Vilanova, Rings of weak dimension one and syzygetic ideals, *Proc. Amer. Math. Soc.* **124** (1996), 3015–3017.

[24] F. Planas-Vilanova, On the module of effective relations of a standard algebra, *Math. Proc. Cambridge Philos. Soc.* **124** (1998), 215–229.

[25] F. Planas-Vilanova, The strong uniform Artin-Rees property in codimension one, *J. Reine Angew. Math.* **527** (2000), 185–201.

[26] F. Planas-Vilanova, Normal transversality and uniform bounds, *J. London Math. Soc.* **62** (2000), 716–728.

[27] K.N. Raghavan, Uniform annihilation of local cohomology and of Koszul homology, *Math. Proc. Cambridge Philos. Soc.* **112** (1992), 487–494.

[28] K.N Raghavan, Powers of ideals generated by quadratic sequences, *Trans. Amer. Math. Soc.* **343** (1994), 727–747.

[29] D. Rees, Two classical theorems of ideal theory, *Proc. Cambridge Philos. Soc.* **52** (1956), 155–157.

[30] J.D. Sally, *Number of Generators of Ideals in Local Rings, Lecture Notes in Pure and Appl. Math.* **35** (1978), Marcel Dekker, New York.

[31] J.D. Sally and W.V. Vasconcelos, Stable rings, *J. Pure Appl. Algebra* **4** (1974), 319–336.

[32] P. Schenzel, Castelnuovo's index of regularity and reduction numbers, *Banach Center Publications* **26** (1990), 201–208.

[33] V. Trivedi, Hilbert functions, Castelnuovo-Mumford regularity and uniform Artin-Rees numbers, *Manuscripta Math.* **94** (1997), 485–499.

[34] N.V. Trung, Absolutely superficial elements, *Math. Proc. Cambridge Philos. Soc.* **93** (1983), 35–47.

[35] N.V. Trung, Reduction exponent and degree bound for the defining equations of graded rings, *Proc. Amer. Math. Soc.* **101** (1987), 229–236.

[**36**] N.V. Trung, The Castelnuovo regularity of the Rees algebra and the associated graded ring, *Trans. Amer. Math. Soc.* **350** (1998), 2813–2832.

[**37**] G. Valla, Elementi independenti rispetto ad un ideale, *Rend. Sem. Mat. Univ. Padova* **44** (1970), 339–354.

[**38**] G. Valla, On the symmetric and Rees algebras of an ideal, *Manuscripta Math.* **30** (1980), 239–255.

[**39**] W.V. Vasconcelos, *Arithmetic of Blowup Algebras*, Cambridge University Press, Cambridge, 1994.

[**40**] W.V. Vasconcelos, *Computational Methods in Commutative Algebra and Algebraic Geometry*, Springer-Verlag, Berlin, 1998.

[**41**] H.J. Wang, Some uniform properties of 2-dimensional local rings, *J. Algebra* **188** (1997), 1–15.

[**42**] H.J. Wang, The Relation-Type Conjecture holds for rings with finite local cohomology, *Comm. Algebra* **25** (1997), 785–801.

[**43**] O. Zariski, A fundamental lemma from the theory of holomorphic functions on an algebraic variety, *Ann. Mat. Pura Appl.* **29** (1949), 187–198.

Castelnuovo-Mumford regularity and finiteness of Hilbert functions

MARIA EVELINA ROSSI
Dipartimento di Matematica, Università di Genova
Via Dodecaneso 35, I-16146 Genova, Italia
E-mail: rossim@dima.unige.it

NGÔ VIÊT TRUNG
Institute of Mathematics, Vien Toan Hoc
18 Hoang Quoc Viet, 10307 Hanoi, Vietnam
E-mail: nvtrung@math.ac.vn

GIUSEPPE VALLA
Dipartimento di Matematica, Università di Genova
Via Dodecaneso 35, I-16146 Genova, Italia
E-mail: valla@dima.unige.it

Introduction

The Castelnuovo-Mumford regularity is a very important invariant of graded modules which arises naturally in the study of finite free resolutions. There have been several results which establish bounds for the Castelnuovo-Mumford regularity of projective schemes in terms of numerical characters. Unfortunately, these invariants are often difficult to handle and the problem of finding good bounds in terms of simpler invariants is a topic featured in much recent research.

The notion of regularity has been used by S. Kleiman in the construction of bounded families of ideals or sheaves with given Hilbert polynomial, a crucial point in the construction of a Hilbert or Picard scheme. In a related direction, Kleiman proved that if I is an equidimensional reduced ideal in a polynomial ring S over an algebraically closed field, then the coefficients of the Hilbert polynomial of $R = S/I$ can be bounded by the dimension and the multiplicity of R (see [**11**, Corollary 6.11]). Srinivas and Trivedi proved that the corresponding result does not hold for a local domain. However, they proved that there exist a finite number of Hilbert functions for a local Cohen-Macaulay ring of given multiplicity and dimension (see [**19**]). The proofs of the above results are very difficult and involve deep results from algebraic geometry.

The aim of this paper is to introduce a unified approach which gives more general results and easier proofs of the above mentioned results. This approach is based on a classical result of Mumford on the behavior of the geometric regularity by hyperplane sections in [**14**], which allows us to bound the regularity once there is a bound for the size of certain component of the first local cohomology modules or a uniform bound for the Hilbert polynomials. The finiteness of the Hilbert functions then follows from the boundness of the regularity. More precisely, we shall see that a class C of standard graded algebras has a finite number of Hilbert functions if and only if there are upper bounds for the regularity and the embedding dimension of the members of C.

In Section 1 we will prepare some preliminary facts on Castelnuovo-Mumford regularity and related notions such as weak regularity and geometric regularity. In Section 2 we will clarify the relationship between the boundedness of Castelnuovo-regularity (resp. geometric regularity) and the finiteness of Hilbert functions (resp. Hilbert polynomials). The main technique is the theory of Gröbner basis.

In Section 3 we will show that the boundedness of the regularity of classes of graded algebras with positive depth can be deduced from the boundedness of the dimension of the zero-graded component of the first local cohomology modules. As a consequence, Kleiman's result follows from the well-known fact that for equidimensional reduced schemes, the dimension of the first sheaf cohomology module is equal to the number of the connected components minus one. We also give an example showing that Kleiman's result does not hold if the graded algebras are not equidimensional.

The local case is studied in Section 4. First we present a local version of Mumford's Theorem, which allows us to reduce the problem of bounding the regularity of the tangent cone to the problem of bounding the Hilbert polynomial. Such a bound for the Hilbert polynomials was already established in [**17**, **22**]. From this we can easily deduce Srinivas and Trivedi's result on the finiteness of the Hilbert functions of Cohen-Macaulay local rings with given dimesion and multiplicity. This approach can be used to prove the more general result that the number of numerical functions which can arise as the Hilbert functions of local rings with given dimension and extended degree is finite (see [**16**] for details).

1 Variations on Castelnuovo-Mumford regularity

Throughout this paper let $S = k[x_1, \dots, x_r]$ be a polynomial ring over a field k. Let $M = \oplus_t M_t$ be a finitely generated graded S-module and let

$$0 \to F_s \longrightarrow \cdots \longrightarrow F_1 \longrightarrow F_0 \longrightarrow M \to 0$$

be a minimal graded free resolution of M as S-module. Write b_i for the maximum of the degrees of the generators of F_i. Following [**6**], Section 20.5, we say that M is *m-regular* for some integer m if $b_j - j \le m$ for all j. The *Castelnuovo-Mumford regularity* $\operatorname{reg}(M)$ of M is defined to be the least integer m for which M is m-regular, that is,

$$\operatorname{reg}(M) = \max\{b_i - i |\ i = 0, \dots, s\}.$$

It is well known that M is m-regular if and only if $\operatorname{Ext}^i_S(M,S)_n = 0$ for all i and all $n \le -m-i-1$ (see [**7**]). This result is hard to apply because in principle infinitely many conditions must be checked. However, in some cases, it suffices to check just a few. We say that M is *weakly m-regular* if $\operatorname{Ext}^i_S(M,S)_{-m-i-1} = 0$ for all $i \ge 0$. Due to a result of Mumford, if $\operatorname{depth} M > 0$ and M is weakly m-regular, then M is m-regular (see [**6**, 20.18]). From this we can easily deduce the following result (see [**16**, Corollary 1.2]).

Proposition 1.1 *Let $R = S/I$ be a standard graded algebra and m a non-negative integer. If R is weakly m-regular, then R is m-regular.*

Using local duality we can characterize these notions of regularity by means of the local cohomology modules of M. Let $\mathfrak{I}$ denote the maximal graded ideal of S and let M be a finitely generated graded S-module. For any integer i we denote by $H^i_{\mathfrak{I}}(M)$ the i-th local cohomology module of M with respect to $\mathfrak{I}$. By local duality (see [**6**], A4.2) we have

$$H^i_{\mathfrak{I}}(M)_m \cong \mathrm{Ext}_S^{r-i}(M,S)_{-m-r}$$

for all i and m. Thus, M is m-regular if and only if $H^i_{\mathfrak{I}}(M)_n = 0$ for all i and $n \geq m-i+1$, and M is weakly m-regular if and only if $H^i_{\mathfrak{I}}(M)_{m-i+1} = 0$ for all i. In particular, reg(M) is the least integer m for which $H^i_{\mathfrak{I}}(M)_n = 0$ for all i and $n \geq m-i+1$. Hence the Castelnuovo-Mumford regularity can be defined for any finite S-module regardless of its presentation. For any integer i we set $a_i(M) := \max\{n |\ H^i_{\mathfrak{I}}(M)_n \neq 0\}$, where $a_i(M) = -\infty$ if $H^i_{\mathfrak{I}}(M) = 0$. Then

$$\mathrm{reg}(M) = \max\{a_i(M) + i |\ i \geq 0\}.$$

A relevant remark is that the Castelnuovo-Mumford regularity controls the behavior of the Hilbert function. We recall that the Hilbert function of M is the numerical function

$$h_M(t) = \dim_k(M_t).$$

The Hilbert polynomial $p_M(X)$ of M is the polynomial of degree $d-1$ such that we have $h_M(t) = p_M(t)$ for $t \gg 0$.

The Hilbert function $h_M(n)$ and the Hilbert polynomial $p_M(n)$ are related by the formula $h_M(n) = p_M(n)$ for $n > \mathrm{reg}(M)$. This is a consequence of the Serre formula which holds for every integer n:

$$h_M(n) - p_M(n) = \sum_{i \geq 0} (-1)^i \dim_k H^i_{\mathfrak{I}}(M)_n.$$

For a proof, see for example [**3**, Theorem 4.4.3].

Inspired by the notion of regularity for sheaves on projective spaces, it is natural to introduce the following weaker notion of regularity. We say that M is *geometrically m-regular* if $H^i_{\mathfrak{I}}(M)_n = 0$ for all $i > 0$ and $n \geq m-i+1$, and we define the *geometric regularity* g-reg(M) of M to be the least integer m for which M is geometrically m-regular. It is clear that

$$\text{g-reg}(M) = \max\{a_i(M) + i \mid i > 0\}.$$

Hence we always have

$$\text{g-reg}(M) = \text{reg}(M/H^0_{\mathfrak{S}}(M)) \le \text{reg}(M).$$

In particular, reg(M) = g-reg(M) if depth$M > 0$. For a standard graded algebra $R = S/I$, a theorem of Gotzmann gives an upper bound for the geometric regularity in terms of an integer which can be computed from the Hilbert polynomial of R.

Theorem 1.2 (see [9]) *Assume that*

$$p_R(n) = \binom{n+a_1}{a_1} + \binom{n+a_2-1}{a_2} + \cdots + \binom{n+a_s-(s-1)}{a_s}$$

with $a_1 \ge a_2 \ge \cdots \ge a_s \ge 0$. *Then*

$$\text{g-reg}(R) = \text{reg}(S/I^{sat}) \le s-1.$$

For example, if R has dimension 1 and multiplicity e, then its Hilbert polynomial is

$$p_R(n) = e = \binom{n}{0} + \binom{n-1}{0} + \cdots + \binom{n-(e-1)}{0}$$

so that g-reg$(R) \le e-1$. In particular, if R is Cohen-Macaulay of dimension 1 and multiplicity e, then reg$(R) \le e-1$.

Unlike the regularity, the geometric regularity does not behave well under generic hyperplane sections. Consider for example the ring $R = k[x,y,z]/(x^2,xy)$. Then g-reg(R) = reg$(R) = 1$ while g-reg$(R/zR) = 0$, reg$(R/zR) = 1$. However the following result of Mumford (see [**14**, page 101, Theorem]) gives us the possibility to control this behavior. It will be the basic result for our further investigation on the Castelnuovo-Mumford regularity of a standard graded algebra R. A ring theoretic proof can be found in [**16**, Theorem 1.4].

Theorem 1.3 *Let* $R = S/I$ *be a standard graded algebra and* $z \in R_1$ *a regular linear form in* R. *If* g-reg$(R/zR) \le m$, *then*

(a) *for every* $s \geq m+1$,

$$\dim_k H^1_{\mathfrak{J}}(R)_m = \dim_k H^1_{\mathfrak{J}}(R)_s + \sum_{j=m+1}^{s} \dim_k H^0_{\mathfrak{J}}(R/zR)_j,$$

(b) $\operatorname{reg}(R) \leq m + \dim_k H^1_{\mathfrak{J}}(R)_m$,

(c) $\dim_k H^1_{\mathfrak{J}}(R)_t = p_R(t) - h_R(t)$ *for every* $t \geq m-1$.

2 Finiteness of Hilbert functions

The aim of this section is to clarify how the finiteness of Hilbert functions for a given class of standard graded algebras is related to the Castelnuovo-Mumford regularity of the members of the class. In the following we let $R = S/I$ where $S = k[x_1,\dots,x_r]$ is a polynomial ring over an infinite field k of any characteristic and I a homogeneous ideal of S. We will denote by $\operatorname{embdim}(R)$ its embedding dimension, that is $h_R(1)$. It is well known that

$$h_{S/I}(t) = h_{S/\operatorname{in}(I)}(t)$$

for all $t \geq 0$ where $\operatorname{in}(I)$ denotes the initial ideal of I with respect to some term order. Therefore, we can go on to our study to Hilbert function of factor rings by initial ideal.

On the other hand, we have the following basic result of Bayer and Stillman on the behavior of the regularity when passing to initial ideals.

Proposition 2.1 (see [1]) *Let* $R = S/I$ *be a standard graded algebra. Then*

$$\operatorname{reg}(R) = \operatorname{reg}(S/\operatorname{gin}(I)),$$

where $\operatorname{gin}(I)$ *is the generic initial ideal of* I *with respect to the reverse lexicographic order.*

Moreover, it follows from a result of Bigatti and Hulett (for $\operatorname{char}(k) = 0$) and of Pardue (for $\operatorname{char}(k) > 0$) that among all the ideals with the same Hilbert functions, the lex-segment ideal has the largest regularity. Recall that the lex-segment ideal $\operatorname{Lex}(I)$ of I is the monomial ideal which is generated in every degree t by the first $h_I(t)$ monomials in the lexicographical order.

Proposition 2.2 (see [2, 10, 15]) *Let $R = S/I$ be a standard graded algebra. Then*

$$\operatorname{reg}(R) \leq \operatorname{reg}(S/\operatorname{Lex}(I)).$$

In the following we will employ the following notations. Let C be a class of standard graded algebras. We say:

- C *is HF-finite* if the number of numerical functions which arise as the Hilbert functions of $R \in C$ is finite,
- C is *HP-finite* if the number of polynomials which arise as the Hilbert polynomials of $R \in C$ is finite,
- C is *reg-bounded* if there exists an integer t such that $\operatorname{reg}(R) \leq t$ for all $R \in C$,
- C is *g-reg-bounded* if there exists an integer t such that $\text{g-reg}(R) \leq t$ for all $R \in C$,
- C is *embdim-bounded* if there exists an integer t such that $\operatorname{embdim}(R) \leq t$ for all $R \in C$.

Moreover, we will denote by $\overline{C}$ the class of graded algebras of the form $\bar{R} := R/H^0_{\mathfrak{F}}(R)$ for $R \in C$. Note that

$$h_R(t) = h_{\bar{R}}(t)$$

for $t \gg 0$ and that

$$\text{g-reg}(R) = \text{g-reg}(\bar{R}) = \operatorname{reg}(\bar{R}).$$

Then C is HP-finite if and only if $\overline{C}$ is HP-finite and C is g-reg-bounded if and only if $\overline{C}$ is reg-bounded.

Theorem 2.3 *Let C be a class of standard graded algebras. Then*

(a) *C is HF-finite if and only if C is reg-bounded and embdim-bounded,*

(b) *C is HP-finite if and only if C is g-reg-bounded and $\overline{C}$ is embdim-bounded.*

Proof: We may assume that the base field is infinite by tensoring it with a transcendental extension.

(a) If C is HF-finite, then C is embdim-bounded. By Proposition 2.2, for every $S/I \in C$ we have $\operatorname{reg}(S/I) \leq \operatorname{reg}(S/Lex(I))$. Since there is a finite number of possible Hilbert functions for S/I, there is also a finite number of lex-segment ideals for I, so that C is reg-bounded by Proposition 2.2. Conversely, assume C is reg-bounded and embdim-bounded. By Proposition 2.1, for all $R = S/I \in C$ we have

$$\operatorname{reg}(S/I) = \operatorname{reg}(S/\operatorname{gin}(I)) \geq D - 1,$$

where D is the maximum degree of the monomials in the standard set of generators of $\operatorname{gin}(I)$. Since the embedding dimension is bounded, the number of monomials of degree smaller than or equal to D in S is finite. Thus, there is only a finite number of possibilities for $\operatorname{gin}(I)$ and hence also for the Hilbert functions of S/I because $h_{S/I}(t) = h_{S/\operatorname{gin}(I)}(t)$.

(b) By Theorem 1.2 we have g-reg$(S/I) \leq s - 1$ where the integer s depends only on the Hilbert polynomial of S/I. Therefore, if C is HP-finite, then C is g-reg-bounded. As remarked above, $\overline{C}$ is reg-bounded. Furthermore, for every $R \in \overline{C}$ we have $h_R(1) \leq h_R(n) = p_R(n)$ for $n = \operatorname{reg}(R)$. Since $\overline{C}$ is HP-finite, there is only a finite number of Hilbert polynomials $p_R(n)$. Let $t = \max\{p_R(n) | n = \operatorname{reg}(R), R \in \overline{C}\}$. Then $h_R(1) \leq t$ for all $R \in \overline{C}$. Hence $\overline{C}$ is embdim-bounded. Conversely, if C is g-reg-bounded, then $\overline{C}$ is reg-bounded. If moreover $\overline{C}$ is embdim-bounded, then $\overline{C}$ is HF-finite by (a). This implies that C is HP-finite. □

Corollary 2.4 *Let C be a class of standard graded algebras. Then C is HP-finite if and only if $\overline{C}$ is HF-finite.*

The following example shows that $\overline{C}$ being embdim-bounded does not imply C is embdim-bounded. Let C be the class of algebras of the form

$$R_n = k[x_1, \ldots, x_n]/(x_1^2, \ldots, x_{n-1}^2, x_1x_n, \ldots, x_{n-1}x_n)$$

for $n > 0$. Then $\overline{R}_n \cong k[x_n]$. Hence $\overline{C}$ is embdim-bounded while C is not because $\operatorname{embdim} R_n = n$.

3 Reg-bounded algebras and Kleiman's Theorem

The aim of this section is to present a relevant class of algebras which are reg-bounded and HF-finite. As an application, we give a proof of a theorem of Kleiman (see [**11**, Corollary 6.11]), which says that the class of graded reduced and equidimensional algebras with given multiplicity and dimension is reg-bounded and HF-finite. The main tool is the aforementioned result of Mumford (Theorem 1.3). For every integer $p \geq 1$ we define recursively the following polynomials $F_p(X)$ with rational coefficients. We let

$$F_1(X) := X$$

and, if $p \geq 2$, then we let

$$F_p(X) := F_{p-1}(X) + X\binom{F_{p-1}(X) + p - 1}{p-1}.$$

Theorem 3.1 *Let C be a class of standard graded algebras with the following properties:*

1. $\operatorname{depth}(R) > 0$ *for every* $R \in C$,
2. *for every* $R \in C$ *with* $\dim R \geq 2$, *there exists some regular linear form* $x \in R$ *such that* $R/(x)^{sat} \in C$,
3. *there exists an integer* t *such that* $\dim_k H^1_{\mathfrak{Z}}(R)_0 \leq t$ *for every* $R \in C$.

Then for every $R \in C$ *with* $d = \dim R$ *we have*

$$\operatorname{reg}(R) \leq F_d(t).$$

The problem of finding the complexity of the regularity bound would be interesting. Explicit bounds can be found in [**16**].

We need the following observation for the proof of the above theorem.

Lemma 3.2 *Let C be a class of standard graded algebras as in the above theorem. Then for all $j \geq 0$ and for all $R \in C$ we have*

$$\dim_k H^1_{\mathfrak{Z}}(R)_j \leq t\binom{j+d-1}{d-1}.$$

Proof: If $j = 0$, the conclusion holds by the assumption. If $d = 1$, then R is a one-dimensional standard graded algebra of positive depth. Hence $H^0_{\mathfrak{M}}(R) = 0$ so that by Serre formula we get for every j,

$$\dim_k H^1_{\mathfrak{M}}(R)_j = p_R(j) - h_R(j) = e - h_R(j),$$

where e denotes the multiplicity of R. In particular, if $R \in C$ and $j \geq 0$, we have

$$\dim_k H^1_{\mathfrak{M}}(R)_j = e - h_R(j) \leq e - 1 = \dim_k H^1_{\mathfrak{M}}(R)_0 \leq t.$$

Hence the conclusion holds for every $R \in C$ of dimension 1.

Now let $j \geq 1$ and $R \in C$ of dimension $d \geq 2$. We may identify R with the graded flat extension R' of Theorem 3.1 (2). Then there exists a regular element $x \in R_1$ such that $B = R/(x)^{sat} \in C$. It is clear that $H^i(R/xR) = H^i(B)$ for every $i > 0$. From the short exact sequence

$$0 \to R(-1) \xrightarrow{x} R \longrightarrow R/xR \to 0$$

we get for every j an exact sequence

$$\cdots \longrightarrow H^1_{\mathfrak{M}}(R)_{j-1} \longrightarrow H^1_{\mathfrak{M}}(R)_j \longrightarrow H^1_{\mathfrak{M}}(R/xR)_j \longrightarrow \cdots.$$

Hence

$$\dim_k H^1_{\mathfrak{M}}(R)_j \leq \dim_k H^1_{\mathfrak{M}}(R)_{j-1} + \dim_k H^1_{\mathfrak{M}}(R/xR)_j = \dim_k H^1_{\mathfrak{M}}(R)_{j-1} + \dim_k H^1_{\mathfrak{M}}(B)_j.$$

Since $\dim B = d - 1$, we can use the inductive assumption and we get

$$\dim_k H^1_{\mathfrak{M}}(R)_j \leq t\binom{j+d-2}{d-1} + t\binom{j+d-2}{d-2} = t\binom{j+d-1}{d-1}.$$

□

Proof of Theorem 3.1: If $R \in C$ is one-dimensional, then R is Cohen-Macaulay. Using Theorem 1.2 and the proof of Lemma 3.2 we have

$$\operatorname{reg}(R) = \text{g-reg}(R) \leq e - 1 = \dim_k H^1_{\mathfrak{M}}(R)_0 \leq t = F_1(t).$$

If $R \in C$ is of dimension $d \geq 2$, we replace R by the graded flat extension R' of (2). Then there exists a regular element $x \in R_1$ such that $B = R/(x)^{sat} \in C$. We have $\operatorname{depth}(B) \geq 1$ and if we let $m := \operatorname{reg}(B)$, then

$$m = \text{g-reg}(B) = \text{g-reg}(R/xR).$$

By Theorem 1.3 and Lemma 3.2, this implies

$$\operatorname{reg}(R) \le m + \dim_k H^1_{\mathfrak{M}}(R)_m \le m + t\binom{m+d-1}{d-1}.$$

Since $\dim B = d-1$, by induction we have $m \le F_{d-1}(t)$. Thus,

$$\operatorname{reg}(R) \le F_{d-1}(t) + t\binom{F_{d-1}(t)+d-1}{d-1} = F_d(t).$$

□

We want to apply now the above result to the class C of reduced equidimensional graded algebras with given multiplicity e and dimension $d \ge 1$. It is clear that every $R \in C$ has positive depth. Moreover, we have the following easy lemma.

Lemma 3.3 *If R is a reduced equidimensional graded algebra over an algebraically closed field with multiplicity e, then* $\dim_k H^1_{\mathfrak{M}}(R)_0 \le e-1$.

Proof: If $\dim R = 1$, then R is Cohen-Macaulay and, as above, $\dim_k H^1_{\mathfrak{M}}(R)_0 = e-1$. If $\dim R \ge 2$, then $\dim_k H^1_{\mathfrak{M}}(R)_0 = N-1$ where N is the number of the connected components of the corresponding scheme [**13**, Theorem 1.2.6 (b)]. Since $N \le e$, we get $\dim_k H^1_{\mathfrak{M}}(R)_0 \le e-1$. □

We can prove now that Kleiman's result is a particular case of our theorem.

Theorem 3.4 (see [11, Corollary 6.11]) *Let C be the class of reduced equidimensional graded algebras over an algebraically closed field with given multiplicity e and dimension $\le d$, $d \ge 1$. Then C is HF-finite.*

Proof: By Bertini theorem [**8**, Corollary 3.4.14], if $\dim R \ge 2$, there exists a regular linear form $x \in R$ such that $R/(x)^{sat}$ is a reduced equidimensional algebra with $\dim R/(x)^{sat} = \dim R - 1$. Therefore, we may apply Theorem 3.1 and Lemma 3.3 to get $\operatorname{reg}(R) \le F_d(e-1)$ for every R in the class C. So C is reg-bounded. Now we need only to prove that C is embdim-bounded since by Theorem 2.1 (a), these conditions will imply that C is finite. Let $R \in C$ be arbitrary. Write $R = S/I$ where S is a polynomial ring over an algebraically closed field and $I = \bigcap_{i=1}^r \mathfrak{p}_i$ is an intersection of equidimensional ideals $\mathfrak{p}_i$. It is obvious that $\operatorname{embdim} R \le \sum_{i=1}^r \operatorname{embdim} S/\mathfrak{p}_i$ and $r \le e$. On the other hand, we know that $\operatorname{embdim} S/\mathfrak{p}_i \le e_i + d - 1$, where e_i is the multiplicity of $S/\mathfrak{p}_i$. Therefore,

$$\operatorname{embdim} R \le \sum_{i=1}^{r} e_i + d - 1 = e + r(d-1) \le ed.$$

□

Theorem 3.1 does not hold if we delete the assumption that every element of the class is reduced. Take for example the class C of the graded algebras

$$R_n := k[x,y,z,t]/(y^2, xy, x^2, xz^n - yt^n) \ (n \ge 1).$$

Note that $(y^2, xy, x^2, xz^n - yt^n)$ is a primary ideal. We have $\dim(R_n) = 2$ and $e(R_r) = 2$. The minimal free resolution of R_r over $S = k[x,y,z,t]$ is given by

$$0 \to S(-n-3) \to S(-3)^2 \oplus S(-n-2)^2 \to S(-2)^3 \oplus S(-n-1) \to S \to R_n \to 0.$$

Hence $\operatorname{reg}(R_n) = n$. Therefore, C is not reg-bounded and hence not HF-finite. Theorem 3.1 does not hold if we consider reduced graded algebras which are not necessarily equidimensional. Let us consider the class of standard graded algebras

$$R_n := k[x,y,z_1,\dots,z_r]/(x) \cap (y, f_n)$$

where $f_n \in k[z_1,\dots,z_r]$ is an irreducible form of degree n. We have $\dim R_n = r+1$, $e(R_n) = 1$, but $\operatorname{reg}(R_n) = n$.

Remark 3.5 The same arguments (almost word by word) can be used to prove that the class of reduced graded algebras over an algebraically closed field with given arithmetic degree and dimension is HF-finite.

4 The local version of Mumford's Theorem

Let $(A,\mathfrak{m})$ be a local ring of dimension d and multiplicity e. Let

$$G = \operatorname{gr}_{\mathfrak{m}}(A) = \oplus_{n \ge 0}(\mathfrak{m}^n/\mathfrak{m}^{n+1})$$

be the associated graded ring of A. The Hilbert function and the Hilbert polynomial of A are by definition the Hilbert function and the Hilbert polynomial of the standard graded algebra G, namely $h_A(t) := h_G(t) = \lambda(\mathfrak{m}^t/\mathfrak{m}^{t+1})$ and $p_A(t) := p_G(t)$. We will need also the first iterated of these functions and polynomials. So we let

$$h_A^1(t) := h_G^1(t) = \sum_{j=0}^{t} h_G(j) = \lambda(A/\mathfrak{m}^{t+1})$$

and we denote by $p_G^1(t)$ the corresponding polynomial, that is, the polynomial which verifies the equality $h_G^1(t) = p_G^1(t)$ for $t \gg 0$. Let x be a superficial element in $\mathfrak{m}$ and let

$$\bar{G} := \mathrm{gr}_{\mathfrak{m}/xA}(A/xA)$$

be the associated graded ring of the local ring A/xA. Let $x^* = \bar{x} \in (\mathfrak{m}/\mathfrak{m}^2)$ be the initial form of x in G. We can consider the standard graded algebra G/x^*G and compare it with $\overline{G}$. These two algebras are not the same, unless x^* is a regular element in G, but they have the same geometric regularity, namely

$$\text{g-reg}(G/x^*G) = \text{g-reg}(\bar{G}).$$

This has been proved in [**16**, Lemma 2.2]. If A has positive depth, then it is well known that x is a regular element in A and furthermore

$$p_G(t) = p_{\overline{G}}^1(t).$$

In that case, one can prove (see [**16**, Lemma 3.2]) that $\mathrm{reg}(G) = \text{g-reg}(G)$ (even when G does not necessarily have positive depth). With the above notations we present now a local version of Mumford's Theorem 1.3.

Theorem 4.1 *Let $(A,\mathfrak{m})$ be a local ring of dimension $d \geq 2$ and positive depth. Let x be a superficial element in $\mathfrak{m}$ and $m := \mathrm{reg}(\overline{G})$. Then*

$$\mathrm{reg}(G) \leq m + h_{\overline{G}}^1(m).$$

Proof: We have

$$m = \mathrm{reg}(\overline{G}) \geq \text{g-reg}(\overline{G}) = \text{g-reg}(G/x^*G) = \text{g-reg}(G/(H_{G_+}^0(G) + x^*G)),$$

where the last equality follows because $(H^0_{G_+}(G)+x^*G)/x^*G)$ has finite length in G/x^*G. Now we remark that x^* is a regular element in $G/H^0_{G_+}(G)$, hence, by using Mumford's theorem, we get

$$\begin{aligned}\operatorname{reg}(G) = \text{g-reg}(G) = \text{g-reg}(G/(H^0_{G_+}(G)) &\leq m + p_{G/H^0_{G_+}(G)}(m)\\ &= m + p_G(m) = m + p^1_{\overline{G}}(m) = m + h^1_{\overline{G}}(m)\end{aligned}$$

where the equality $p^1_{\overline{G}}(m) = h^1_{\overline{G}}(m)$ is a consequence of the fact that $m = \operatorname{reg}(\overline{G})$. □

Using this local version of Mumford's theorem we can easily deduce the result of Srinivas and Trivedi which says that the number of Hilbert functions of Cohen-Macaulay local rings with fixed dimension and multiplicity is finite (see [**19**]). For that we need the following inequality proved in [**17**] and [**22**].

Proposition 4.2 *Let $(A,\mathfrak{m})$ be a local ring of dimension $d \geq 1$ and J an ideal generated by a system of parameters in $\mathfrak{m}$. Then*

$$H_A(n) \leq \ell(A/J)\binom{n+d-2}{d-1} + \binom{n+d-2}{d-2}.$$

If A is a d-dimensional Cohen-Macaulay ring of multiplicity e, then from the above proposition we immediately get the inequality

$$H_A(n) \leq e\binom{n+d-2}{d-1} + \binom{n+d-2}{d-2}.$$

For every $d \geq 1$ we define recursively the following polynomials $Q_d(X)$ with rational coefficients. We let

$$Q_1(X) := X - 1$$

and, if $d \geq 2$, then we let

$$Q_d(X) := Q_{d-1}(X) + X\binom{Q_{d-1}(X)+d-2}{d-1} + \binom{Q_{d-1}(X)+d-2}{d-2}.$$

Theorem 4.3 *Let $(A,\mathfrak{m})$ be a Cohen-Macaulay local ring of dimension $d \geq 1$ and multiplicity e. Then*

$$\operatorname{reg}(G) \leq Q_d(e).$$

Proof: If $d = 1$, then $\text{reg}(G) \leq e - 1 = Q_1(e)$. Let $d \geq 2$ and x be a superficial element in $\mathfrak{m}$. Then A/xA is a Cohen-Macaulay local ring of dimension $d-1$ and multiplicity e. By Proposition 4.2 we get

$$H^1_{\overline{G}}(n) \leq e\binom{n+d-2}{d-1} + \binom{n+d-2}{d-2}.$$

By Theorem 4.1 it follows that, if $m = \text{reg}(\overline{G})$, then

$$\text{reg}(G) \leq m + h^1_{\overline{G}}(m) \leq m + e\binom{m+d-2}{d-1} + \binom{m+d-2}{d-2}.$$

By induction we have $m \leq Q_{d-1}(e)$, so that

$$\text{reg}(G) \leq Q_{d-1}(e) + e\binom{Q_{d-1}(e)+d-2}{d-1} + \binom{Q_{d-1}(e)+d-2}{d-2} = Q_d(e).$$

□

Corollary 4.4 *The number of numerical functions which can arise as the Hilbert functions of Cohen-Macaulay local rings with given dimension and multiplicity is finite.*

Proof: By a classical result of Abhyankar we have $v(\mathfrak{m}) \leq e+d-1$, where $v(\mathfrak{m})$ is the embedding dimension of G. Now we need only to apply Proposition 2.3 and the above theorem. □

The analogous of Kleiman result does not hold in the local case. Srinivas and Trivedi gave the following example showing that classes of local domains of fixed dimension and multiplicity need not be HF-finite. Let

$$A_r := k[\![x,y,z,t]\!]/(z^r t^r - xy, x^3 - z^{2r}y, y^3 - t^{2r}x, x^2t^r - y^2z^r).$$

It is easy to see that A_r is a local domain and the associated graded ring of A_r is the standard graded algebra

$$G_r = k[x,y,z,t]/(xy, x^3, y^3, x^2t^r - y^2z^r).$$

We have $\text{reg}(G_r) = r+1$ and

$$H_{A_r}(n) = \begin{cases} 5n-1 & \text{for} \quad n \leq r, \\ 4n+r & \text{for} \quad n > r. \end{cases}$$

Finally, we remark that the above approach can be used to prove that the number of numerical functions which can arise as the Hilbert functions of local rings with given dimension and extended degree is finite. Note that extended degree coincides with the usual multiplicity for Cohen-Macaulay local rings. We refer to [**16**] for details. Furthermore, one can prove similar results for Hilbert functions of finitely generated modules over local rings with respect to $\mathfrak{m}$-primary ideals (see [**20**, **21**, **12**]).

Acknowledgments

The first and third authors are partially supported by MIUR of Italy. The second author is partially supported by the National Basic Research Program of Vietnam.

References

[**1**] D. Bayer and D. Stillman, A criterion for detecting m-regularity, *Invent. Math.* **87** (1987), 1–11

[**2**] A.M. Bigatti, Upper bounds for the Betti numbers of a given Hilbert function, *Comm. Algebra* **21** (1993), 2317–2334.

[**3**] W. Bruns and J. Herzog, *Cohen-Macaulay Rings*, Cambridge Studies in Advanced Mathematics **39**, Cambridge University Press, Cambridge, 1993.

[**4**] M. Brodmann and R. Sharp, *Local Cohomology: An Algebraic Introduction with Geometric Applications*, Cambridge Studies in Advanced Mathematics **60**, Cambridge University Press, Cambridge, 1998.

[**5**] L.R. Doering, T. Gunston and W.V. Vasconcelos, Cohomological degrees and Hilbert functions of graded modules, *Amer. J. Math.* **120** (1998), 493–504.

[**6**] D. Eisenbud, *Commutative Algebra. With a view toward Algebraic Geometry*, Graduate Texts in Mathematics **150**, Springer, 1995.

[**7**] D. Eisenbud and S. Goto, Linear free resolutions and minimal multiplicities, *J. Algebra* **88** (1984), 89–133.

[**8**] H. Flenner, L. O'Carroll and W. Vogel, *Joins and Intersections*, Springer, 1999.

[**9**] G. Gotzmann, Eine Bedingung für die Flachheit und das Hilbertpolynom eines graduierten Ringes, *Math. Z.* **158** (1978), 61–70.

[**10**] H.A. Hulett, Maximal Betti numbers with a given Hilbert function, *Comm. Algebra* **21** (1993), 2335–2350.

[**11**] S. Kleiman, Théorie des intersections et théorème de Riemann-Roch, in: SGA 6, *Lect. Notes in Math.* **225**, Springer, 1971.

[**12**] C.H. Linh, Upper bounds for the Castelnuovo-Mumford regularity of associated graded modules, *Comm. Algebra*, to appear.

[**13**] J. Migliore, *Introduction to Liaison Theory and Deficiency Modules*, Progress in Mathematics **165**, Birkhäuser, Boston, MA, 1998.

[**14**] D. Mumford, *Lectures on Curves on an Algebraic Surfaces*, Princeton Univ. Press, Princeton, 1966.

[**15**] K. Pardue, Deformation of classes of graded modules and maximal Betti numbers, *Illinois J. Math.* **40** (1996), 564–585.

[**16**] M.E. Rossi, N.V. Trung and G. Valla, Cohomological degree and Castelnuovo-Mumford regularity, *Trans. Amer. Math. Soc.* **355** (2003), 1773–1786.

[**17**] M.E. Rossi, G. Valla, and W. Vasconcelos, Maximal Hilbert functions, *Results in Math.* **39** (2001), 99–114.

[**18**] V. Srinivas and V. Trivedi, A finiteness theorems for the Hilbert functions of complete intersection local rings, *Math. Z.* **225** (1997), 543–558.

[**19**] V. Srinivas and V. Trivedi, On the Hilbert function of a Cohen-Macaulay ring, *J. Algebraic Geom.* **6** (1997), 733–751.

[**20**] V. Trivedi, Hilbert functions, Castelnuovo-Mumford regularity and uniform Artin-Rees numbers, *Manuscripta Math.* **94** (1997), 485–499.

[**21**] V. Trivedi, Finiteness of Hilbert fuctions for generalized Cohen-Macaulay modules, *Comm. Algebra* **29** (2001), 805–813.

[**22**] N.V. Trung, Absolutely superficial sequence, *Math. Proc. Camb. Philos. Soc.* **93** (1983), 35–47.

[**23**] W.V. Vasconcelos, The homological degree of a module, *Trans. Amer. Math. Soc.* **350** (1998), 1167–1179.

[**24**] W.V. Vasconcelos, *Cohomological Degrees of Graded Modules. Six Lectures on Commutative Algebra* (Bellaterra, 1996), 345–392, Progress in Mathematics **166**, Birkhäuser, Basel, 1998.

Differential idealizers and algebraic free divisors

ARON SIMIS
Departmento de Matemática, Universidade Federal de Pernambuco, Recife, Brazil
E-mail: aron@dmat.ufpe.br

1 Introduction

The original goal of this paper was to translate the classical results of Saito ([**7**]) on logarithmic vector fields to the polynomial case and subsequently prove them anew by pure ring theory. As it turned out, the project landed on very basic module theory. Moreover, by just going slightly more nonsensical one found a general frame for other questions as well. This explains why parts of the present version look so far apart from the original purpose, for which we expect the reader's indulgence.

Going polynomial in Saito's theory means primevally to replace the ring of germs of analytical functions by a polynomial ring over a field. Thus, Saito's definition of a free divisor ought to be transcribed in terms of a certain module being locally free (i.e., projective) rather than free. If this transcription had been made years ago, then it would naturally stand at that. However, we are now aware for quite some time that projective modules over a polynomial ring over a field are in fact (globally) free. Therefore, to update the theory we decided on a definition in which the module is globally free. Of course, logically it does not matter whether one takes it to be projective or free. From the viewpoint of actually writing down a free basis the situation becomes more challenging than in Saito's local context, since the module makes its début with rather many "natural" generators. To get a free basis, one might resort to the classical theorems of Eisenbud-Evans on generating modules efficiently. Luckily, the subjacent theory makes life a bit easier.

We intentionally stayed away from hastily transcribing those analytical methods depending on analytical transformations — such as the theory of local quasi-homogenous divisors. Of course a transcription is not at all impossible and we would expect some more-versed people to take it up.

Among other things we realized that free divisors which are not smooth can be characterized as those having a codimension two perfect Jacobian ideal (i.e., both the equation of the divisor and its partial derivatives are the maximal minors of a Hilbert-Burch matrix). That may help explain why non-smooth free divisors are rather rare in some sense. After this paper was ready for submission, we found that this had previously been shown by H. Terao in the differential context. We decided to keep our result in the paper for the sake of completeness.

We also transcribed a recent notion established by the "school of Seville," that of a Koszul free divisor. The notion allows for an encore of the (now classical) theory of Cohen-Macaulay symmetric algebras and ideals of linear type. These "old" concepts as applied to the Jacobian ideal of a polynomial may be of renewed interest. Moreover, they provide a fairly simple computational framework to test whether a given free divisor is Koszul free, which may also have some interest.

As to the content of the paper, we divided the presentation into three sections.

In the first section we focus on some basic module theory which yields a general framework, thus providing some general exact sequences that apply in the context and may be of further use in different problems.

In the second section we make the appropriate transcription of the theory of free divisors. At the beginning of the section we also establish other exact sequences tagged to the Jacobian matrix that may be useful in other contexts.

The last section is a short pointer to a curious aspect of the theory, namely, that of classifying free and Koszul free divisors that are irreducible and homogeneous. To our knowledge this has yet not been considered in depth, while it may lead to interesting geometric considerations.

2 Basic module theory

Let R be a commutative ring. Recall that the concept of quotient $I : I'$ of two ideals I, I' of R can be extended to the case of submodules E, E' of a module F, by setting $E : E' = \{h \in \mathrm{End}_R(F) \mid h(E') \subset E\}$. As in the case of ideals, we omit the colon subscript $\mathrm{End}_R(F)$ and, when E' is a cyclic module generated by the element z, we simplify the notation by writing simply $E : z$.

Given a map $G \xrightarrow{\varphi} F$ of finitely generated R-modules, with F free, let there be given an ideal $\mathfrak{A} \subset R$ and consider the composition $\varphi_{[\mathfrak{A}]}$ of φ and the canonical map $F \twoheadrightarrow F/\mathfrak{A}F$. The following easy result is nearly a tautology but it is surprisingly useful.

Lemma 2.1 *Suppose that* $\mathrm{rank}\, F = m$ *and fix a basis* $\{\varepsilon_1, \ldots, \varepsilon_m\}$ *of* F. *Let there be given an element* $z \in F$ *and let* $\mathfrak{A}$ *be the ideal generated by its coordinates in the fixed basis (the "content ideal" of* z*). Then there is an* R*-module isomorphism*

$$\ker \varphi_{\mathfrak{A}} / \ker \varphi \simeq (\mathrm{Im}(\varphi) : z)/(0 : z).$$

Proof: If $y \in \ker \varphi_{[\mathfrak{A}]}$ then $\varphi(y) \in \mathfrak{A}F$, so there are elements $b_{jk} \in R\,(1 \leq j, k \leq m)$ such that $\varphi(y) = \sum_{j,k} b_{jk} a_j \varepsilon_k$, where $\{a_1, \ldots, a_m\}$ are the coordinates of z in the fixed basis. Thinking of the matrix (b_{jk}) as an element $h \in \mathrm{End}_R(F)$ this means that $h \in \mathrm{Im}(\varphi) : z$. Clearly, the class of h modulo $(0 : z)$ is uniquely defined, hence a well defined map $\ker \varphi_{[\mathfrak{A}]} \to (\mathrm{Im}(\varphi) : z)/(0 : z)$ which is an R-homomorphism. By construction the map is surjective and its kernel is $\ker \varphi$. □

The above lemma is a general template for different situations (cf. Proposition 3.1). In its simplest form, namely, when G is also free and $\mathrm{rank}\, F = 1$, it gives part of the next proposition — of course, the proof of the latter could be given directly. Below we will take $F = R$ with its natural basis $\{1\}$. Moreover we focus on the special case where $\mathfrak{A} = (z)$ and z is a regular element of R. In this case, we will shorten $\varphi_{[(f)]}$ (resp. $Z_{[(f)]}$) to $\varphi_{[f]}$ (resp. $Z_{[f]}$).

Proposition 2.2 *Let R be a Noetherian commutative ring, let $f \in R$ be a regular element and let $\varphi\colon G \to R$ be an R-homomorphism, with G a free module of finite rank. Set $I = \mathrm{Im}(\varphi)$, $Z = \ker\varphi$ and $Z_{[f]} = \ker\varphi_{[f]}$.*

(i) *There are exact sequences of R-modules*

$$0 \to Z \to Z_{[f]} \to I\colon (f) \to 0, \tag{1}$$

and of $R/(f)$-modules

$$0 \to Z/fZ \to Z_{[f]}/fG \to \frac{I\colon (f)}{I} \to 0. \tag{2}$$

(ii) *If $\psi\colon = \varphi \oplus f\colon G \oplus R \longrightarrow (I, f)$ then $Z_{[f]} \simeq \ker(\psi)$ and $Z_{[f]}$ is reflexive.*

Proof: (i) The first exact sequence follows immediately from Lemma 2.1 with $z = f$. The second exact sequence follows from the first. Indeed, since we have that $\ker(Z/fZ \to Z_f/fG) = (Z \cap fG)/fZ$ and f is a regular element in R, it follows easily that $\ker(Z/fZ \to Z_f/fG) = 0$. On the other hand, the homomorphism $Z_{[f]} \to I\colon (f)$ maps an element $\mathfrak{z} \in Z_{[f]}$ to $b \in I$, where $\varphi(\mathfrak{z}) = bf$. In particular, if $\mathfrak{z} = fe \in fG \subset Z_{[f]}$, with $e \in G$, then $\varphi(\mathfrak{z}) = \varphi(e)f$, hence $\mathfrak{z}$ is mapped to $\varphi(e)$. Therefore, the restriction to fG maps onto I. This explains the second exact sequence.

(ii) Let $\widetilde{Z_{[f]}} = \ker(\psi)$. It is easy to see that, since f is a regular element, projection of $G \oplus R$ onto the first summand induces a bijection of $\widetilde{Z_{[f]}}$ onto $Z_{[f]}$.

Since f is a regular element, the ideal (I, f) has grade at least one, hence it has a well defined rank (one) as an R-module. Therefore, $\widetilde{Z_{[f]}}$ being a first syzygy of such a module, it too has a well defined rank. But a finitely generated second syzygy with this property is reflexive. □

Corollary 2.3 *With the notation of* Proposition 2.2 *one has:*

(a) *If $f \in I$ then $Z_{[f]} \simeq Z \oplus R$.*

(b) *Suppose that projective R-modules of finite rank are free. If the ideal (I, f) is proper and has grade at least two then $Z_{[f]}$ is a free module if and only (I, f) is a codimension two perfect ideal.*

Proof: (a) The first assertion is obvious since R is free. Note, for future reference, that a splitting map ρ will map 1 to a vector $\mathfrak{z}$ such that $\varphi(\mathfrak{z}) = f$.

(b) If $Z_{[f]}$ is free then (I, f) has projective dimension one, hence it must be a codimension two perfect ideal because it has grade at least two. Conversely, if (I, f) is a codimension two perfect ideal then it has projective dimension one. Since $Z_{[f]}$ is a first syzygy thereof, it must be a projective module (of finite rank) by a well known device, hence it is free. □

For the next result we denote $\mathrm{Hom}_R(E, R) = E^*$, where E is an R-module.

Proposition 2.4 *Keeping the previous notation, set moreover $J = (I, f)$.*

(i) *If* $\mathrm{grade} J \geq 2$, *there is an exact sequence of R-modules*

$$0 \to (G^* \oplus R) \Big/ (\varphi, f)R \longrightarrow (Z_{[f]})^* \longrightarrow \mathrm{Ext}^2_R(R/J, R) \to 0, \qquad (3)$$

where we identify $J^* = R$.

(ii) *If R is a domain and $\{f, f_i\}$ is a regular sequence for every $1 \leq i \leq n$, then there is an exact sequence of R-modules*

$$0 \to G^* + \frac{1}{f}\varphi \cdot R \to (Z_{[f]})^* \longrightarrow \mathrm{Ext}^2_R(R/J, R) \to 0.$$

Proof: (i) This follows by applying $\mathrm{Hom}(\_, R)$ to the exact sequence obtained from identifying $Z_{[f]}$ with $\ker(G \oplus R \to J)$ as in Proposition 2.2 (ii), and using that $\mathrm{grade} J \geq 2$.

(ii) Consider the R-module map $\eta \colon G^* \oplus R \longrightarrow S^{-1}G^*$ (module of fractions with respect to the set of powers of f), where η restricted to G^* is the natural localization map and $\eta((0,1)) = -\frac{1}{f}\varphi$. Let $\{e_1, \ldots, e_n\}$ denote a basis of G such that $\varphi(e_i) = f_i$ and let $\{e_1^*, \ldots, e_n^*\}$ denote its dual basis. Then $\varphi = \sum_i f_i e_i^*$. Writing up the condition for an element of the form $\sum_i g_i e_i^* - g\frac{1}{f}\sum_i f_i e_i^*$ to be zero in G_f^* and using the hypothesis on the pairs f_i, f it is straightforward to see that $\ker(\eta) = (\varphi, f)R$. This shows that $(G^* \oplus R)/(\varphi, f)R \simeq G^* + \frac{1}{f}\varphi \cdot R$. Now apply the previous exact sequence. □

Corollary 2.5 *Assuming the conditions of* Proposition 2.4 (ii), *the following conditions are equivalent:*

(a) $\mathrm{grade}\, J \geq 3$.

(b) $(Z_{[f]})^* = G^* + \frac{1}{f}\varphi \cdot R$.

(c) $G^* + \frac{1}{f}\varphi \cdot R$ *is a reflexive module.*

(d) *There is a natural exact sequence* $0 \to G^* \longrightarrow (Z_{[f]})^* \longrightarrow R/(f) \to 0$.

Furthermore, if one (any) of these conditions holds, then $(Z_{[f]})^*$ *is generated by* $n+1$ *elements and has projective dimension at most one.*

Proof: To see the stated equivalences, note that the 0th Fitting ideal of $(G^* \oplus R)/(\varphi, f)R$ is J. Therefore, this module is reflexive if and only if $\mathrm{grade}\, J \geq 3$.

(a) $\Rightarrow$ (b) If $\mathrm{grade}\, J \geq 3$ then $\mathrm{Ext}^2_R(R/J, R) = 0$, hence $(Z_{[f]})^* = G^* + \frac{1}{f}\varphi \cdot R$ by the exact sequence of Proposition 2.4 (ii).

(b) $\Rightarrow$ (c) This is obvious since $(Z_{[f]})^*$ is reflexive.

(c) $\Rightarrow$ (d) As remarked above, if $G^* + \frac{1}{f}\varphi \cdot R$ is a reflexive module then its 0th Fitting J has grade at least three. Therefore, $\mathrm{Ext}^2_R(R/J, R) = 0$. But, taking duals in the natural exact sequence $0 \to (f) \longrightarrow J \longrightarrow J/(f) \to 0$ yields an exact sequence $0 \to R/(f) \longrightarrow \mathrm{Ext}^1_R(J/(f), R) \longrightarrow \mathrm{Ext}^1_R(J, R) \simeq \mathrm{Ext}^2_R(R/J, R) \to 0$. On the other hand, taking duals in the exact sequence $0 \to Z_{[f]} \longrightarrow G \longrightarrow J/(f) \to 0$ yields an exact sequence

$$0 \to G^* \longrightarrow (Z_{[f]})^* \longrightarrow \mathrm{Ext}^1_R(J/(f), R) \to 0.$$

Therefore, we get an exact sequence as required.

(d) $\Rightarrow$ (a) Since all maps are natural, the reverse argument to the one in the previous implication yields the vanishing of $\mathrm{Ext}^2_R(R/J, R)$. Since we are assuming that $\mathrm{grade}\, J \geq 2$ then it must be the case that $\mathrm{grade}\, J \geq 3$.

The last statement of the proposition follows immediately from the exact sequence (3) or from condition (d). □

3 Differential idealizers

In this section we apply the previous results to a basic differential situation.

3.1 The basic exact sequence of the differential idealizer

Fix a perfect field k of high enough characteristic and let R denote a polynomial ring over k in $n \geq 1$ indeterminates. (Presumably most of the present material could be discussed, without substantial change, over a regular domain essentially of finite type over k.)

We consider the module $\Omega(R) = \Omega(R/k)$ of k-differentials of R, which is a free R-module, and its R-dual, the module $\mathrm{Der}(R) = \mathrm{Der}_k(R)$ of k-derivations of R.

For a nonzero ideal $I \subset R$, define $\mathrm{Der}_I(R) :\, = \{\delta \in \mathrm{Der}(R) \mid \delta(I) \subset I\}$, where $\delta(I)$ means of course that we apply δ to every element of I. This is called the *differential idealizer* of I. Clearly, $\mathrm{Der}_I(R)$ is an R-submodule of $\mathrm{Der}(R)$, hence it is torsion free. It is immediately seen to have rank $n = \dim R$ as it contains the rank n submodule $I\mathrm{Der}(R)$.

Let $\{dx_1, \ldots, dx_n\}$ be the free basis of $\Omega(R/k)$ induced by the universal derivation $d\colon R \to \Omega(R/k)$ and let $\{\frac{\partial}{\partial x_1}, \ldots, \frac{\partial}{\partial x_n}\}$ denote the dual basis to it, a basis of $\mathrm{Der}(R)$. Then

$$\mathrm{Der}_I(R) = \left\{ \sum_{i=1}^{n} g_i \frac{\partial}{\partial x_i} \in \mathrm{Der}(R) \;\middle|\; \sum_{i=1}^{n} g_i \frac{\partial f}{\partial x_i} \in I, \forall f \in I \right\}. \tag{4}$$

The module $\mathrm{Der}_I(R)$ admits important submodules, that is, for any choice of a set of generators $\mathbf{f} = \{f_1, \ldots, f_m\}$ of I, one defines

$$\mathrm{Der}_{\mathbf{f}}(R)^0 = \left\{ \sum_{i=1}^{n} g_i \frac{\partial}{\partial x_i} \in \mathrm{Der}(R) \;\middle|\; \sum_{i=1}^{n} g_i \frac{\partial f_j}{\partial x_i} = 0,\ 1 \leq j \leq m \right\}.$$

In order to describe these modules in more detail, let us fix a set of generators $\mathbf{f} = \{f_1, \ldots, f_m\}$ of I and consider the Jacobian matrix $\Theta = \Theta(\mathbf{f})$ of these generators. Denote by the same symbol the associated R-module homomorphism $\mathrm{Der}(R) \to R^m$ with respect to the canonical basis $\{\varepsilon_1, \ldots, \varepsilon_m\}$ of R^m and the dual basis $\{\frac{\partial}{\partial x_1}, \ldots, \frac{\partial}{\partial x_n}\}$ of $\mathrm{Der}(R)$. In terms of this choice, we can translate the two modules as $\mathrm{Der}_{\mathbf{f}}(R)^0 = \ker(\Theta) \subset \mathrm{Der}_I(R) = \ker(\overline{\Theta})$, where $\overline{\Theta}$ is the composite map of Θ and the natural surjection $R^m \twoheadrightarrow (R/I)^m$. The first of these is a well-known module, namely, the first syzygy of the so-called *Jacobian module* $\mathrm{Im}(\Theta(\mathbf{f}))$ of $\mathbf{f}$, which will be denoted $Z(\Theta)$.

The *Jacobian content module* of $\mathbf{f}$ is the R-module $\mathrm{Im}(\Theta) : \mathbf{f}'$, where $\mathbf{f}' = f_1\varepsilon_1 + \cdots + f_m\varepsilon_m$. Let $Z(\mathbf{f})$ denote the first syzygy module of $\mathbf{f}$. Write $Z(\mathbf{f})^{\oplus m} =$

$Z(\mathbf{f}) \oplus \cdots \oplus Z(\mathbf{f})$ (m-times) viewed as an R-submodule of $\mathrm{End}_R(R^m) \simeq R^m \oplus \cdots \oplus R^m$.

Proposition 3.1 *There is an exact sequence of R-modules*

$$0 \longrightarrow Z(\Theta) \to \mathrm{Der}_I(R) \to \frac{\mathrm{Im}(\Theta) : \mathbf{f}^t}{Z(\mathbf{f})^{\oplus m}} \to 0, \tag{5}$$

and an exact sequence of R/I-modules

$$0 \longrightarrow \frac{Z(\Theta) + I\mathrm{Der}_k(R)}{I\mathrm{Der}_k(R)} \to \mathrm{Der}_k(R/I) \to \frac{\mathrm{Im}(\Theta) : \mathbf{f}^t}{I\,\mathrm{Im}(\Theta) : \mathbf{f}^t} \to 0. \tag{6}$$

Proof: The first sequence follows immediately from Lemma 2.1 with $\varphi = \Theta(\mathbf{f})$, $z = \mathbf{f}^t$ and $\mathfrak{A} = I$. The second exact sequence follows from the first one by taking $\mathrm{Der}_I(R)$ and $Z(\Theta)$ modulo $I\mathrm{Der}_k(R)$. □

It may be an interesting problem to understand the right-most modules in the above exact sequences in the case where $m \leq n$, e.g., when the ideal I is generated by a regular sequence. If the ideal is principal, we give an answer below. For a homogeneous isolated complete intersection singularity, it would be interesting to check whether the right-most module of the second exact sequence is cyclic as a positive answer would imply a result of Kersken ([**6**], also [**13**]).

3.2 Algebraic version of Saito's logarithmic vector fields

We proceed to prove some algebraic analogues of Saito's results in [**7**].

As above, R is a ring of polynomials in n indeterminates over a perfect field k of high enough characteristic. Let $f \in R$ be nonzero, $I = (\frac{\partial f}{\partial x_1}, \ldots, \frac{\partial f}{\partial x_n})$ (the *critical ideal* or *gradient ideal* of f) and $J = (I, f)$ (the *Jacobian ideal* of f). We denote $\mathrm{Der}_f(R) = Z_{[f]}$ and $\Omega_f(R) = (Z_{[f]})^*$. The first module of syzygies of $\frac{\partial f}{\partial x_1}, \ldots, \frac{\partial f}{\partial x_n}$ will be denoted by $Z(\partial)$.

First we isolate the two above exact sequences in this setup, since they have apparently not been noted before in this generality in the polynomial context. They can also be directly derived from Proposition 2.2.

The Jacobian content module becomes now $I : (f)$ — the *Euler conductor* of f and f is called *Eulerian* if $f \in I$.

Proposition 3.2 *There are exact sequences of R-modules*

$$0 \to Z(\partial) \to \mathrm{Der}_f(R) \to I\colon (f) \to 0, \qquad (7)$$

and of $R/(f)$*-modules*

$$0 \to \frac{Z(\partial)}{fZ(\partial)} \to \mathrm{Der}_k(R/(f)) \to \frac{I\colon (f)}{I} \to 0. \qquad (8)$$

In particular, when f *is Eulerian there is a decomposition of R-modules* $\mathrm{Der}_f(R) = Z(\partial) \oplus R\varepsilon$*, where*

$$\varepsilon = \frac{\partial f}{\partial x_1}\frac{\partial}{\partial x_1} + \cdots + \frac{\partial f}{\partial x_n}\frac{\partial}{\partial x_n}$$

is the Euler derivation of R.

The following gives a slightly souped-up version of the differential nature of $\Omega_f(R)$ and also one of the equivalences in [**7**, Theorem 2.9], for the case where f is irreducible.

Proposition 3.3 ([7, Lemma 1.6 (ii)]) *Let* $f \in R$ *be a square-free polynomial. Then*

(i) $\mathrm{Der}_f(R)$ *and* $\Omega_f(R)$ *are reflexive R-modules and dual to each other.*

(ii) *There is an exact sequence of R-modules*

$$0 \to (\Omega(R) \oplus R) \Big/ R.(df, f) \longrightarrow \Omega_f(R) \longrightarrow \mathrm{Ext}^2_R(R/J, R) \to 0. \qquad (9)$$

(iii) *If* $R/(f)$ *is a domain the following are equivalent:*

(a) $R/(f)$ *is normal.*

(b) $\Omega_f(R) = \Omega(R) + \frac{df}{f}R$ *(as an R-submodule of* $\frac{1}{f}\Omega(R)$*).*

Moreover, any one of these conditions implies that $\Omega_f(R)$ *is* $n+1$*-generated and has projective dimension at most one.*

Proof: (i) Apply Proposition 2.2 (ii) with $f_i = \frac{\partial f}{\partial x_i}$.

(ii) This follows from Proposition 2.4 (i), again applying it with $f_i = \frac{\partial f}{\partial x_i}$ and noticing that $\mathrm{height} J \geq 2$ because $R/(f)$ is reduced.

(iii) It follows from Corollary 2.5 (ii), after sufficient translation: $R/(f)$ a domain implies that $\gcd(f_i, f) = 1$ for every $1 \leq i \leq n$. On the other hand, as $R/(f)$ obviously satisfies (S_2) then it is normal if and only if $\mathrm{height} J \geq 3$. □

Remark 3.4 Saito proves an equivalence such that the one in Proposition 3.3 (iii) in the case where f is only assumed to be square-free, in terms of the divisor configuration structure of f. It ought to be possible to deduce this more general result by the previous method as soon as the configuration conditions are sufficiently translated in algebraic terms.

Definition 3.5 (According to Saito) f is said to be a *free divisor* if f is square-free and $\mathrm{Der}_f(R)$ is a free R-module.

Remark 3.6 Any smooth f (i.e., $J = R$) is a free divisor: this follows trivially from the above characterization of $\mathrm{Der}_f(R)$ as a syzygy module of J. (Of course, no homogeneous f of degree ≥ 2 is smooth since it is a cone.) Any quasi-homogeneous divisor is Eulerian. A smooth f is Eulerian if and only if $I = R$. Also, if f is Eulerian but not smooth then $f + \alpha$ is smooth non-Eulerian for every $\alpha \in k, \alpha \neq 0$. If $f \in R$ is an irreducible homogeneous free divisor of degree ≥ 2 then the singular locus of the associated projective hypersurface has codimension 2 — this will easily follow from the proposition below.

Apart from these easy general considerations, and away of the case $n = 2$, the theory of free divisors has by and large a collection of remarkable particular results (cf. [**1**], [**4**], [**9**], [**10**], [**12**]). The next result explains a common feature of all these examples. When this work was ready for submission, we realized after a more careful reading of the previous literature that Terao [**9**, Proposition 2.4] had earlier shown this result for hyperplane arrangements. His proof is in fact valid for any free divisor — he actually quotes the result in its generality in another work ([**11**, Proposition 3]). However, since the present result follows from the general considerations of the previous section, we have decided to state it anyway.

Proposition 3.7 *Let $R/(f)$ be reduced.*

(i) *f is a free divisor if and only if either f is smooth or else the Jacobian ideal J is a codimension two perfect ideal.*

(ii) *If f is a non-smooth Eulerian divisor then f is a free divisor if and only if I is a codimension two perfect ideal.*

Proof: (i) Since $R/(f)$ is reduced, height$J \geq 2$. Therefore the result is an immediate consequence of Corollary 2.3 (b).

(ii) This is a particular case of (i). □

Supplement (to Proposition 3.7): Explicitly, the previous result says that if f is a non-smooth free divisor then, as a submodule of R^n (identified with $\mathrm{Der}_k(R)$), $\mathrm{Der}_f(R)$ is freely generated by the column vectors of an $n \times n$ matrix

$$M = \begin{pmatrix} g_{11} & \cdots & g_{1n} \\ \vdots & & \vdots \\ g_{n1} & \cdots & g_{nn} \end{pmatrix},$$

while its "lifting" $\widetilde{\mathrm{Der}_f(R)} = \ker(R^{n+1} \to J)$ as in Proposition 2.2 (ii), is generated by the column vectors of the $(n+1) \times n$ matrix obtained from M by stacking to it the row vector $(-h_1, \ldots, -h_n)$, where $\sum_i g_{ij} \frac{\partial f}{\partial x_i} = h_j f$ for $1 \leq j \leq n$ (cf. the proof of Proposition 2.2). Then the other $n \times n$ signed minors of the latter Hilbert-Burch matrix are the partial derivatives of f. On the other hand, the result of part (i) of Proposition 3.7 reads to the effect that $\mathrm{Der}_f(R)$ is freely generated by the column vectors of an $n \times n$ matrix such as M, where one of the columns can be taken to have as coordinates the coefficients of a Euler vector and the remaining columns can be taken to be syzygies of the gradient ideal I. Then the $(n-1) \times (n-1)$ signed minors of these $n \times (n-1)$ columns will be the partial derivatives while the total determinant gives back f, as is seen by expansion along the Euler column.

A piece of notation: given an R-module E, an element $\mathfrak{z} \in E$ (resp. a submodule $Z \subset E$) gives rise to a 1-form $\mathfrak{z}(1)$ (resp. to an ideal $Z(1)$ generated by 1-forms) in the symmetric algebra $\mathcal{S}_R(E)$.

The following notion appears in [**2**] (cf. also [**3**]) in the local analytic setup. With slightly appropriate language changes, we have:

Definition 3.8 Let $f \in R$ be such that $R/(f)$ is reduced. One says that f is a *Koszul free divisor* if

1. f is a free divisor (i.e., $\mathrm{Der}_f(R)$ is free), and
2. Locally everywhere on R, the ideal $\mathrm{Der}_f(R)(1) \subset \mathcal{S}_R(\mathrm{Der}(R))$ has grade n.

Since $\mathrm{Der}_f(R)$ is n-generated when it is free, then $\mathrm{Der}_f(R)(1)$ is globally generated by n elements when f is a free divisor. If, moreover, f is Koszul free then this ideal is locally everywhere on $\mathcal{S}_R(\mathrm{Der}(R))$ generated by a regular sequence of n elements. Since in a Noetherian ring (global) grade is attained locally somewhere, it follows that a free divisor f is a Koszul free divisor if and only if the ideal $\mathrm{Der}_f(R)(1)$ is globally generated by a regular sequence.

Example 3.9 The easiest example of a (reduced, non-smooth) Koszul free divisor is the square-free monomial $f = x_1 \cdots x_n$.

Note that the cokernel of the inclusion $\mathrm{Der}_f(R) \subset \mathrm{Der}(R)$ is $J/(f)$. Finally, since grade and height coincide in the polynomial ring $\mathcal{S}_R(\mathrm{Der}(R))$, f is Koszul free if and only if f is free and $\dim \mathcal{S}_R(J/(f)) = n$. We then arrive at the following characterization of a Koszul free divisor.

Proposition 3.10 *Let $f \in R$ be a free divisor. Then f is Koszul free if and only if $\dim \mathcal{S}_R(J) = n+1$ and $f(1)$ avoids all minimal primes of $\mathcal{S}_R(J)$ of maximal dimension.*

Proof: We note that since $\mathrm{Der}(R)/\mathrm{Der}_f(R) \simeq J/(f)$, our subject fits as the cokernel of the exact sequence $0 \to (f(1)) \longrightarrow \mathcal{S}_R(J) \longrightarrow \mathcal{S}_R(J/(f)) \to 0$, where $f(1)$ is f viewed as a 1-form in $\mathcal{S}_R(J)_1 = J$. Now, we have that $\dim \mathcal{S}_R(J) \geq n+1$ and $\dim \mathcal{S}_R(J/(f)) \geq n$ — these are the minimum possible dimensions of these symmetric algebras. Clearly then, $\dim \mathcal{S}_R(J/(f)) = n$ if and only if $\dim \mathcal{S}_R(J) = n+1$ and $f(1)$ is not contained in any minimal prime of $\mathcal{S}_R(J)$ of maximal dimension. Therefore, we are through. □

For non-smooth divisors one derives the following version.

Proposition 3.11 *Let $f \in R$ be a non-smooth reduced divisor. The following conditions are equivalent:*

(i) *f is Koszul free.*

(ii) *The Jacobian ideal J is a codimension two perfect ideal with $\mathcal{S}_R(J)$ a Cohen-Macaulay ring and $f(1)$ a non-zero divisor.*

Proof: It is known that, over a Cohen-Macaulay ring, the symmetric algebra of a codimension two perfect ideal is of minimal dimension if and only it is Cohen–Macaulay as well (cf., e.g., [**5**]). □

The proposition has the following consequence, which is similar to a result of Calderón-Narváez ([**3**, Proposition 3.2]).

Corollary 3.12 *Let f be a square-free divisor. If J is a codimension two perfect ideal of linear type then f is Koszul free. In particular, if f is Eulerian then it is Koszul free if I is a codimension two perfect ideal of linear type.*

Proof: Since J is of linear type, the symmetric algebra $\mathcal{S}_R(J)$ is a Cohen–Macaulay domain (see [**5**, Corollary 10.8]). □

Remark 3.13 It was shown in [**3**, Theorem 5.6] in the local analytic setup that the Jacobian ideal of a locally quasi-homogeneous free divisor is of linear type. It is possible to introduce a notion of a family of free divisors (relative to the base of the family). In this extended context there are examples of families of free divisors whose (relative) Jacobian ideal is not of linear type (cf. [**8**]).

4 Irreducible homogeneous free divisors

In this short section we briefly discuss homogeneous free divisors. It is a provocative question hunting for such divisors which are irreducible homogeneous in the least possible number of variables. For low degrees, one of the first examples to our knowledge is a Cayley sextic (cf. [**8**]).

In dimension three, we have the following special result.

Proposition 4.1 *Let $f \in R = k[x,y,z]$ be a square-free homogeneous polynomial of degree ≥ 3. The following conditions are equivalent:*

(i) *f is a free divisor.*

(ii) *f is a Koszul free divisor.*

(iii) $\mathrm{depth}(R/J) > 0$.

Proof: The equivalence of (i) and (iii) follows from Proposition 3.7. It suffices to show that (i) implies (ii). For that, since f is Eulerian, according to Proposition 3.2, the Euler vector can always be taken as part of a free basis of $\mathrm{Der}_f(R)$. Since f is further homogeneous, the coordinates of the Euler vector generate an ideal of codimension three. This easily implies that $\mathrm{Der}(R)/\mathrm{Der}_f(R) \simeq J/(f)$ satisfies the property called $\mathcal{F}_0$, hence its symmetric algebra has minimal dimension. ([**5**]). □

Remark 4.2 It is clear that the equivalence of conditions (i) and (ii) above hold also for a quasi-homogeneous divisor.

It is natural to look for irreducible homogeneous free divisors of higher dimension.

By a divisor of "tangential nature" is meant a hypersurface whose definition is given in terms of tangent behavior of other varieties, such as the tangential surface to a curve and a non-deficient dual variety. Replacing a variety by a regular map, one obtains yet another class of divisors reminiscent of some tangential behavior such as discriminants and bifurcation varieties. The latter have been the subject of focused work by several authors: K. Saito, H. Terao, J.W. Bruce, J. Damon, D. Mond, D. Van Straten, among others.

Question 4.3 *Let $C_n \subset \mathbb{P}^n$ be a rational normal curve. Is the dual to C a Koszul free divisor of linear type?*

Note that the dual is non-deficient, hence a hypersurface. This would show the existence of irreducible homogeneous Koszul free divisors in any number ≥ 4 of variables as well.

We can show that the answer to the question is affirmative for $n \leq 5$, but the proof we know of is (as of present) a rather uninspiring syzygy calculation.

Remark 4.4 There is a way to get back the tangential surface to C_3 which is reminiscent of the present theory. Namely, let $I \subset R = k[x_0, \ldots, x_3]$ denote the homogeneous ideal of the rational normal cubic in $\mathbb{P}^3$. Consider the R/I-module of k-derivations of the ring R/I. A computation yields that this module is minimally generated by 4 generators. The lifting of these generators to the free R-module $\mathrm{Der}_k(R)$ generates a free R-module of rank 4 (beware: this is not the entire I-idealizer $\mathrm{Der}_I(R)$ of $\mathrm{Der}_k(R)$ as defined in the first section). Its determinant Δ is a degree four Koszul free divisor as the partial derivatives generate a codimension two perfect ideal presented by a 4×3 linear matrix which satisfies $\mathcal{F}_0$. One can also show that Δ is the equation of the tangential surface to the rational normal cubic, hence, in particular, this determinant is irreducible.

As a last side remark, the dual to either a *non*-smooth arithmetically Cohen-Macaulay curve or to a smooth *non*-arithmetically Cohen-Macaulay curve in $\mathbb{P}^3$ fails even to be a free divisor in general.

Acknowledgments

This work was partially supported by a grant from CNPq (Brazilian Research Council), by PRONEX (Brazilian Group in Commutative Algebra and Algebraic Geometry) and by a grant from the Brazil-France International Cooperation Agreement, for which we are indebted. We also thank the referee for numerous remarks and corrections in the submitted version.

References

[1] J.W. Bruce, Vector fields on discriminants and bifurcation varieties, *Bull. London Math. Soc.* **17** (1985), 257–262.

[2] F.J. Calderón-Moreno, Logarithmic differential operators and logarithmic de Rham complexes relative to a free divisor, *Ann. Sci. E. N. S.* **32** (1999), 577–595.

[3] F.J. Calderón-Moreno and L. Narváez-Macarro, The module $\mathcal{D}f^s$ for locally quasi-homogeneous free divisors, *Compositio Math.* **134** (2002), 59–74.

[4] J. Damon, On the legacy of free divisors: Discriminants and Morse type singularities, *Amer. J. Math.* **120** (1998), 453–492.

[5] J. Herzog, A. Simis and W.V. Vasconcelos, Koszul homology and blowing-up rings, in *Commutative Algebra*, Proceedings: Trento 1981 (S. Greco and G. Valla, Eds.), 79–169, Lecture Notes in Pure and Applied Mathematics **84**, Marcel-Dekker, New York, 1983.

[6] M. Kersken, Der Residuenkomplex in der lokalen algebraischen und analytischen Geometrie, *Math. Ann.* **265** (1983) 423–455.

[7] K. Saito, Theory of logarithmic differential forms and logarithmic vector fields, *J. Fac. Sci. Univ. Tokyo Sect. 1A Math.* **27** (1980), 265–291.

[**8**] A. Simis, The depth of the Jacobian ring of a homogeneous polynomial in three variables, *Proc. Amer. Math. Soc*, to appear.

[**9**] H. Terao, Arrangements of hyperplanes and their freeness I, II, *J. Fac. Sci. Univ. Tokyo Sect. Math.* **27** (1980), 293–320.

[**10**] H. Terao, The bifurcation set and logarithmic vector fields, *Math. Ann.* **263** (1983), 313–321.

[**11**] H. Terao, The exponents of a free hypersurface, *Singularities*, *Proc. Symp. Pure Math.* **40** (1983), 561–566.

[**12**] D. Van Straten, A note on the discriminant of a space curve, *Manuscripta Math.* **87** (1995), 167–177.

[**13**] J. Wahl, Derivation, automorphisms and deformations of quasihomogeneous singularities, *Singularities*, *Proc. Symp. Pure Math.* **40** (1983), 613–624.

Gorenstein rings call the tune

ANNE-MARIE SIMON
Service d'Algèbre, C.P. 211, Université Libre de Bruxelles
1050 Bruxelles, Belgique
E-mail: `amsimon@ulb.ac.be`

JAN R. STROOKER
Mathematisch Instituut, Universiteit Utrecht
Postbus 80010, 3508 TA Utrecht, Nederland
E-mail: `strooker@math.uu.nl`

1 Introduction

The "Homological Conjectures" in local algebra date back to Serre's beautiful 1957-58 course at the Collège de France [**19**]. How to count the multiplicities of components in an intersection of two algebraic varieties was reduced to questions in local algebra involving the celebrated Tor-formula. Several important results were proven, other questions remained. These led to related but also different conjectures by other mathematicians, notably M. Auslander, Bass [**2**], Vasconcelos [**26**], Peskine-Szpiro [**15**], and Hochster [**10**], [**12**]. Some of these were proved for all noetherian local rings, others up to a certain dimension or in equal characteristic; others, again, remain virtually untouched. A masterly survey of results and of the many interconnections between these Homological Conjectures was presented in [**11**]. This was updated in [**16**]. Progress depending on De Jong's theory of alterations is discussed in [**3**] and [**17**]. Finally, a substantial part of these conjectures is treated more leisurely in the monographs [**24**] and [**5**].

Our aim in this note is far less ambitious. We focus on two of Hochster's conjectures: the Monomial Conjecture (MC) and the Canonical Element Conjecture (CEC). We point out that these are equivalent with certain statements about Gorenstein rings, in a way to be made precise as we go along. Thus Gorenstein rings "call the tune:" It is their module theory, or if one likes, representation theory, which controls MC and CEC for all noetherian local rings. To us, this suggests that the Gorenstein property, besides the several "symmetries" and "dualities" known, harbors further ones awaiting discovery.

It is not customary to write a survey of material already in print. However, we think that a brief presentation of selected results from a series of papers [**25**], [**20**], [**21**], [**22**] and [**23**], highlighting the Gorenstein connection, will help the reader to understand our point of view and perhaps continue this line of research. No proofs are given, rather remarks about ideas and results used along the way. We are grateful to the editors for this opportunity.

2 The conjectures

We use the word "ring" for unital, commutative, noetherian local rings, since no others shall be needed, and all modules will be finitely generated unless otherwise stated. As a standard notation we use $(A, \mathfrak{m}, k)$ for the ring A, its maximal ideal and its residue class field. Let A be a d-dimensional ring, and $x_1, \ldots, x_d = \mathbf{x}$ a system of parameters (sop). Let t be a positive integer. Is the product $x_1^t \ldots x_d^t$ in the ideal generated by $x_1^{t+1}, \ldots, x_d^{t+1}$?

The ring is said to satisfy MC provided for every choice of $\mathbf{x}$ and of t, one has $x_1^t \ldots x_d^t \notin (x_1^{t+1}, \ldots, x_d^{t+1})$ [**10**, Conj. 2]. It is often advantageous to translate this to saying that the maps $x_1^t \ldots x_d^t : A/(x_1, \ldots, x_d) \to A/(x_1^{t+1}, \ldots, x_d^{t+1})$ are never null. Since the direct limit of this directed system is the d-th local cohomology module $H^d_{\mathfrak{m}}(A)$, we can also state:

Definition 2.1 A ring A is said to satisfy the Monomial Conjecture if the induced map $\mu_{\mathbf{x}}(A) : A/(x_1,...,x_d) \to H^d_{\mathfrak{m}}(A)$ is nonnull for every sop $\mathbf{x}$.

To prepare for the Canonical Element Conjecture, let M be a finitely generated A-module and consider the natural maps $\theta^i_A(M) : \mathrm{Ext}^i_A(k,M) \to H^i_{\mathfrak{m}}(k,M)$ for $i \geq 0$. This Ext is a finite dimensional vector space over k, whose dimension is often called the i-th Bass number $\mu^i_A(M)$ of M; we call the k-dimension of the image of $\theta^i_A(M)$ the reduced Bass number $\nu^i_A(M)$ of M. Thus $\nu^i_A(M) \leq \mu^i_A(M)$ and both are nonnegative integers.

Let $\mathbf{F}$ be a resolution of k in terms of free A-modules. Let $\mathbf{x}$ again be a sop, and $\mathbf{K}(\mathbf{x},A)$ its Koszul complex. Then the surjection $A/(x_1,...,x_d) \to k$ lifts to a map of complexes $\phi : \mathbf{K}(\mathbf{x},A) \to \mathbf{F}$.

Definition 2.2 A ring A is said to satisfy the Canonical Element Conjecture if the map ϕ_d is nonnull for every lifting belonging to every sop $\mathbf{x}$.

Notice that the source of ϕ_d is just a copy of A. This was Hochster's original approach [**12**, Def. 2.1]. He and others have put forward several equivalent formulations of CEC. One of these led to [**12**, Ths. 3.7 and 4.3]; see also [**22**, 5.1]:

Theorem 2.3 *Write $S \subset F_d$ for the d-th syzygy of k. Then the following are equivalent:*

(i) The ring A satisfies the Canonical Element Conjecture;

(ii) $\nu^d_A(S) > 0$*;*

(iii) $\nu^d_A(M) > 0$ *for some M, not necessarily finitely generated.*

In case A possesses a canonical module K, this is also equivalent with

(iv) $\nu^d_A(K) > 0$.

Here K is called a canonical module of A when $K^\vee = H^d_{\mathfrak{m}}(A)$, where $-^\vee$ stands for the Matlis dual, e.g., [**22**, 1.6]. Both MC and CEC need only be proved for rings which are complete in their $\mathfrak{m}$-adic topology, and a ring which satisfies CEC satsifies MC. In case the sop $\mathbf{x}$ forms a regular sequence, i.e., A is a Cohen-Macaulay ring, both conjectures are true. Better even, if there is an

A-module M, not necessarily finitely generated, such that $x_1, \ldots, x_d$ form a regular sequence on M and $M \neq (x_1, \ldots, x_d)M$, then both remain true for $\hat{M}$, the $\mathfrak{m}$-adic completion of M, and hold for all sops $\mathbf{x}$ [**24**, Th. 9.1.1]. Such, possibly "Big," Cohen-Macaulay modules provide one of the proofs of the conjectures in equal characteristic. All this, and more, is in the papers [**10**] and [**12**]. Thanks to Heitmann's recent advance in dimension 3 [**9**] and [**18**], MC and CEC are known in mixed characteristic for $d \leq 3$ but not beyond. It should be mentioned that the two conjectures are closely interwoven with others, in particular Hochster's Direct Summand Conjecture, but we don't deal with these.

Here we wish to emphasize that it is the connection with local cohomology, and then the description in terms of reduced Bass numbers, which shall allow us to tie up the conjectures with properties of Gorenstein rings.

3 Why Gorenstein?

Since we need only prove MC for rings complete in their maximal ideal topology, let A be such a ring. According to Cohen, one can write $A = S/\mathfrak{b}$ where b is an ideal in the regular ring S. By judiciously dividing out a maximal regular sequence $z_1, \ldots, z_n$ in $\mathfrak{b}$, one can prove [**25**, Prop. 1], [**22**, 5.5] the following:

Lemma 3.1 *There exists a d-dimensional complete intersection ring R with d =* $\dim A$ *which maps onto* $A = R/\mathfrak{a}$*. Moreover, sops in R map to sops in A, and any sop* $\mathbf{x}$ *in A can be lifted to a sop* $\mathbf{y}$ *in R.*

In the situation of this Lemma, $\mu_{\mathbf{x}}(A) = \mu_{\mathbf{y}}(R) \otimes_R A$ and $\mu_{\mathbf{x}}(A) = 0$ if and only if its Matlis dual $\mu_{\mathbf{x}}(A)^{\vee}$ is. Working out this dual, using the adjointness of tensor product and Hom and a few facts from local Grothendieck duality regarding the Gorenstein ring R, one obtains [**25**, Prop 2]:

Proposition 3.2 *In the situation just described,* $\mu_{\mathbf{x}}(A) = 0$ *if and only if* $\mathrm{Ann}_R\, \mathfrak{a} \subset (y_1, \ldots, y_d)$.

Definition 3.3 We say that a Gorenstein ring R and a ring A are in our favorite position if $A = R/\mathfrak{a}$ where $\mathfrak{a}$ in a non-null ideal in R consisting of zero divisors. We standardly put $\mathfrak{b} = \mathrm{Ann}_R\, \mathfrak{a}$.

With the help of the above and further reasoning, one proves

Theorem 3.4 *Let R and A be in our favorite position. Then A satisfies* MC *if and only if the ideal* $\mathfrak{b}$ *is not contained in any parameter ideal of R.*

It is well known that in a Gorenstein ring the ideals $\mathfrak{b} = \operatorname{Ann} \mathfrak{a}$, where $\mathfrak{a}$ is a nonnull ideal consisting of zero divisors, are just the unmixed nonnull ideals consisting of zero divisors. So one has:

Corollary 3.5 *Every ring satisfies the Monomial Conjecture if and only if in no Gorenstein ring is an unmixed nonnull ideal of zero divisors contained in a parameter ideal.*

Remark 3.6 In our favorite situation, if R is equicharacteristic, so is the ring A, which therefore satisfies MC. It follows that the above statement is true for every equicharacteristic Gorenstein ring. It only remains an open question in mixed characteristic for (Heitmann) dimension > 3.

Remark 3.7 Furthermore, in the corollary we may replace the word Gorenstein by complete intersection. At face value, the statement becomes stronger in one direction, weaker in the other. We have not managed so far to utilize properties of complete intersections which are not shared by all Gorenstein rings to prove MC or CEC in cases yet unknown. Also our feeling is that the dualities and symmetries pertaining to Gorenstein rings and their module theory are just "right." So here, and in following sections, we shall stick to Gorenstein and leave it to the reader to realize that if one can obtain certain results for complete intersections, this is enough.

For details and further observations we refer to [**25**].

4 Auslander-Buchweitz theory

The theory which M. Auslander and R.-O. Buchweitz developed in [**1**] is quite abstract and covers various special cases, of which we treat only one: (finitely generated) modules over (local) Gorenstein rings. For the sake of brevity, we recall this in a manner which is not quite that of [**1**], but was worked out later by various authors and is systematized in the first four sections of [**22**]. In this section we fix a Gorenstein ring R and only consider modules and several invariants over this. We therefore drop the R from notation.

Let M be a module over the Gorenstein ring R and C a maximal Cohen-Macaulay module (MCM). An R-homomorphism $f: C \to M$ is called an MCM-precover of M if, for every MCM-module D, the induced map $\mathrm{Hom}(D,C) \to \mathrm{Hom}(D,M)$ is surjective. In other words, every map from D to M can be factored through C. If, in addition, the map f is right minimal, then it is called a cover. Here right minimal means that if $f \circ g = f$ for any endomorphism g of C, then g is an automorphism.

Dually, one defines a homomorphism $h: M \to J$, with J a module of finite injective dimension (FID), to be an FID-preenvelope (resp. -envelope) of M. Very briefly, the basic result of [**1**] can in our case be summarized [**22**, section 3] as:

Theorem 4.1 *Over a Gorenstein ring, there exist* MCM-*covers and* FID-*envelopes. In these cases f is a surjection and h an injection with an* FID-*kernel resp.* MCM-*cokernel. Covers and envelopes are determined up to isomorphisms over M.*

Though we shall stick with our Gorenstein environment, it should be pointed out that "Auslander-Buchweitz contexts" turn out to be relevant in different areas of algebra [**8**], beyond what was originally envisaged by these authors.

By the f-rk of a module X, we mean the maximal rank of a free direct summand of X. For a module M, by the above theorem, the f-rk of its MCM-cover C and its FID-envelope J are well defined. In [**1**, p. 8] it is suggested to examine these. The first one is called Auslander's delta invariant $\delta(M)$ and has been investigated by several of his last students and others. The one on the FID-side so far has hardly been looked at. We hope to convince you that this is unjustified, and that these invariants are most profitably studied in tandem.

Proposition 4.2 *Let J be an* FID-*envelope of M. Then f-rk$(J) = \nu^d(M)$.*

This result, [**21**, Prop. 6], [**22**, Prop. 3.10], connects the second invariant with local cohomology. What about the δ-invariant? Well, [**21**, Prop. 7] and [**22**, Prop. 4.1 (iv)]:

Proposition 4.3 *Let $p: R^t \to M$ be a surjection of a free module onto M. This induces a map* $\mathrm{Ext}^d(k,R^t) \to \mathrm{Ext}^d(k,M)$. *The first* Ext *being simply k^t, its image is a finite dimensional vector space V over k. Then* $\delta(M) = \dim V$.

While this description of δ does not involve local cohomology, it does bring into play the d-th Ext whose relation with d-th local cohomology is crucial for the reduced Bass number. We shall exploit this affinity in the next section, returning to our main theme. There are many interactions between the two invariants, which suggest a formal kind of duality. Perhaps more accurately one should speak of orthogonality, stemming from the fact that if X is an MCM and Y an FID, then $\mathrm{Ext}^i(X,Y) = 0$ for $i > 0$. Of course, the yoga becomes even more powerful when one realizes that over a Gorenstein ring, FID-modules are exactly the ones of finite projective dimension [**24**, Th. 10.1.9]. For our purposes, we collect a few useful facts from this wondrous world:

Lemma 4.4 *For any module M, there exists a map $s : M \to R$ such that* $\mathrm{Ext}^d(k,s)$ *is nonnull if and only if* $\nu^d(M) > 0$.

Indeed, bearing in mind that the canonical module of the Gorenstein ring R is just R itself, one concludes this from [**22**, Cor. 3.11]. A more precise statement, which can be regarded as a counterpart on the envelope side of Proposition 4.3, can be found in [**21**, Th. 3] or [**22**, Cor. 3.12].

The next lemma is [**22**, Prop. 4.1 (iii), Prop. 4.5]:

Lemma 4.5 *For any module M, $\delta(M) \leq \beta(M)$, where $\beta(M)$ means the minimal number of generators of the module M. For* FID*-modules this is actually an equality. Also $\delta(M) = 0$ precisely when M is a homomorphic image of an* MCM*-module without a free direct summand.*

5 Harvesting from favorite position

Here is one more reason why our favorite position is so effective. In the set-up of Definition 3.3, the ideal $\mathfrak{b}$, being annihilated by $\mathfrak{a}$, is also an A-module. In fact, it is a canonical module of the ring A [**22**, 6.9 and 1.6].

Theorem 5.1 *In our favorite position consider the statements*

(i) The ring A satisfies the Canonical Element Conjecture;

(ii) $\nu_R^d(\mathfrak{b}) > 0$;

(iii) $\delta(R/\mathfrak{b}) = 0$;

(iv) $\mathfrak{b}$ *is not contained in any* FID-*ideal;*

(v) $\mathfrak{b}$ *is not contained in any parameter ideal;*

(vi) The ring A satisfies the Monomial Conjecture;

where items (ii)-(v) all refer to ideals of the Gorenstein ring R. Then (i) $\Rightarrow$ *(ii)* $\Leftrightarrow$ *(iii)* $\Rightarrow$ *(iv)* $\Rightarrow$ *(v)* $\Leftrightarrow$ *(vi).*

Sketch of proof: *(i)* $\Rightarrow$ *(ii)*: Theorem 2.3 tells us that *(i)* and $\nu^d_A(\mathfrak{b}) > 0$ are equivalent statements. Since the map $\theta^d_A(\mathfrak{b})$ (section 2) factors through $\theta^d_R(\mathfrak{b})$, one has *(ii)*. *(ii)* $\Leftrightarrow$ *(iii)*: The embedding $\mathfrak{b} \to R$ yields an exact sequence $\mathrm{Ext}^d_R(k,\mathfrak{b}) \to \mathrm{Ext}^d_R(k,R) \to \mathrm{Ext}^d_R(k,R/\mathfrak{b})$. Since the middle Ext is just a single copy of k for the Gorenstein ring R, and any map $\mathfrak{b} \to R$ lands in $\mathrm{Ann}\,\mathfrak{a} = \mathfrak{b}$, Lemma 4.4 and Prop. 4.3 allow one to conclude. *(iii)* $\Rightarrow$ *(iv)*: If $\mathfrak{c}$ is an FID-ideal containing $\mathfrak{b}$, then $\delta(R/\mathfrak{c}) = 1$ by Lemma 4.5, which contradicts the last statement of this lemma, because $R/\mathfrak{c}$ is a homomorphic image of $R/\mathfrak{b}$. *(iv)* $\Rightarrow$ *(v)*: In the Gorenstein ring R, every sop is a regular sequence which generates an ideal of finite projective dimension. This then also has FID. *(v)* $\Leftrightarrow$ *(vi)*: Our Theorem 3.4.

We now recall one of the principal results in [**12**], its Theorem 2.8.

Theorem 5.2 *If the Monomial Conjecture is true for all rings, then all rings satisfy the Canonical Element Conjecture.*

Well, this is not quite what is stated there, but it is if one takes into account the equivalence of Direct Summand Conjecture and Monomial Conjecture from the earlier paper [**10**, Th. 1].

In a more precise statement of Hochster's Theorem 5.2 one may replace "all rings" both times by "all rings of a particular dimension and a fixed residue characteristic." To a ring $(A,\mathfrak{m},k)$ we attach the pair of integers (u,v) when $u = \mathrm{char}\,A$ and v = char k. In the cases (0,0) and (p,p), p a prime, we speak of equal characteristic, while (0,p) is mixed. For fixed u and v we speak of a particular characteristic. In our favorite position, if R is equicharacteristic, so is A of the same particular characteristic. Now contemplate the statements and implications in Theorem 5.1, bearing in mind that CEC is known in equal characteristic and in the other case for $d \leq 3$.

Corollary 5.3 *All rings of a particular characteristic and dimension satisfy* CEC *(or* MC, *for that matter) if and only if statements (ii)-(v) hold for all nonnull unmixed ideals of zero divisors* $\mathfrak{b}$ *in all such Gorenstein rings.*

This gives new information about Gorenstein rings of equal characteristic and for mixed characteristic in dimension ≤ 3. We believe that this should be exploited and further structure will come to light. In the remaining cases, the burden of the Conjectures is borne by the Gorenstein specimens.

Using several of these results one can prove [**22**, Cor. 7.2].

Proposition 5.4 *The Canonical Element Conjecture is equivalent with the following statement. For any system of parameters in a Gorenstein ring the kernels of all maps* ϕ_d *of Definition 2.2 are contained in the nilradical of the ring.*

This again suggests some kind of symmetry, at least for the minimal primes in a Gorenstein ring. For further results and refinements see [**21**] and [**22**].

We finish this section with a tribute to the late Maurice Auslander. Once again ideas of his, this time in collaboration with Buchweitz, which started out as a kind of abstract "Spielerei," revealed themselves over the years as relevant to more concrete problems investigated by others.

6 Back to linear algebra: stiffness

Let A be a ring and $f : A^m \to A^n$ a homomorphism between two free A-modules. By choosing bases, one describes f as an $n \times m$ matrix with coefficients in A. For $r \leq \min(m,n)$ consider the ideal in A which is generated by all the $r \times r$ minors of the matrix. It is well known that this ideal $I_r(f)$ does not depend on the choice of bases; the largest r for which $I_r(f)$ does not vanish is called rk f, the rank of f.

Let

$$\mathbf{F} \quad = \quad 0 \to F_s \to \cdots \to F_i \stackrel{d_i}{\to} F_{i-1} \to \cdots \to F_0$$

be a complex of free modules. The by now classical Buchsbaum-Eisenbud criterion [**6**], as expressed in [**4**, Th. 1], tells us precisely when this complex is exact. Recall that the grade of an ideal I is the longest length of a regular sequence contained in I; this is equal to $\mathrm{ext}_A^-(A/I,A)$, the smallest degree for which this Ext is $\neq 0$. The grade of the improper ideal is taken to be ∞.

Theorem 6.1 *Put $f_i = \mathrm{rk}\, F_i, i = 0, \ldots, s$ and $r_i = f_i - f_{i+1} + \ldots \pm f_s$. The complex* **F** *is acyclic precisely when $\mathrm{gr}\, I_{r_i}(d_i) \geq i$ for all i. In this case $r_i = \mathrm{rk}\, d_i$.*

Instead of minors of a matrix attached to a map between free modules, let us look at column ideals, i.e., ideals generated by all elements in one particular column. Easy examples show that the grade of such ideals depends on choice of bases. This motivates the following definition, where we suppose that the complex **F** is exact and minimal in the sense that all boundary maps in the complex $k \otimes_A \mathbf{F}$ are null. The latter condition is not essential, but it makes the treatment a bit easier.

Definition 6.2 Let **F** now be acyclic and minimal. We say that this complex is "stiff" if, regardless of base choice, $\mathrm{gr}\, \mathfrak{c} \geq i$ for every column ideal $\mathfrak{c}$ belonging to d_i, $i = 1, \ldots, s$. We call the ring A stiff if every such complex over it is stiff.

Theorem 6.3 *Every ring of equal characteristic is stiff.*

This result [**23**, Th. 1] has some overlap with earlier work of Evans-Griffith [**7**] and Hochster-Huneke [**13**], [**14**] as explained in the paper. Our proof uses Big Cohen-Macaulay modules.

Remark 6.4 An immediate question springs to mind: what have these two theorems to do with one another? The Buchsbaum-Eisenbud criterion is by no means obvious. Yet its proof employs techniques and arguments which are rather common in this area of local algebra, and are not concerned with characteristic. Is stiffness really restricted to equal characteristic? The two statements involving grade appear to be in league, but how? For a more detailed discussion, see [**23**, section 6].

What use is stiffness? Well, for a consistent class of rings (if A belongs to this class, so does $A/(x)$ for every non-zero divisor $x \in \mathfrak{m}$) stiffness is controlled by first syzygies. In other words, one only needs to check that $\mathrm{gr}\, \mathfrak{c} \geq 1$ for every column ideal $\mathfrak{c}$ belonging to d_1 in each **F** over every ring in this class [**23**, Prop. 8]. This makes stiffness more accessible, because it is easily seen to mean that $\mathrm{Ann}\, z = 0$ for every minimal generator z of a syzygy of finite projective dimension. For Gorenstein rings, this turns out to be equivalent with a condition we have already encountered. A little juggling with FID-envelopes yields a proof of [**23**, Prop. 10]:

Proposition 6.5 *Let R be a Gorenstein ring. Then* $\delta(R/\mathfrak{b}) = 0$ *for every unmixed non-null ideal of zero divisors* $\mathfrak{b}$ *if and only if* $\operatorname{Ann} z = 0$ *for every minimal generator* z *of every syzygy of finite projective dimension.*

Tying this together with Corollary 5.3 we obtain [**23**, Th. 2] which brings us back to our main theme.

Theorem 6.6 *The Canonical Element Conjecture holds for all rings of a particular characteristic and dimension if and only if all such Gorenstein rings are stiff.*

Only the mixed characteristics in dimensions ≥ 4 remain undecided. In all other cases, Gorenstein rings are stiff. However, in equal characteristic Theorem 6.3 asserts a stronger result. Several obvious questions remain, inviting further exploration.

References

[1] M. Auslander and R.-O. Buchweitz, The homological theory of maximal Cohen-Macaulay approximations, *Mémoire* **38**, Soc. Math. France (1989), 5–37.

[2] H. Bass, On the ubiquity of Gorenstein rings, *Math. Z.* **82** (1963), 8–28.

[3] P. Berthelot, Altérations de variétes algébriques [d'après A.J. de Jong], Sém. Bourbaki, *Astérisque* **241**, Soc. Math. France (1997), 273–311.

[4] W. Bruns, The Evans-Griffith syzygy theorem and Bass numbers, *Proc. Amer. Math. Soc.* **115** (1992), 939–946.

[5] W. Bruns and J. Herzog, *Cohen-Macaulay Rings*, Cambridge Studies in Advanced Mathematics **39**, revised edition, Cambridge University Press, Cambridge, 1998.

[6] D.A. Buchsbaum and D. Eisenbud, What makes a complex exact?, *J. Algebra* **25** (1973), 259–268.

[7] E.G. Evans and P. Griffith, Order ideals, in *Commutative Algebra*, 212–225, MSRI Publ. **15**, Springer, New York, 1989.

[8] M. Hashimoto, Auslander-Buchweitz approximations of equivariant modules, *London Mathematical Society Lecture Notes* **282**, Cambridge University Press, Cambridge, 2000.

[9] R.C. Heitmann, The direct summand conjecture in dimension 3, *Ann. Math.* **156** (2002), 695–712.

[10] M. Hochster, Contracted ideals from integral extensions of regular rings, *Nagoya Math. J.* **51** (1973), 25–43.

[11] M. Hochster, *Topics in the Homological Theory of Modules over Commutative Rings*, CBMS Regional Conference Series in Mathematics **24**, American Mathematical Society, Providence, 1975.

[12] M. Hochster, Canonical elements in local cohomology modules and the Direct Summand Conjecture, *J. Algebra* **84** (1983), 503–553.

[13] M. Hochster and C. Huneke, Tight closure, in *Commutative Algebra*, 305–324, MSRI Publ. **15**, Springer, New York, 1989.

[14] M. Hochster and C. Huneke, Tight closure, invariant theory, and the Briançon-Skoda theorem, *J. Amer. Math. Soc.* **3** (1990), 31–116.

[15] C. Peskine and L. Szpiro, Dimension projective finie et cohomologie locale, *Publ Math. Inst. Hautes Étud. Scient.* **42** (1973), 77–119.

[16] P.C. Roberts, The Homological Conjectures, in *Free Resolutions in Commutative Algebra and Algebraic Geometry*, 121–132, Research Notes in Mathematics **2**, Jones and Bartlett, Boston-London, 1992.

[17] P.C. Roberts, Recent developments on Serre's multiplicity conjectures: Gabber's proof of the nonnegativity conjecture, *Enseignem. Math.* (2e série) **44** (1998), 305–324.

[18] P.C. Roberts, Heitmann's proof of the direct summand conjecture in dimension 3, arXiv:math.AC/0212073.

[19] J.-P. Serre, *Algèbre Locale-Multiplicités*, Lecture Notes in Mathematics **11**, Springer, Berlin-New York, 1965.

[20] A.-M. Simon and J.R. Strooker, Monomial conjecture and Auslander's δ-invariant, in *Commutative Algebra and Algebraic Geometry*, F. van Oystaeyen, Ed., 265–273, Lecture Notes in Pure Applied Mathematics **206**, Marcel Dekker, New York, 1999.

[21] A.-M. Simon and J.R. Strooker, Connections between reduced Bass numbers and Auslander's δ-invariant, *J. Pure Appl. Algebra* **152** (2000), 303–309.

[22] A.-M. Simon and J.R. Strooker, Reduced Bass numbers, Auslander's δ-invariant and certain homological conjectures, *J. Reine Angew. Math.* **551** (2002), 173–218.

[23] A.-M. Simon and J.R. Strooker, Stiffness of finite free resolutions and the canonical element conjecture, *Bull. Sci. Math.* **127** (2003), 251-260.

[24] J.R. Strooker, *Homological Questions in Local Algebra*, London Mathematical Society Lecture Note Series **145**, Cambridge University Press, Cambridge, 1990.

[25] J.R. Strooker and J. Stückrad, Monomial conjecture and complete intersections, *Manuscripta Math.* **79** (1993), 153–159.

[26] W.V. Vasconcelos, *Divisor theory in Module Categories*, Mathematics Studies **14**, North-Holland, Amsterdam-Oxford, 1973.

On free integral extensions generated by one element

IRENA SWANSON
Department of Mathematical Sciences
New Mexico State University, Las Cruces, NM 88003, USA
E-mail: `iswanson@nmsu.edu`

ORLANDO VILLAMAYOR
Departamento de Matemáticas
Facultad de Ciencias, C-XV, Universidad Autónoma de Madrid
ctra. de Colmenar Viejo, Km. 15, 28049 Madrid, Spain
E-mail: `villamayor@uam.es`

1 Introduction

Let R be a commutative integral domain with unity, and θ an element of an extension domain satisfying the relation

$$\theta^d = a_1\theta^{d-1} + a_2\theta^{d-2} + \cdots + a_{d-1}\theta + a_d,$$

with $a_i \in R$. We assume throughout that $R[\theta] \cong R[X]/(X^d - \sum_{i=1}^d a_i X^{d-i})$, where X is an indeterminate over R.

Suppose that R is a normal domain with quotient field K, and $K \subset L$ an algebraic extension. Let $\overline{R}$ be the integral closure of R in L, and fix $\theta \in \overline{R}$. There is information on the element θ encoded in the coefficients a_i. The first example arises when characterizing if θ belongs to the integral closure of the extended ideal $I\overline{R}$, for some ideal I in R. The objective of this paper is to study more precisely what information about θ is encoded in the coefficients a_i.

In a first approach, in Section 2, we show that for an ideal I in R, $a_i \in I^i$ for all i implies that $\theta^n R[\theta] \cap R \in I^n$ for all n, but that the converse fails. Thus contractions of powers of $\theta^n R[\theta]$ to R contain some information, but not enough.

We turn to a different approach in Sections 3 and 4, where we replace contractions by the trace functions (the image of $\theta^n R[\theta]$ in R by the trace function), and it turns out that if θ is separable over K, then the trace codes more information.

The main results in this paper are:

(a) Propositions 4.6 and 4.8 with conditions that assert that θ belongs to the integral closure of an extended ideal, and

(b) Propositions 4.12 and 4.14 with conditions that assert that θ belongs to the tight closure of an extended ideal.

In all these propositions we fix an ideal $I \subset R$ and consider the extended ideal $I \cdot R[\theta]$. It should be pointed out that normally the condition for θ to belong to the integral closure of $I \cdot R[\theta]$ is expressed in terms of a polynomial with coefficients in the ring $R[\theta]$; whereas we will express the same fact but in terms of a polynomial with coefficients in R, furthermore, in terms of the minimal polynomial of θ over R in case R is normal.

We also point out that we start with an ideal I in R, and an element θ in $\overline{R}$, and we study if θ belongs to integral or tight closure of the extended ideal, but only for the extension $R \subset R[\theta]$. This situation is however quite general, at least if I is a parameter ideal. In fact, given a complete local reduced ring (B, M) of dimension d containing a field, and with residue field k, and given a system of parameters $\{x_1, \ldots, x_d\}$, then B is finite over the subring $R = k[\![x_1, \ldots, x_d]\!]$. Furthermore an element $\theta \in B$ is in the integral closure (in the tight closure) of the parameter ideal $< x_1, \ldots, x_d > B$, if and only if it is so in $< x_1, \ldots, x_d > R[\theta]$.

Throughout the previous argumentation there is a difference between characteristic zero and positive characteristic. The point is that our arguments will rely on properties of the subring of symmetric polynomials in a polynomial ring.

The relation of symmetric polynomials with our problem will arise and be discussed in the paper. We will show that the properties of θ that we are considering can be expressed in terms of symmetric functions on the roots of the minimal polynomial of θ, and hence as functions on the coefficients a_i of the minimal polynomial.

If k is a field of characteristic zero and S is a polynomial ring over k, the subring of symmetric polynomials of S can be generated in terms of the trace; however this is not so if k is of positive characteristic. In Section 4 we address the pathological behavior in positive characteristic, and we give an example in which R is a k-algebra, k a field of positive characteristic, and the k-subalgebra generated by all the $Tr(\theta^n)$, as n varies, is not finitely generated.

We try to develop our results in maximal generality, in order to distinguish properties that hold under particular conditions (e.g., on the characteristic of R, separability of θ over K, etc.).

Our arguments rely on a precise expression of the powers θ^n of θ in terms of the natural basis $\{1,\theta,\theta^2,\ldots,\theta^{d-1}\}$ of $R[\theta]$ over R. This is done in Section 1 by using compositions, that is, ordered tuples of positive integers. Similarly, we also develop a product formula for elements of $R[\theta]$ in terms of the natural basis.

2 Power and product formula

Every element of $R[\theta]$ can be written uniquely as an R-linear combination of $1,\theta,\theta^2, \ldots, \theta^{d-1}$. In this section we develop formulas for the R-linear combinations for all powers of θ, and for linear combinations of products.

Definition 2.1 Let e be a positive integer. A **composition** of e is an ordered tuple $(e_1,\ldots,e_k)$ of positive integers such that $\sum e_i = e$. Let $\mathcal{E}_e$ denote the set of all compositions of e.

For example, $\mathcal{E}_1 = \{(1)\}$, $\mathcal{E}_2 = \{(2),(1,1)\}$, $\mathcal{E}_3 = \{(3),(2,1),(1,2),(1,1,1)\}$.

We will express θ^n in terms of these compositions. Without loss of generality we may use the following notation:

Notation 2.2 For $i > d$, set $a_i = 0$.

Definition 2.3 Set $C_0 = 1$, and for all positive integers e set

$$C_e = \sum_{(e_1,\dots,e_k)\in\mathcal{E}_e} a_{e_1}a_{e_2}\cdots a_{e_k}.$$

Remark 2.4 It is easy to see that for all $e > 0$, $C_e = C_0a_e + C_1a_{e-1} + \cdots + C_{e-1}a_1$.

Proposition 2.5 *For all* $e \geq 0$,

$$\theta^{d+e} = \sum_{i=0}^{d-1} (C_0a_{d+e-i} + C_1a_{d+e-i-1} + C_2a_{d+e-i-2} + \cdots + C_ea_{d-i})\,\theta^i.$$

Proof: The proof follows by induction on e. When $e = 0$, the coefficient of θ^i in the expression on the left above is $C_0a_{d-i} = a_{d-i}$, so the proposition holds for the base case by definition.

Now let $e > 0$. Then

$$\begin{aligned}
\theta^{d+e} &= \theta^{d+e-1}\theta \\
&= \sum_{i=0}^{d-1} (C_0a_{d+e-i-1} + C_1a_{d+e-i-2} + C_2a_{d+e-i-3} + \cdots + C_{e-1}a_{d-i})\,\theta^{i+1} \\
&= \sum_{i=0}^{d-2} (C_0a_{d+e-i-1} + C_1a_{d+e-i-2} + C_2a_{d+e-i-3} + \cdots + C_{e-1}a_{d-i})\,\theta^{i+1} \\
&\quad + (C_0a_e + C_1a_{e-1} + C_2a_{e-2} + \cdots + C_{e-1}a_1)\,\theta^d \\
&= \sum_{i=1}^{d-1} (C_0a_{d+e-i} + C_1a_{d+e-i-1} + C_2a_{d+e-i-2} + \cdots + C_{e-1}a_{d-i+1})\,\theta^i \\
&\quad + C_e\sum_{i=0}^{d-1} a_{d-i}\theta^i = \sum_{i=0}^{d-1}\sum_{j=0}^{e} C_ja_{d+e-i-j}\theta^i.
\end{aligned}$$

□

Recall that $a_i = 0$ if $i > d$. Thus in the expression for θ^{d+e} in the proposition above, many of the terms $C_ja_{d+e-i-j}$ are trivially zero.

We similarly determine the product formula:

Let $f = \sum_{i=0}^{d-1} f_i\theta^i$, $g = \sum_{i=0}^{d-1} g_i\theta^i$ be two elements in $R[\theta]$. Write fg as an R-linear combination of $1, \theta, \ldots, \theta^{d-1}$. (Here, $f_i = g_i = 0$ if $i < 0$ or $i \geq d$.)

$$\begin{aligned}
fg &= \sum_{i=0}^{2d-2}\sum_{k=0}^{d-1} f_k g_{i-k}\theta^i \\
&= \sum_{i=0}^{d-1}\sum_{k=0}^{d-1} f_k g_{i-k}\theta^i + \sum_{i=d}^{2d-2}\sum_{k=0}^{d-1} f_k g_{i-k}\theta^i \\
&= \sum_{i=0}^{d-1}\sum_{k=0}^{d-1} f_k g_{i-k}\theta^i + \sum_{e=0}^{d-2}\sum_{k=0}^{d-1} f_k g_{d+e-k}\theta^{d+e} \\
&= \sum_{i=0}^{d-1}\sum_{k=0}^{d-1} f_k g_{i-k}\theta^i + \sum_{e=0}^{d-2}\sum_{k=0}^{d-1} f_k g_{d+e-k}\sum_{i=0}^{d-1}\sum_{j=0}^{e} C_j a_{d+e-i-j}\theta^i \\
&= \sum_{i=0}^{d-1}\sum_{k=0}^{d-1}\left(f_k g_{i-k} + \sum_{e=0}^{d-2} f_k g_{d+e-k}\sum_{j=0}^{e} C_j a_{d+e-i-j}\right)\theta^i \\
&= \sum_{i=0}^{d-1}\sum_{k=0}^{d-1} f_k\left(g_{i-k} + \sum_{e=0}^{d-2} g_{d+e-k}\sum_{j=0}^{e} C_j a_{d+e-i-j}\right)\theta^i.
\end{aligned}$$

We will use this expression mainly for the cases when $fg \in R$. Then the coefficients of θ^i in the expression above, for $i > 0$, are 0, and the constant coefficient is

$$\begin{aligned}
&\sum_{k=0}^{d-1} f_k\left(g_{-k} + \sum_{e=0}^{k-1} g_{d+e-k}(C_0a_{d+e} + C_1a_{d+e-1} + \cdots + C_ea_d)\right) \\
&= f_0g_0 + \sum_{k=0}^{d-1} f_k \sum_{e=0}^{k-1} g_{d+e-k}C_ea_d.
\end{aligned}$$

3 Contractions

In this section we examine implications between $a_i \in I^i$ for all i, and $\theta^n R[\theta] \cap R \in I^n$ for all n, where I is an ideal of R. In case R is an $\mathbb{N}$-graded ring with $R = R_0[R_1]$ and $I = R_1R$, then $a_i \in I^i$ is equivalent to saying that $\deg(a_i) \geq i$. (The two statements are not equivalent in general.)

We examine how under some $\mathbb{N}$-gradings on R, the degrees of the a_i affect and are affected by the degrees of the elements of $\theta^n R[\theta] \cap R$.

Proposition 3.1 *With set-up on R, $a_1, \ldots, a_d$, and θ as in the introduction, if I is any ideal of R and $a_i \in I^i$ for all i, then $\theta^n R[\theta] \cap R \in I^n$ for all n.*

Similarly, if R is an $\mathbb{N}$-graded regular ring with a_i an element of R of degree at least i, then for all $n \geq 0$, $\theta^n R[\theta] \cap R$ is an ideal all of whose elements lie in degrees at least n.

Proof: First let $n < d$. Let $g = \sum_{i=0}^{d-1} g_i \theta^i$ be such that $\theta^n g \in R$. By the product formula from the previous section, the constant coefficient of $\theta^n g$ is

$$\delta_{n0} g_0 + \sum_{k=0}^{d-1} \delta_{kn} \sum_{e=0}^{k-1} g_{d+e-k} C_e a_d,$$

where δ_{ij} is the Kronecker delta function. If $n = 0$, the proposition follows trivially, and if $n > 0$, $\theta^n g$ is a multiple of a_d, so it is in $I^d \subseteq I^n$.

Now let $n \geq d$. Write $n = d + e$. Let $g \in R[\theta]$ such that $\theta^{d+e} g \in R$. Write $\theta^{d+e} = \sum_{i=0}^{d-1} f_i \theta^i$. By assumption each a_i is in I^i, so that each $a_{e_1} a_{e_2} \cdots a_{e_k}$ lies in I raised to the power $\sum e_i$. Thus each C_e is in I^e. It follows that the coefficient f_i of θ^i in the expression of θ^{d+e} above is in I^{d+e-i}. Then by the product formula the constant part of $\theta^{d+e} g$ is in I raised to the power

$$\begin{aligned}
&\min\{\deg f_0, \deg(f_k C_e a_d) | k = 0, \ldots, d-1; e = 0, \ldots, k-1\} \\
&\geq \min\{d+e, d+e-k+e+d | k = 0, \ldots, d-1; e = 0, \ldots, k-1\} = d+e,
\end{aligned}$$

which equals n. This proves the proposition. □

However, the converse does not hold in general:

Proposition 3.2 *Let R be a regular local ring with maximal ideal m, and let $a_1, \ldots, a_d$ be a regular sequence. Then for all $n \geq 0$, $\theta^n R[\theta] \cap R \subseteq m^n$ (yet the a_i need not be in progressively higher powers of m).*

Proof: Let $n \geq 0$ and f a non-zero element of $\theta^n R[\theta] \cap R$. Write $f = \theta^n(s_0 + s_1\theta + \cdots + s_{d-1}\theta^{d-1})$ for some $s_i \in R$. Let $s = s_0 + s_1\theta + \cdots + s_{d-1}\theta^{d-1}$.

For each non-negative integer n, repeatedly rewrite each occurrence of θ^d in $\theta^n \cdot s$ as $\sum_{i=1}^d a_i\theta^{d-i}$ until $\theta^n s$ is in the form $\sum_{i=0}^{d-1}\sum_{j=0}^{d-1} b_{ij}s_j\theta^i$ for some $b_{ij} \in R$. In other words, $\sum_{i=0}^{d-1}\sum_{j=0}^{d-1} b_{ij}s_j\theta^i$ is the reduction of $\theta^n \cdot s$ with respect to the polynomial $\theta^d - \sum_{i=1}^d a_i\theta^{d-i}$. Set B_n to be the $d \times d$ matrix (b_{ij}).

Note that if $\theta^n s$ reduces to $\sum_{i=0}^{d-1}\sum_{j=0}^{d-1} b_{ij}s_j\theta^i$, then $\theta^{n+1}s$ reduces to the same polynomial as $\sum_{i=0}^{d-1}\sum_{j=0}^{d-1} b_{ij}s_j\theta^{i+1}$. But this is

$$\sum_{i=0}^{d-2}\sum_{j=0}^{d-1} b_{ij}s_j\theta^{i+1} + \sum_{j=0}^{d-1} b_{d-1,j}s_j \sum_{i=1}^{d} a_i\theta^{d-i}.$$

Thus the first row of B_{n+1} is a_d times the last row of B_n, and row i of B_{n+1}, with $i > 1$, equals row $i-1$ of B_n plus a_{d-i+1} times row d of B_n.

Note that B_0 is the identity matrix. Then by induction on n one can easily prove that for all $n \geq 0$, $\det B_n = \pm a_d^n$.

Now let C_n be the submatrix of B_n obtained from B_n by removing the first row and the first column. We claim that for all $n \geq 1$, $\det C_n = \pm a_{d-1}^{n-1} + p_n$ for some $p_n \in (a_1, \ldots, a_{d-2}, a_d)$.

As B_0 is the identity matrix, then C_1 is the identity matrix, and the claim holds for $n = 1$. Suppose that the claim holds for $n \geq 1$. Let R_i be the ith row of B_n after deleting the first column. Then

$$C_{n+1} = \begin{bmatrix} R_1 + a_{d-1}R_d \\ R_2 + a_{d-2}R_d \\ \vdots \\ R_{d-2} + a_2R_d \\ R_{d-1} + a_1R_d \end{bmatrix}.$$

Then modulo $(a_1, \ldots, a_{d-2}, a_d)$, as R_1 is a multiple of a_d,

$$\det(C_{n+1}) \equiv \det \begin{bmatrix} a_{d-1}R_d \\ R_2 \\ \vdots \\ R_{d-2} \\ R_{d-1} \end{bmatrix} = \pm a_{d-1} \det \begin{bmatrix} R_2 \\ R_3 \\ \vdots \\ R_{d-1} \\ R_d \end{bmatrix} = \pm a_{d-1} \det C_n,$$

so that the claim holds by induction.

We have proved that $\det(B_n) = \pm a_0^n \neq 0$. As $B_n(s_0, s_1, \ldots, s_{d-1})^T = (f, 0, \ldots, 0)^T$, by Cramer's rule $s_0 = \pm f \det(C_n)/a_d^n$. But $\det(C_n)$ and a_d are relatively prime, so that as $s_0 \in R$, necessarily f is a multiple of a_d^n. Thus $f \in m^n$. □

4 Trace

In the previous section we showed that $a_i \in I^i$ for all i implies that $\theta^n R[\theta] \cap R \in I^n$ for all n, but that the converse fails. In this section we analyze the situation when the contraction is replaced with the trace function. Namely, we prove that the condition $a_i \in I^i$ for all i implies that $Tr(\theta^n) \in I^n$ for all n, that the converse fails in general, but holds in several cases, for example in characteristic 0; see Proposition 4.6. Other special cases of the converse assume that θ is separable over R.

We start by proving the positive results. We first introduce some more notation. Throughout this section let k be a ring; in our applications it will be either the ring of integers, or a field, and R will be a k-algebra. (This imposes no condition on R if k is the ring of integers.) Let Y_i, $i = 1, \ldots, d$ and Z be variables over k. Consider the polynomial

$$(Z - Y_1) \cdots (Z - Y_d) = Z^d - s_1 \cdot Z^{d-1} + \cdots + (-1)^d s_d$$

in $k[Y_1, \ldots, Y_d, Z]$, where $s_i = s_i(Y_1, \ldots, Y_d)$ denotes the elementary symmetric polynomials. It is well known that $k[s_1, \ldots, s_d] \subset k[Y_1, \ldots, Y_d]$ is the subring of invariants by permutations, that the extension is finite, and hence that $k[s_1, \ldots, s_d]$ is also a polynomial ring over k.

Since each s_i is homogeneous of degree i in the graded ring $k[Y_1, \ldots, Y_d]$, a natural weighted homogeneous structure is defined in the polynomial ring $k[s_1, \ldots, s_d]$ by setting $\deg(s_i) = i$, which makes the inclusion an homogeneous morphism of graded rings.

Remark 4.1 Set $v_i = Y_1^i + Y_2^i + \cdots + Y_d^i$, for $i \geq 0$. Then $k[v_1, v_2, \ldots] \subset k[s_1, \ldots, s_d]$, and since each v_i is homogeneous of degree i in $k[Y_1, \ldots, Y_d]$, the inclusion is homogeneous by setting $\deg(v_i) = i$. In other words, $v_i = v_i(s_1, \ldots, s_d)$ is weighted homogeneous of degree i in $k[s_1, \ldots, s_d]$. Let us finally recall that when k is a field of characteristic zero, then $k[v_1, \ldots, v_d] = k[s_1, \ldots, s_d]$.

Remark 4.2 The ring $k[s_1, \ldots, s_d][\Theta] = k[s_1, \ldots, s_d][Z] / < Z^d - s_1 \cdot Z^{d-1} + \cdots + (-1)^d \cdot s_d >$ is a free module of rank d over $k[s_1, \ldots, s_d]$. The trace of the endomorphism, on this finite module, defined by multiplication by Θ^i, is the weighted homogeneous polynomial $v_i \in k[s_1, \ldots, s_d]$ mentioned above.

In fact there are d different embeddings $\sigma_i : k[s_1,\ldots,s_d][\Theta] \to k[Y_1,\ldots,Y_d]$ of $k[s_1,\ldots,s_d]$-algebras, each defined by $\sigma_i(\Theta) = Y_i$, and the trace (of the endomorphism) of any element $\Gamma \in k[s_1,\ldots,s_d][\Theta]$ is $\sum \sigma_i(\Gamma)$.

Remark 4.3 Any primitive extension over a ring R, say

$$R[\theta] = R[Z]/ < Z^d - a_1 \cdot Z^{d-1} + \cdots + (-1)^d.a_d >$$

is

$$k[s_1,\ldots,s_d][Z]/ < Z^d - s_1 \cdot Z^{d-1} + \cdots + (-1)^d.s_d > \otimes_{k[s_1,\ldots,s_d]} R,$$

where k denotes here the ring of integers, and $\phi : k[s_1,\ldots,s_d] \to R$ defined by $\phi(s_i) = a_i$. By change of base rings it follows that the trace of the endomorphism of R modules defined by $\theta^i : R[\theta] \to R[\theta]$ is $\phi(v_i(s))$. When R is a normal domain with quotient field K, and θ is an algebraic element over K with minimal polynomial $Z^d - a_1 \cdot Z^{d-1} + \cdots + (-1)^d.a_d \in R[Z]$, then the trace of the endomorphism $\theta^i : R[\theta] \to R[\theta]$ is $Tr(\theta^i)$, where Tr denotes the trace of the field extension $K \subset K[\theta]$. In what follows, for an arbitrary ring R, we abuse notation and set $Tr(\theta^i) = \phi(v_i(s))$.

Remark 4.4 Fix an ideal I in a k-algebra R. Suppose that a weighted homogeneous structure on the polynomial ring $k[T_1,\ldots,T_d]$ is defined by setting $\deg(T_i) = m_i$, and let $G(T_1,\ldots,T_d)$ be a weighted homogeneous element of degree m. If $\phi : k[T_1,\ldots,T_d] \to R$ is a morphism of k-algebras and $\phi(T_i) \in I^{m_i}$, then $\phi(G) \in I^m$.

Now we can finally prove that the analog of Proposition 3.1 holds also for the Trace function:

Proposition 4.5 *Let I be an ideal of R. Assume that for each $i = 1,\ldots,d$, $a_i \in I^i$. Then $Tr(\theta^n) \in I^n$ for all positive integers n.*

Proof: The polynomial $Z^d - \sum_{i=0}^{d-1} a_i Z^i$ is the image of $Z^d - \sum_{i=0}^{d-1} (-1)^{i+1} s_i Z^i$ by the morphism $\phi : k[s_1,\ldots,s_d] \to R$, $\phi(s_i) = (-1)^i a_i \in I^i$, so we may apply Remark 4.4. □

The converse holds easily when k is a field of characteristic zero:

Proposition 4.6 *If the ring R contains a field, say k, of characteristic zero then $a_i \in I^i$ for $i = 1,\ldots,d$ if and only if $Tr(\theta^n) \in I^n$ for $1 \leq n \leq d$.*

Proof: The proof follows from the proof of the previous proposition and the second assertion in Remark 4.1. □

Furthermore, the converse holds in a much greater generality; see Proposition 4.8 below. We first introduce some conditions, and show some implications among them, culminating in Proposition 4.8.

Let R be an excellent normal domain, and K the quotient field of R. Normality asserts that if θ is a root of a polynomial $Z^n + b_1 \cdot Z^{n-1} + \cdots + b_n \in R[Z]$, then the minimal polynomial of θ over K is also in $R[Z]$. For an ideal I in R we study the following conditions:

Condition 1): The minimal polynomial of θ, $Z^d + a_1 \cdot Z^{d-1} + \cdots + a_d$, is such that $a_i \in I^i$.

Condition 2): The minimal polynomial of θ, $Z^d + a_1 \cdot Z^{d-1} + \cdots + a_d$, is such that $a_i \in \overline{I^i}$, the integral closure of I^i.

Condition 3): The element θ satisfies a polynomial equation $Z^n + b_1 \cdot Z^{n-1} + \cdots + b_n$, for some n, all $b_i \in I^i$.

Condition 4): θ is separable over K and $Tr_{K[\theta]/K}(\theta^i) \in I^i$.

It is clear that 1) implies both 2) and 3).

Proposition 4.7 Condition 3) *implies* Condition 2).

Proof: (Case I principal) If $I = <t>$ is a principal ideal and Condition 3) holds, it follows that θt^{-1} is an integral element over the ring R. If $Z^m + c_1 Z^{m-1} + \cdots + c_m \in R[Z]$ denotes the minimal polynomial of θt^{-1}, it is easy to check that $Z^m + tc_1 Z^{m-1} + t^2 c_2 Z^{m-2} + \cdots + t^m c_m$ is the minimal polynomial of θ over R. Hence, even Condition 1) holds in this case.

(The general case) Assume that, for some n, the element θ satisfies a polynomial equation $Z^n + b_1 Z^{n-1} + \cdots + b_n$, all $b_i \in I^i$. Let $Z^d + a_1 Z^{d-1} + \cdots + a_d$ denote the minimal polynomial of θ. We claim that $a_i \in \overline{I^i}$. Let S be the integral closure of the Rees algebra $R[It, t^{-1}]$ of I. Here t is a variable over R. As R is excellent, S is still Noetherian, excellent, normal. Its quotient field is $K(t)$. The minimal polynomial of θ over $K(t)$ is the same as the minimal polynomial of θ over K. Also, θ satisfies the polynomial equation $Z^n + b_1 Z^{n-1} + \cdots + b_n$, all $b_i \in I^iS = (It)^i t^{-i}S$, so that θ is integral over the principal ideal $t^{-1}S$. By the principal ideal case then all $a_i \in \overline{t^{-i}S} \cap R = \overline{I^n}$. $\square$

Proposition 4.8 *If θ is separable over K, and $Tr(\theta^r) \in I^r$ for all r big enough, then* Condition 3) *holds. In particular,* Condition 2) *holds.*

Proof: Let R be a normal ring with quotient field K, and set $L = K[\theta]$, where θ has minimal polynomial $f = Z^d + a_1 Z^{d-1} + \cdots + a_d$ with coefficients in R. So $\{1, \theta, \ldots, \theta^{d-1}\}$ is a basis of $R[\theta]$ over R.

For each index $j = 0, 1, \ldots, d-1$ we define $Tr(\theta^j.V)$ as a K-linear function on the variable V, say $Tr(\theta^j.V) : L \to K$. In addition $\{Tr(\theta^j.V) \mid j = 0, 1, \ldots, d-1\} \subset Hom_R(R[\theta], R)$ is a subset of the R-dual of the free module $R[\theta]$.

We will assume that the extension $K \subset L$ is separable, namely, that the discriminant Δ_f of the minimal polynomial f is non-zero in K (actually $\Delta_f \in R$), and we now argue as in [**3**] (Prop 11, page 40). Recall that setting $N = (n_{i,j})$ the $d \times d$ matrix where $n_{i,j} = Tr(\theta^i.\theta^j)$, then $\Delta_f = det(N)$. Since $\Delta_f \neq 0$ and $\{1, \theta, \ldots, \theta^{d-1}\}$ is a basis of $L = K[\theta]$ over K, it follows that $\{Tr(\theta^j.V), j = 0, 1, \ldots, d-1\}$ is a basis of $L^* = Hom_K(L, K)$.

Let T denote the free R-submodule in L^* generated by $\{Tr(\theta^j.V) \mid j = 0, 1, \ldots, d-1\}$. So $T \subset Hom_R(R[\theta], R)$ is an inclusion of two free R submodules in L^*. Since the functor $Hom_R(-, R)$ reverses inclusions,

$$R[\theta] = Hom_R(Hom_R(R[\theta], R), R) \subset Hom_R(T, R) \subset L.$$

Let $\{\omega_i, i = 0, 1, \ldots, d-1\}$ be the dual basis of $\{Tr(\theta^j.V), j = 0, 1, \ldots, d-1\}$ over the field K; it is also a basis of the R-module $Hom_R(T, R)$. Furthermore, for any element $\beta \in L$,

$$\beta = \sum_i Tr(\theta^i.\beta)\omega_i$$

is the expression of β as a K-linear combination in the basis $\{\omega_i, i = 0, 1, \ldots, d-1\}$. Note also that if $\beta \in R[\theta]$, all $Tr(\theta^i.\beta)$ are elements in R.

Set $R[\theta] = R^d$ by choosing basis $\{1, \theta, \ldots, \theta^{d-1}\}$, and $Hom_R(T, R) = R^d$ with basis $\{\omega_i, i = 0, 1, \ldots, d-1\}$, so the inclusion $R[\theta] \subset Hom_R(T, R)$ defines a short exact sequence

$$0 \to R^d \to R^d \to C \to 0$$

where C denotes the cokernel of the morphism given by the square matrix $N = (n_{i,j})$ mentioned above. Since $\Delta_f = det(N)$ it follows that $\Delta_f.Hom_R(T,R) \subset R[\theta]$; in fact $\Delta_f \in Ann(C)$.

Assume that for some ideal $I \subset R$, $Tr(\theta^r) \in I^r$ and all r big enough. In order to prove that Condition 3) holds we first note that

$$\theta^r = \sum_i Tr((\theta)^{i+r}).\omega_i \in I^r.Hom_R(T,R).$$

In fact, for r big enough,

$$J_r =< Tr(\theta^r), Tr(\theta^{r+1}), \ldots, Tr(\theta^{r+d-1}) > \subset I^r$$

in R. But then,

$$\Delta_f \theta^r \in I^r \cdot \Delta_f \cdot Hom_R(T,R) \subset I^r R[\theta]$$

for all r big enough. This already shows that θ is in the integral closure of $IR[\theta]$ (integral closure in the ring $R[\theta]$). That means that θ satisfies a polynomial equation $Z^n + b_1.Z^{n-1} + \cdots + b_n \in R[\theta][Z]$ with $b_i \in J^i$, $J = IR[\theta]$. As in [**4**] (page 348), this is equivalent to the existence of a finitely generated $R[\theta]$ submodule, say Q, in the field L, such that $\theta \cdot Q \subset J \cdot Q$. In fact Q can be chosen as the ideal $(J + \theta \cdot R[\theta])^{n-1}$ in $R[\theta]$. Finally, since Q is a finitely generated $R[\theta]$-module, it is also a finitely generated R-module. On the other hand note that $J \cdot Q = I \cdot Q$, and Condition 3) follows now from the determinant trick applied to $\theta \cdot Q \subset I \cdot Q$. □

Corollary 4.9 *If θ is separable over a local regular ring (R,m), then $Tr(\theta^n) \in m^n$ for all n big enough if and only if $a_i \in m^i$ for all $i = 1, \ldots, d$.* □

However, this equivalence fails in general for arbitrary rings and arbitrary ideals. The converse fails, for example, if θ is not separable over R:

Example 4.10 Let k be a field of characteristic 2, $d = 2$, $a_1 = 0$. Then $Tr(\theta^n) = 0$ for all n, but a_2 need not be in I^2.

Another failure of the converse is if the powers of I are not integrally closed:

Example 4.11 Let $R = k[X,Y]$ be a polynomial ring in two variables X and Y over a field k of characteristic 2. Let I be the ideal generated by $X^8, X^7Y, X^6Y^2, X^2Y^6, XY^7, Y^8$, and the minimal equation for θ being

$$\theta^2 - X^8\theta - X^{11}Y^5.$$

Note $a_1 = X^8 \in I$, $a_2 = X^{11}Y^5 \notin I^2$, but $X^{11}Y^5 \cdot I \subseteq I^3$. Hence

$$Tr(\theta) = X^8 \in I,$$
$$Tr(\theta^2) = X^8\,Tr(\theta) + Tr(X^{11}Y^5) = X^{16} \in I^2,$$

and for $n \geq 3$,

$$Tr(\theta^n) = X^8\,Tr(\theta^{n-1}) + X^{11}Y^5\,Tr(\theta^{n-2}) \in I^n.$$

Set as before the ideals $J_r = < Tr(\theta^r), Tr(\theta^{r+1}), \ldots, Tr(\theta^{r+d-1}) >$ in R. Note that $\{\theta^r, \theta^{r+1}, \cdots, \theta^{r+d-1}\}$ generate the ideal $\theta^r R[\theta]$ as R-module, so that J_r is the image of this ideal by the trace map.

If R is of characteristic $p > 0$, and $I = < f_1, \cdots, f_l >$, then $I^{[p^e]}$ denotes the ideal $< f_1^{p^e}, \cdots, f_l^{p^e} > \subset R$.

Proposition 4.12 *Let θ be separable over a local regular ring (R,m) of characteristic p. If $J_{p^r} \subset m^{[p^r]}$ for all r big enough, then θ is in the tight closure of the parameter ideal $m.R[\theta]$.*

Proof: We apply the same argument as in the previous Proposition. Note that in this case

$$\theta^{p^r} = \sum_{0 \leq i \leq d-1} Tr((\theta)^{i+p^r}) \cdot \omega_i \in m^{[p^r]} \cdot Hom_R(T,R).$$

But then,

$$\Delta_f \theta^{p^r} \in m^{[p^r]} \cdot \Delta_f \cdot Hom_R(T,R) \subset m^{[p^r]} R[\theta]$$

for r big enough. This already shows that θ is in the tight closure of $mR[\theta]$ (tight closure in the ring $R[\theta]$). □

Example 4.13 Consider $R = k[y,z]$ where k is a field of odd characteristic, and set $R[\theta]$, $\theta^2 - a_2 = 0$, where $a_2 = y^3 + z^n$, $n \geq 7$, n some integer. We will prove that $J_r \subseteq < y^{p^r}, z^{p^r} >$. Here $\{1, \theta\}$ is a basis of $R[\theta]$ over R. $Tr(1) = 2$ (invertible in k), and $Tr(\theta) = 0$. Since the trace is compatible with Frobenius, $Tr(\theta^{p^r}) = Tr(\theta)^{p^r} = 0$, so it suffices to check that $Tr(\theta^{p^r+1}) \in < y^{p^r}, z^{p^r} >$. Set $p^r + 1 = 2k$, so $(\theta)^{p^r+1} = a_2^k$, and $Tr(\theta^{p^r+1}) = 2a_2^k$. We finally refer to [**1**], page 14, Example 1.6.5, for a proof that $a_2^k \in < y^{p^r}, z^{p^r} >$ if $n \geq 7$ and r is sufficiently large.

Proposition 4.14 *Assume that θ is separable over a local regular ring (R,m) of characteristic p, and let Δ denote the discriminant. If θ is in the tight closure of the parameter ideal $mR[\theta]$ (in a ring containing $R[\theta]$), then $\Delta.J_{p^r} \subset m^{[p^r]}$ (in R) for all r.*

Proof: Let $f(X) \in R[X]$ denote the minimal polynomial of θ. Recall that the resultant $\Delta \in < f(X), f'(X) > \cap R$ (in $R[X]$), and hence $\Delta \in < f'(\theta) >$ in $R[\theta]$. Since $f'(\theta)$ is a test element, Δ is a test element, and

$$\Delta.(\theta)^{p^r} \in m^{[p^r]}R[\theta]$$

for all r.

Note that $R[\theta] \subset Hom_R(T,R)$ (hence $m^{[p^r]}R[\theta] \subset m^{[p^r]}Hom_R(T,R)$), and that, choosing as before the basis $\{\omega_0, \omega_1, \ldots, \omega_{d-1}\}$ in $Hom_R(T,R)$:

$$\Delta\theta^{p^r} = \sum_{0 \leq i \leq d-1} \Delta.Tr((\theta)^{i+p^r}).\omega_i \in m^{[p^r]}.Hom_R(T,R),$$

which shows that $\Delta.J_{p^r} \subset m^{[p^r]}$ in the ring R. □

5 The subalgebra of R generated by $Tr\theta^n, n \geq 0$

Let R and θ be as before, so that $R[\theta] \cong R[X]/(X^d + \Sigma_{i=1}^d (-1)^i a_i X^{d-i})$. Assume now that R is an algebra over a field k. It follows from Remarks 4.1 and 4.2 that if k is of characteristic zero, the k-subalgebra generated by the traces $Tr\theta^n$ for all n, is $k[a_1, \cdots, a_d] (\subset R)$. In particular it is finitely generated. This subalgebra need not be finitely generated over a field of positive characteristic, as we show below.

First we recall some notation. Let B_n be the matrix as in the proof of Proposition 3.2. The trace of θ^n is exactly the trace of B_n.

Remark 5.1 In the proof of Proposition 3.2 we showed that the first row of B_{n+1} is a_d times the last row of B_n, and row i of B_{n+1}, with $i > 1$, equals row $i-1$ of B_n plus a_{d-i+1} times row d of B_n.

We determine the entries of B_n more precisely:

Lemma 5.2 *For $n \leq d$,*

$$(B_n)_{ij} = \begin{cases} \delta_{i,j+n} & \text{if } j \leq d-n, \\ \sum_{k=d-n+1}^{j-1} a_{j-k}(B_n)_{ik} + a_{n-i+j} & \text{if } j > d-n. \end{cases}$$

Furthermore, for all $j > d-n$,

$$(B_n)_{ij} = (B_d)_{i,j-d+n}.$$

Proof: We proceed by induction on n. The formulation is correct for $n = 0$. Thus we assume that $n > 0$. By Remark 5.1 the formulations of the entries of B_n in the first $d-n+1$ columns are correct: in the first $d-n$ columns, the entries are $\delta_{i,j+n}$, and $(B_n)_{i,d-n+1} = a_{d-i}$.

Now let $i = 1$, $j > d-n+1$. Then

$$\begin{aligned}
(B_n)_{1j} &= a_d(B_{n-1})_{dj} \\
&= a_d\left(\sum_{k=d-(n-1)+1}^{j-1} a_{j-k}(B_{n-1})_{dk} + a_{n-1-d+j}\right) \\
&= \sum_{k=d-n+2}^{j-1} a_{j-k}a_d(B_{n-1})_{dk} + a_d a_{n-1-d+j} \\
&= \sum_{k=d-n+2}^{j-1} a_{j-k}(B_n)_{1k} + (B_n)_{1,d-n+1}a_{j-(d-n+1)} \\
&= \sum_{k=d-n+1}^{j-1} a_{j-k}(B_n)_{1k} \\
&= \sum_{k=d-n+1}^{j-1} a_{j-k}(B_n)_{1k} + a_{n-1+j}
\end{aligned}$$

as $n-1+j>d$ so that $a_{n-1+j}=0$. Now let $i>1,\ j>d-n+1$. Then

$$\begin{aligned}
(B_n)_{ij} &= (B_{n-1})_{i-1,j}+a_{d-i+1}(B_{n-1})_{dj} \\
&= \sum_{k=d-(n-1)+1}^{j-1} a_{j-k}(B_{n-1})_{i-1,k}+a_{(n-1)-(i-1)+j} \\
&\quad + a_{d-i+1}\left(\sum_{k=d-(n-1)+1}^{j-1} a_{j-k}(B_{n-1})_{dk}+a_{(n-1)-d+j}\right) \\
&= \sum_{k=d-n+1}^{j-1} a_{j-k}(B_{n-1})_{i-1,k}+a_{n-i+j}+a_{d-i+1}\sum_{k=d-n+1}^{j-1} a_{j-k}(B_{n-1})_{dk} \\
&\quad \text{(because for } k=d-n+1,\ (B_{n-1})_{i-1,k}=0 \text{ and } (B_{n-1})_{dk}=1) \\
&= \sum_{k=d-n+1}^{j-1} a_{j-k}(B_n)_{ik}+a_{n-i+j}.
\end{aligned}$$

Observe that the last statement is true for $j=d-n+1$. Then by induction on $j>d-n+1$,

$$\begin{aligned}
(B_n)_{ij} &= \sum_{k=d-n+1}^{j-1} a_{j-k}(B_n)_{ik}+a_{n-i+j} \\
&= \sum_{k=d-n+1}^{j-1} a_{j-k}(B_d)_{i,k-d+n}+a_{n-i+j} \\
&= \sum_{l=1}^{j-d+n-1} a_{j-l-d+n}(B_d)_{il}+a_{n-i+j} \\
&= (B_n)_{i,j-d+n}.
\end{aligned}$$

□

It then follows:

Corollary 5.3 *Whenever* $1\le n\le d$,

$$Tr(\theta^n)=\sum_{i=1}^{n-1} a_{n-i}Tr(\theta^i)+na_n,$$

and $Tr(\theta^n)$ *is a polynomial in* $a_1,\ldots,a_n$, *homogeneous of degree n under the weights* $\deg(a_i)=i$.

Proof: By definition, $Tr(\theta^n) = Tr(B_n) = \Sigma_{i=1}^{d}(B_n)_{ii}$, and by Lemma 5.2 this equals

$$Tr(\theta^n) = \sum_{i=d-n+1}^{d} (B_n)_{ii} = \sum_{i=d-n+1}^{d} (B_d)_{i,i-d+n} = \sum_{j=1}^{n} (B_d)_{d-n+j,j},$$

i.e., this is the sum of the elements of B_d on the nth diagonal, counting from the bottom leftmost corner. Hence,

$$\begin{aligned} Tr(\theta^n) &= \sum_{j=1}^{n}\left(\sum_{k=1}^{j-1} a_{j-k}(B_d)_{d-n+j,k} + a_{d-(d-n+j)+j}\right) \\ &= \sum_{j=1}^{n}\sum_{k=1}^{j-1} a_{j-k}(B_d)_{d-n+j,k} + na_n. \end{aligned}$$

Now we change the double summation: c sums over the differences $j-k$, and k keeps the same role:

$$\begin{aligned} Tr(\theta^n) &= \sum_{c=1}^{n-1}\sum_{k=1}^{n-c} a_c(B_d)_{k+c+d-n,k} + na_n \\ &= \sum_{c=1}^{n-1} a_c \sum_{k=1}^{n-c} (B_d)_{k+d-(n-c),k} + na_n \\ &= \sum_{c=1}^{n-1} a_c Tr(\theta^{n-c}) + na_n. \end{aligned}$$

□

For $n \geq 0$ let C_n be as in Definition 2.3. We adopt the notation that for $n < 0$, $C_n = 0$. Then for $n \geq 0$, let P_n be the row matrix $[C_n, C_{n-1}, \ldots, C_{n-d+1}]$, and for each $n = 1, \ldots, d$, let

$$F_n = \sum_{i=0}^{d-1} a_{d+n-1-i} Tr(\theta^i).$$

Let $\vec{F}$ be the vector $(F_1, \ldots, F_d)$. With this we can give another formulation of the trace of powers of θ:

Lemma 5.4 *For each $e \geq 0$,*

$$Tr(\theta^{d+e}) = P_e \cdot \vec{F}.$$

Proof: By Proposition 2.5,

$$\begin{aligned} Tr(\theta^{d+e}) &= \sum_{i=0}^{d-1}\sum_{j=0}^{e} C_j a_{d+e-i-j} Tr(\theta^i) = \sum_{j=0}^{e} C_j \sum_{i=0}^{d-1} a_{d+e-i-j} Tr(\theta^i) \\ &= \sum_{j=e-d+1}^{e} C_j \sum_{i=0}^{d-1} a_{d+e-i-j} Tr(\theta^i) = \sum_{j=e-d+1}^{e} C_j F_{e-j+1} \\ &= P_e \cdot \vec{F}. \end{aligned}$$

□

Now we can give an example of a k algebra R, and θ as before, where k is a field of positive characteristic, and the subalgebra of R generated over k by $Tr(\theta^n)$ as n varies is not a finitely generated algebra (compare with Remark 4.1):

Example 5.5 Let k be a field of positive prime characteristic p, $d = p$, and $a_1, \ldots, a_d$ indeterminates over k, $R = k[a_1, \ldots, a_d]$. Let $A = k[Tr\theta, Tr\theta^2, \ldots]$. It follows from Remark 4.2 and Remark 4.1 that $A \subseteq R$. But this A is not finitely generated over k, as we prove below.

For each $n \geq 1$, let $A_n = k[Tr\theta, Tr\theta^2, \ldots, Tr\theta^n]$.

Claim: For each $n \geq 0$ and $l \in \{0, \ldots, d-1\}$:

$$A_{dn+l} = k[a_i a_d^j \mid \text{either } j < n \text{ or else } j = n \text{ and } i \leq l].$$

We will prove this by induction on n. It holds for $n = 0$ by Corollary 5.3. Thus by the definition of the F_i and by Corollary 5.3, all F_i are in all $A_{(n+1)d+l}$. Furthermore, each F_i is linear in a_d.

By Lemma 5.4, $Tr(\theta^{(n+1)d+l})$ equals

$$C_{nd+l}F_1 + \cdots + C_{nd+1}F_l + C_{nd}F_{l+1} + C_{nd-1}F_{l+1} + \cdots + C_{nd-(d-i-1)}F_d.$$

By the structure of the C_i, a_d appears in C_i with exponent at most i/d. Thus the summand $C_{nd-1}F_{l+1} + \cdots + C_{nd-(d-i-1)}F_d$ lies in $A_{(n+1)d+l-1}$. Also, in the expansion of the summand $C_{nd+l}F_1 + \cdots + C_{nd+1}F_l$, in each term a_d either appears with exponent n or smaller, or else it appears with exponent exactly $n+1$ and is multiplied by one of the variables $a_1, \ldots, a_{l-1}$. Thus also this summand lies in $A_{(n+1)d+l-1}$. Thus

$$A_{(n+1)d+l} = A_{(n+1)d+l-1}[C_{nd}F_{l+1}].$$

F_{l+1} is linear in a_d with leading coefficient $Tr(\theta^l)$. C_{nd} equals a_d^n plus terms of lower a_d-degree, so that similarly, by Corollary 5.3,

$$A_{(n+1)d+l} = A_{(n+1)d+l-1}[a_d^n a_d Tr(\theta^l)] = A_{(n+1)d+l-1}[a_d^{n+1} l a_l].$$

This proves the claim. As $a_1, \dots, a_d$ are variables over k, this means that A is not finitely generated over k.

As an almost immediate corollary we can give another proof of Proposition 4.8 in a special case:

Proposition 5.6 *Let $d = p$, i.e., the degree of the extension is the same as the characteristic of the ring. Assume that $X^d - \sum_{i=1}^d a_i X^{d-i}$ is a separable polynomial. Let v be any valuation $v : R \to \mathbb{N} \cup \{\infty\}$ such that $v(x) = \infty$ if and only if $x = 0$. Then $v(Tr(\theta^n)) \geq n$ for all n if and only if $v(a_i) \geq i$ for all i.*

Proof: With notation as above, one can prove by induction on $nd + l$ that $v(a_d^n a_l) \geq nd + l$. In particular, for $l = 1, \dots, d-1$, $v(a_l) \geq l$. Also,

$$v(a_d) = \frac{1}{n}(v(a_d^n a_l) - v(a_l)) \geq \frac{1}{n}(nd + l - v(a_l))$$

for all n and l. As at least one $v(a_l)$ is finite (by the separability assumption), it follows that $v(a_d) \geq d$. □

References

[1] C. Huneke, Tight closure and its applications. With an appendix by Melvin Hochster, *CBMS Regional Conference Series in Mathematics* **88**, American Mathematical Society, Providence, RI, 1996.

[2] J. Lipman and A. Sathaye, Jacobian ideals and a theorem of Briancon-Skoda, *Michigan Math. J.* **28** (1981), 199–222.

[3] J.-P. Serre, *Local Algebra*, Springer Monographs in Mathematics, Springer-Verlag, Berlin, 2000.

[4] O. Zariski and P. Samuel, *Commutative Algebra II*, D. Van Nostrand Co., Inc., Princeton-Toronto-London-New York, 1960.

Degeneration and G-dimension of modules

YUJI YOSHINO
Department of Mathematics, Okayama University, Okayama 700-8530, Japan
E-mail: yoshino@math.okayama-u.ac.jp

1 Introduction

The aim of this paper is to show that the property for a module having G-dimension 0 is an "open" property. To be more precise, let R be a Noetherian commutative algebra over a field k and let M and N be finitely generated R-modules. Suppose that N is a degeneration of M in some sense. Then we shall prove the inequality G-dim$M \leq$ G-dim$N$ in Theorem (3.2). In particular, that G-dim$N = 0$ implies that G-dim$M = 0$ in this case. We infer from this that if there are some moduli spaces of finitely generated R-modules, then the set of modules with G-dimension 0 should form an open subset.

If the ring R is a Gorenstein local ring, then an R-module has G-dimension zero if and only if it is a maximal Cohen-Macaulay module. And maximal Cohen-Macaulay modules have been studied by many authors and a great deal of properties are known (cf. [**2**]). It will be natural to expect that many of the properties of maximal Cohen-Macaulay modules over a Gorenstein local ring are satisfied as well for modules of G-dimension zero over a general local ring. This is a main reason for us to consider modules of G-dimension zero in this paper. Note, however, that the category of modules of G-dimension zero is not a contravariantly finite subcategory in the category of finitely generated modules for a certain kind of local ring, as we have recently shown in [**4**].

The above inequality of G-dimension could be proved from the definition of degeneration and G-dimension. But in this paper we will prove it by using a general result about degeneration (Theorem 2.2), from which the inequality follows easily. Indeed, M degenerates to N if and only if there is a short exact sequence of the following type:

$$0 \to Z \xrightarrow{\binom{\phi}{\psi}} M \oplus Z \to N \to 0$$

such that the endomorphism ψ on Z is nilpotent. This equivalence has been proved in our previous work [**5**]. We are inspired very much by the work of G. Zwara [**6**] for this. Actually, he proved it for a finite dimensional (noncommutative) algebra over a field. In that case the condition that ψ is nilpotent is not necessary. Many of algebraic properties of degeneration could follow from this exact sequence as well as the inequality of G-dimension.

We shall give a precise definition of degeneration and discuss it to give the above necessary and sufficient condition in Section 2. Using this result, we shall give a proof of the inequality in Section 3.

2 Preliminaries for degenerations of modules

In this section k always denotes a field and R is a Noetherian commutative k-algebra.

Definition 2.1 For finitely generated R-modules M and N, we say that M degenerates to N along a discrete valuation ring, or N is a degeneration of M along a DVR, if there is a discrete valuation ring (V, tV, k) that is a k-algebra (where t is a prime element) and a finitely generated $R \otimes_k V$-module Q which satisfies the following conditions:

(1) Q is flat as a V-module.

(2) $Q/tQ \cong N$ as an R-module.

(3) $Q[\frac{1}{t}] \cong M \otimes_k V[\frac{1}{t}]$ as an $R \otimes_k V[\frac{1}{t}]$-module.

We have proved the following theorem in the previous paper [**5**], which gives a perfect condition for the degeneration along a DVR.

Theorem 2.2 ([5, Theorem 2.2]) *The following conditions are equivalent for finitely generated R-modules M and N:*

(1) *N is a degeneration of M along a DVR.*

(2) *There is a short exact sequence of finitely generated R-modules*

$$0 \to Z \xrightarrow{\binom{\phi}{\psi}} M \oplus Z \to N \to 0,$$

such that an endomorphism ψ *on Z is nilpotent, i.e.,* $\psi^n = 0$ *for* $n \gg 1$.

Remark 2.3 As we remarked in the paper [**5**], G. Zwara has shown in [**6**] that if R is an Artinian k-algebra, then the following conditions are equivalent for finitely generated R-modules M and N:

(1) N is a degeneration of M along a DVR.

(2) There is a short exact sequence of finitely generated R-modules

$$0 \to Z \xrightarrow{\binom{\phi}{\psi}} M \oplus Z \to N \to 0.$$

Note here that we need not the nilpotency assumption for ψ. Actually, suppose that there is such an exact sequence as in the condition (2). Then, since $\mathrm{End}_R(Z)$ is an Artinian ring, we can decompose Z as $Z' \oplus Z''$ and according to this decomposition we can describe ψ as

$$\begin{pmatrix} \psi' & 0 \\ 0 & \psi'' \end{pmatrix} : Z' \oplus Z'' \to Z' \oplus Z'',$$

where ψ' is an isomorphism and ψ'' is nilpotent (Fitting theorem). Therefore we obtain an exact sequence of the type

$$0 \to Z'' \xrightarrow{\binom{\phi''}{\psi''}} M \oplus Z'' \to N \to 0,$$

such that ψ'' is nilpotent.

We collect several remarks here that are proved in the paper [**5**].

Remark 2.4

(1) Our Theorem 2.2 is valid for finitely generated left modules over a general k-algebra which may be noncommutative and non-Noetherian. See [**5**, Theorem 2.2].

(2) We have considered a different kind of degeneration in the paper [**3**], mainly for maximal Cohen-Macaulay modules. To distinguish it from the degeneration along a DVR, we call it a degeneration along an affine line. We have shown the implication: "degeneration along a DVR $\Longrightarrow$ degeneration along an affine line." However, the converse implication does not hold. See [**5**, Section 3].

(3) Suppose that M degenerates to N along a DVR. Then as a discrete valuation ring V we can always take the ring $k[t]_{(t)}$. See [**5**, Corollary 2.4].

3 G-dimension of modules

In this section, R denotes a Noetherian commutative ring without any other assumption. First of all, let us briefly recall the definition of G-dimension.

Definition 3.1 ([1]) Let M be a finitely generated R-module. We say that M has G-dimension zero if it satisfies the following conditions:

(1) The natural morphism $M \to \mathrm{Hom}_R(\mathrm{Hom}_R(M,R),R)$ is an isomorphism.

(2) $\mathrm{Ext}^i_R(M,R) = 0$ for all $i > 0$.

(3) $\mathrm{Ext}^i_R(\mathrm{Hom}_R(M,R),R) = 0$ for all $i > 0$.

For an integer n, we say that M has G-dimension at most n, denoted by $\text{G-dim}_R M \leq n$, if an n-th syzygy module $\Omega^n_R(M)$ of M has G-dimension zero. If there is no such integer n, we denote $\text{G-dim}_R M = \infty$.

Now we state our main theorem of this paper.

Theorem 3.2 *Assume that there is an exact sequence of finitely generated R-modules*

$$0 \to Z \to M \oplus Z \to N \to 0.$$

Then we have an inequality $\text{G-dim}_R M \leqq \text{G-dim}_R N$.

In order to prove the theorem, it will be convenient to prepare the following lemma.

Lemma 3.3 *Let X and Y be finitely generated R-modules and let* $f : X \to X \oplus Y$ *be an R-homomorphism. If f is surjective, then* $Y = 0$ *and f is an isomorphism.*

Proof: Let $p : X \oplus Y \to X$ be the natural projection. Since $p \cdot f$ is a surjective endomorphism on X, it is an automorphism. Likewise, $f \cdot p$ is an automorphism on $X \oplus Y$. Thus f is an isomorphism. In general, it is easy to see that $X \cong X \oplus Y$ implies $Y = 0$. □

Now we proceed to the proof of the theorem. Of course we may assume $\text{G-dim}_R N < \infty$ to prove the inequality. As the first step of the proof, we shall show that we may assume that $\text{G-dim}_R N = 0$.

For this, suppose $\text{G-dim}_R N > 0$. Consider any first syzygy modules ΩM, ΩN, ΩZ of M, N, Z repectively, and we have an exact sequence of the type $0 \to \Omega Z \to \Omega M \oplus \Omega Z \to \Omega N \to 0$. Note here that $\text{G-dim}_R \Omega N = \text{G-dim}_R N - 1$ from the definition, and that $\text{G-dim}_R M - 1 = \text{G-dim}_R \Omega M \leqq \text{G-dim}_R \Omega N$ by induction on $\text{G-dim}_R N$. Therefore, we can reduce the proof to the case where $\text{G-dim}_R N = 0$.

Now in the rest of the proof, we assume that there is an exact sequence

$$0 \to Z \to M \oplus Z \to N \to 0, \tag{1}$$

where $\text{G-dim}_R N = 0$. We want to show $\text{G-dim}_R M = 0$.

Denote $X^* = \text{Hom}_R(X, R)$ for any R-module X. Since $\text{Ext}_R^i(N, R) = 0$ $(i > 0)$, we have a short exact sequence and isomorphisms:

$$\begin{cases} 0 \to N^* \to M^* \oplus Z^* \to Z^* \to 0, \\ \text{Ext}_R^i(M, R) \oplus \text{Ext}_R^i(Z, R) \cong \text{Ext}_R^i(Z, R) \quad (i > 0). \end{cases} \tag{2}$$

It follows from Lemma 3.3 that

$$\text{Ext}_R^i(M, R) = 0 \quad (i > 0). \tag{3}$$

Furthermore, since $\mathrm{Ext}^i_R(N^*,R)=0$ for $i>0$, we have from (2) the following exact sequence and isomorphisms:

$$\begin{cases} 0\to Z^{**}\to M^{**}\oplus Z^{**}\to N^{**} \\ \qquad \to \mathrm{Ext}^1_R(Z^*,R)\xrightarrow{f}\mathrm{Ext}^1_R(M^*,R)\oplus\mathrm{Ext}^1_R(Z^*,R)\to 0, \\ \mathrm{Ext}^i_R(Z^*,R)\cong\mathrm{Ext}^i_R(M^*,R)\oplus\mathrm{Ext}^i_R(Z^*,R)\quad (i>1). \end{cases} \tag{4}$$

Similarly to the above, using Lemma 3.3, we have that

$$\mathrm{Ext}^i_R(M^*,R)=0\quad (i>0), \tag{5}$$

and the mapping f in the diagram (4) is an isomorphism. As a consequence, there is an exact sequence $0\to Z^{**}\to M^{**}\oplus Z^{**}\to N^{**}\to 0$.

Now denote by ν_X the natural homomorphism $X\to X^{**}$ for any R-module X. We notice that there is a commutative diagram with exact rows:

$$\begin{array}{ccccccccc} 0 & \longrightarrow & Z & \longrightarrow & M\oplus Z & \longrightarrow & N & \longrightarrow & 0 \\ & & \nu_Z\downarrow & & \begin{pmatrix}\nu_M & 0\\ 0 & \nu_Z\end{pmatrix}\downarrow & & \nu_N\downarrow & & \\ 0 & \longrightarrow & Z^{**} & \longrightarrow & M^{**}\oplus Z^{**} & \longrightarrow & N^{**} & \longrightarrow & 0 \end{array} \tag{6}$$

Since ν_N is an isomorphism, we see that $\mathrm{Coker}\nu_Z\cong\mathrm{Coker}\nu_M\oplus\mathrm{Coker}\nu_Z$ and $\ker\nu_Z\cong\ker\nu_M\oplus\ker\nu_Z$. Again by Lemma 3.3 we finally have that $\mathrm{Coker}\nu_M=0$ and $\ker\nu_M=0$. Hence,

$$\nu_M : M\to M^{**} \text{ is an isomorphism.} \tag{7}$$

From (3), (5) and (7) we have shown that M has G-dimension 0. □

Finally, as a direct consequence of Theorem 3.2 together with Theorem 2.2, we have the following result as a corollary, which shows the openness of the "G-dimension 0" property.

Corollary 3.4 *Assume that R is a Noetherian algebra over a field k and let M and N be finitely generated R-modules. Suppose that N is a degeneration of M along a DVR. Then we have an inequality* $\text{G-dim}_R M\leqq\text{G-dim}_R N$.

References

[1] M. Auslander and M.Bridger, *Stable Module Theory*, Mem. Amer. Math. Soc. **94**, 1969.

[2] Y. Yoshino, *Cohen-Macaulay Modules over Cohen-Macaulay Rings*, Lecture Note Series **146**, Cambridge University Press, Cambridge, 1990.

[3] Y. Yoshino, On degenerations of Cohen-Macaulay modules, *J. Algebra* **248** (2002), 272–290.

[4] Y. Yoshino, Modules of G-dimension zero over local rings with the cube of maximal ideal being zero, in *Commutative Algebra, Singularities and Computer Algebra*, J. Herzog and V.Vuletescu, Eds., 255–273, NATO Sci. Ser. II, *Math. Phys. Chem.* **115**, Kluwer Academic Publishers, Dordrecht, 2003.

[5] Y. Yoshino, On degenerations of modules, *J. Algebra* **278** (2004), 217–226.

[6] G. Zwara, Degenerations of finite dimensional modules are given by extensions, *Compositio Math.* **121** (2000), 205–218.